Molecular Insight into Plant Bioactive Compounds

Molecular Insight into Plant Bioactive Compounds

Guest Editor

Karolina Wojtunik-Kulesza

MDPI

Basel • Beijing • Wuhan • Barcelona • Belgrade • Novi Sad • Cluj • Manchester

Guest Editor
Karolina Wojtunik-Kulesza
Department of Inorganic
Chemistry
Medical University of Lublin
Lublin
Poland

Editorial Office
MDPI AG
Grosspeteranlage 5
4052 Basel, Switzerland

This is a reprint of the Special Issue, published open access by the journal *International Journal of Molecular Sciences* (ISSN 1422-0067), freely accessible at: https://www.mdpi.com/journal/ijms/special_issues/9K03J1WA88.

For citation purposes, cite each article independently as indicated on the article page online and as indicated below:

Lastname, A.A.; Lastname, B.B. Article Title. *Journal Name* **Year**, *Volume Number*, Page Range.

ISBN 978-3-7258-7264-0 (Hbk)
ISBN 978-3-7258-7265-7 (PDF)
https://doi.org/10.3390/books978-3-7258-7265-7

Contents

About the Editor

Karolina Wojtunik-Kulesza

Karolina Wojtunik-Kulesza is a researcher at the Medical University of Lublin, Poland, specializing in the neuroprotective potential of natural compounds. In 2025, she attained the degree of habilitated doctor following the completion of a distinguished research series "Monoterpenes – old compounds, a new look at their multidirectional biological activity and potential neuroprotective applications." Their academic career, which includes a PhD focused on terpenoids in neurodegeneration (2017), has consistently bridged the gap between phytochemistry and clinical prevention. The research centers on the bioactive properties of secondary plant metabolites, specifically monoterpenes. Utilizing both in vitro spectrophotometric analyses and in vivo mouse models, Dr. hab. Wojtunik-Kulesza investigates antioxidant activity, free radical scavenging, enzyme inhibition, and metal ion chelation. This work aims to identify novel therapeutic pathways for treating neurodegenerative and lifestyle-related diseases. A prolific project leader and collaborator, she has spearheaded several major initiatives, including the National Science Centre "Miniatura 5" project on monoterpene reaction kinetics and various internal grants at the Medical University of Lublin. Wojtunik-Kulesza also serves as a contractor in projects funded by the Medical Research Agency (2024–2026) and the National Centre for Research and Development (project: LIDER), focusing on innovative functional foods and drug development targeting neuronal receptors through molecular modeling.

International Journal of *Molecular Sciences*

MDPI

Editorial

Special Issue "Molecular Insight into Plant Bioactive Compounds"

Karolina Anna Wojtunik-Kulesza

Department of Inorganic Chemistry, Faculty of Pharmacy, Medical University of Lublin, W. Chodźki St 4a, 20-093 Lublin, Poland; karolina.wojtunik-kulesza@umlub.edu.pl

1. Introduction

Plants are an inexhaustible source of natural bioactive compounds. To date, thousands of secondary plant metabolites have been identified. In addition to their various roles in plants, including allelopathy and protection against abiotic stress or pathogens, the compounds have a wide range of biological activities that are significant for human health. Vegetables, fruits, and species rich in polyphenols, terpenoids, or alkaloids are characterized by diverse biological effects, including antioxidants, antibacterial and antiviral effects, metal ion reduction and chelation, and a positive impact on lifestyle-related diseases. An adequate amount of these extremely valuable compounds contributes to the prevention of many diseases, supports faster recovery, and helps maintain a youthful appearance [1,2].

The issue has deepened our understanding of secondary plant metabolites, including polyphenols, flavonoids, lignins, alkaloids, and terpenoids. The multidirectional scope of these published papers focuses on health-promoting properties, such as antioxidant properties, antimicrobial, enzyme-inhibitory, and procognitive activities, as well as providing a toxicity evaluation of these compounds. Various advanced techniques and methodologies, applied in both in vitro and in vivo studies, are also presented.

2. Overview of Published Articles

This Special Issue brings together a diverse collection of studies addressing the complex chemistry, biology, pharmacy, and therapeutic potential of plants as a source of bioactive compounds. This Special Issue accepted eight papers, including both original research articles and comprehensive reviews, highlighting key advances in the field. Collectively, they provide valuable molecular insights and highlight the importance of interdisciplinary collaboration among pharmacy, chemistry, and biology in the field of natural products research.

The research articles provide a detailed characterization of secondary plant metabolites, along with quality and quantity analyses, using advanced analytical methods. Granda-Santos et al. (contribution 1) presented a characterization of terpenoids in aromatic plants using Raman spectroscopy and GC-MS. The basis of the analyses was medicinal plants: *Tagetes filifolia*, *Aloysia citrodora*, *Cymbopogon citratus*, *Eucalyptus globulus*, *Chamaemelum nobile*, *Piper aduncum*, *Minthostachys mollis*, and *Rosmarinus officinalis*. The studies revealed differences in the proportion of oxygenated compounds compared with aromatic hydrocarbons and their influence on the differences between species; these were observed using Raman spectroscopy and GC-MS. Cis,cis-nepetalactone was identified in *M. mollis* for the first time. Comprehensive studies based on European *Phellinus* species were presented by Świderski et al. (contribution 2). The authors presented a multi-directional evaluation of the chemical potential and biological properties of samples collected from

22 *Phellinus* species. The most valuable biological activity was observed for *Phellinus igniarius*, *Fomitiporia robusta*, and *Porodaedalea pini*, which were the subject of detailed analyses. Interesting results are also presented by Valdivia-Padilla et al. (contribution 3), who focused on metabolomics' characterization and bioinformatics studies of the bioactive compounds of *Psidium guajava* L. Studies based on principal component analysis (PCA) showed that β-caryophyllene was the predominant secondary plant metabolite in both genotypes of the tropical plant. Advanced analyses, including computational approaches, demonstrated significant interactions with target proteins, such as CB2, PPARα, BAX, BCL2, and AKT1. The results underscore the pharmacological promise of guava leaf metabolites in the context of natural product research and drug development. An evaluation of the pro-cognitive properties of citral was presented by Wojtunik-Kulesza et al. (contribution 4). A comprehensive evaluation based on in vitro, in vivo and ex vivo studies revealed the positive influence of citral (50 mg/kg; mice model) on the memory processes associated with acquisition. Additionally, GC-MS analysis revealed the presence of monoterpene in the blood and brain after the subchronic administration of citral and its accumulation in hippocampus. Li et al. (contribution 5) focused polyphenols from *Nymphaea candida* and their preventative activity against LPS-induced septic acute lung injury (ALI) in a mouse model. The polyphenols markedly alleviated lung damage, reduced the wet/dry ratio in the lung and reduced MPO activity. Additionally, the normalization of leukocyte profiles, lower level of pro-inflammatory cytokines, and suppression of TLR4/NF-κB and NLRP3 were observed. These findings indicate that NCTP exerts preventive effects against septic ALI, likely through the modulation of inflammatory signaling pathways.

The review articles collectively highlight the growing scientific interest in plant-derived bioactive compounds as complementary or alternative strategies for the prevention and treatment of major human diseases, particularly cancer, metabolic disorders, neurological conditions, and viral infections.

The review presented by Gökçen et al. (contribution 6) summarizes evidence that extracts from different parts of *Ficus carica* exhibit antiproliferative, pro-apoptotic and antioxidant effects in various in vitro cancer models. However, the findings vary depending on the plant part, extraction method and dosage, while in vivo and clinical data remain limited.

ElShamey et al. (contribution 7) explored the genetic basis of key bioactive compounds—GABA, resistant starch and alkaloids—in rice and barley, highlighting important genes and QTLs involved in their biosynthesis. These compounds contribute to cardiovascular health, glycemic control, gut microbiota modulation, and neuroprotection.

Bukowska (contribution 8) decided to review the therapeutic applications of cannabinoids from *Cannabis sativa*, particularly THC and CBD, which interact with the endocannabinoid system and exhibit anti-inflammatory, anticancer, and antiviral properties. Approved formulations are already used for epilepsy and multiple sclerosis, while emerging studies suggest broader potential in oncology and other chronic diseases.

3. Conclusions

This Special Issue brings together original research and review articles highlighting the chemical diversity and biological potential of plant secondary metabolites. Research papers provide detailed qualitative and quantitative characterization using advanced analytical techniques such as Raman spectroscopy, GC-MS, metabolomics, and computational approaches. Several studies demonstrated novel compounds, species-specific metabolic profiles, and significant biological activities. The review articles complement these findings by summarizing the current knowledge on plant-derived bioactive compounds in cancer, metabolic, neurological, and viral diseases. Collectively, the contributions underscore the

https://doi.org/10.3390/ijms27052340

translational potential of natural products while highlighting the need for further in vivo and clinical validation.

Informed Consent Statement: Not applicable.

Data Availability Statement: Data are available upon request.

Conflicts of Interest: There is no conflict of interest.

List of Contributions:

1. Granda-Santos, M.; Reyna-Gonzales, K.; Torrejón-Valqui, L.; Valle-Epquín, M.G.; Caetano, A.C.; Díaz-Valderrama, J.R.; Castro-Alayo, E.M.; Cayo-Colca, I.S.; Maicelo, J.L.; Balcázar-Zumaeta, C.R. Characterization of Terpenoids in Aromatic Plants Using Raman Spectroscopy and Gas Chromatography–Mass Spectrometry (GC–MS). *Int. J. Mol. Sci.* **2025**, *26*, 11254. https://doi.org/10.3390/ijms262311254.
2. Świderski, G.; Kalinowska, M.; Zapora, E.; Wołkowycki, M.; Stocki, M.; Ciszkowicz, E.; Bocian, A.; Jaromin, M.; Tyrka, M.; Lecka-Szlachta, K.; et al. Bioactive Properties of Selected European *Phellinus* Species: A Comprehensive Study. *Int. J. Mol. Sci.* **2025**, *26*, 8013. https://doi.org/10.3390/ijms26168013.
3. Valdivia-Padilla, A.V.; Sharma, A.; Zegbe, J.A.; Morales-Domínguez, J.F. Metabolomic Characterization and Bioinformatic Studies of Bioactive Compounds in Two Varieties of *Psidium guajava* L. Leaf by GC–MS Analysis. *Int. J. Mol. Sci.* **2025**, *26*, 2530. https://doi.org/10.3390/ijms26062530.
4. Wojtunik-Kulesza, K.; Rudkowska, M.; Klimek, K.; Agacka-Mołdoch, M.; Mołdoch, J.; Michalak, A. Expanding Knowledge about the Influence of Citral on Cognitive Functions—In Vitro, In Vivo and Ex Vivo Studies. *Int. J. Mol. Sci.* **2024**, *25*, 6866. https://doi.org/10.3390/ijms25136866.
5. Li, C.; Qi, X.; Xu, L.; Sun, Y.; Chen, Y.; Yao, Y.; Zhao, J. Preventive Effect of the Total Polyphenols from *Nymphaea candida* on Sepsis-Induced Acute Lung Injury in Mice via Gut Microbiota and NLRP3, TLR-4/NF-κB Pathway. *Int. J. Mol. Sci.* **2024**, *25*, 4276. https://doi.org/10.3390/ijms25084276.
6. Gökçen, E.N.; Gezici, S.; Raposa, B.L.; Szép, D.; Budán, F.; Ağagündüz, D. Effects of Common Fig (*Ficus carica* L.) and Its Extracts on Certain Cancer Types: Focusing on the Mechanism of Action. *Int. J. Mol. Sci.* **2026**, *27*, 56. https://doi.org/10.3390/ijms27010056.
7. ElShamey, E.; Yang, J.; Yang, X.; Hasan, M.M.; Yang, T.; Zeng, Y. Genetic Divergence and Functional Significance of Bioactive Compounds in Rice and Barley: Implications for Biofortification and Human Health. *Int. J. Mol. Sci.* **2025**, *26*, 7374. https://doi.org/10.3390/ijms26157374.
8. Bukowska, B. Current and Potential Use of Biologically Active Compounds Derived from *Cannabis sativa* L. in the Treatment of Selected Diseases. *Int. J. Mol. Sci.* **2024**, *25*, 12738. https://doi.org/10.3390/ijms252312738.

References

1. Zielińska-Błajet, M.; Feder-Kubis, J. Monoterpenes and Their Derivatives-Recent Development in Biological and Medical Applications. *Int. J. Mol. Sci.* **2020**, *21*, 7078. [CrossRef] [PubMed] [PubMed Central]
2. Chen, X.; Pan, S.; Li, F.; Xu, X.; Xing, H. Plant-Derived Bioactive Compounds and Potential Health Benefits: Involvement of the Gut Microbiota and Its Metabolic Activity. *Biomolecules* **2022**, *12*, 1871. [CrossRef] [PubMed] [PubMed Central]

International Journal of
Molecular Sciences

MDPI

Article

Characterization of Terpenoids in Aromatic Plants Using Raman Spectroscopy and Gas Chromatography–Mass Spectrometry (GC–MS)

Milagros Granda-Santos [1,2,3], Katherine Reyna-Gonzales [2], Llisela Torrejón-Valqui [2], Marvin G. Valle-Epquín [2], Aline C. Caetano [3], Jorge R. Díaz-Valderrama [3], Efraín M. Castro-Alayo [2], Ilse S. Cayo-Colca [1], Jorge L. Maicelo [4] and César R. Balcázar-Zumaeta [2,*]

1 Programa Doctoral en Ciencias para el Desarrollo Sustentable, Escuela de Posgrado, Universidad Nacional Toribio Rodríguez de Mendoza de Amazonas, Chachapoyas 01001, Peru; milagros.granda@untrm.edu.pe (M.G.-S.); icayo.fizab@untrm.edu.pe (I.S.C.-C.)
2 Instituto de Investigación, Innovación y Desarrollo para el Sector Agrario y Agroindustrial (IIDAA), Facultad de Ingeniería y Ciencias Agrarias, Universidad Nacional Toribio Rodríguez de Mendoza de Amazonas, Chachapoyas 01001, Peru; 7702961920@untrm.edu.pe (K.R.-G.); llisela.torrejon@untrm.edu.pe (L.T.-V.); marvin.valle@untrm.edu.pe (M.G.V.-E.); efrain.castro@untrm.edu.pe (E.M.C.-A.)
3 Instituto de Investigación para el Desarrollo Sustentable de Ceja de Selva, Universidad Nacional Toribio Rodríguez de Mendoza de Amazonas, Calle Universitaria N° 304, Chachapoyas 01001, Peru; aline.caetano@untrm.edu.pe (A.C.C.); jorge.diaz@untrm.edu.pe (J.R.D.-V.)
4 Facultad de Ingeniería Zootecnista, Biotecnología, Agronegocios y Ciencia de Datos, Universidad Nacional Toribio Rodríguez de Mendoza de Amazonas, Chachapoyas 01001, Peru; jmaicelo@untrm.edu.pe
* Correspondence: cesar.balcazar@untrm.edu.pe; Tel.: +51-935987326

Abstract: The study characterized the essential oils of eight aromatic medicinal plants (*Tagetes filifolia, Aloysia citrodora, Cymbopogon citratus, Eucalyptus globulus, Chamaemelum nobile, Piper aduncum, Minthostachys mollis*, and *Rosmarinus officinalis*) using Raman spectroscopy and gas chromatography–mass spectrometry (GC–MS). Raman spectra allowed the identification of bands associated with C–H, C=C, C–O, and C=O bonds, evidencing the presence of monoterpenes, sesquiterpenes, and oxygenated compounds. GC–MS analysis confirmed these results, detecting 224 compounds, predominantly terpenoids. Among the major compounds, cis,cis-nepetalactone (30.16%), β-caryophyllene (up to 18.26%), citronellol (10.92%), citral, and linalool were found. The combination of both techniques made it possible to relate the chemical composition to the molecular structure. This showed that the differences between species are mainly due to the proportion of oxygenated compounds (citral, linalool, geraniol) compared with aromatic hydrocarbons (β-caryophyllene, D-limonene, β-pinene). Additionally, the presence of cis,cis-nepetalactone in M. mollis was reported for the first time, representing a significant chemical finding.

Keywords: bioactivity; essentials oils; herbal medicines; linalool; terpenoids

1. Introduction

Currently, the food, pharmaceutical, and cosmetic industries show a growing interest in the use of natural compounds, driven by the demand for safer, more sustainable products with proven bioactive properties [1,2]. In this context, aromatic and medicinal plants represent a direct source of essential oils—volatile mixtures mainly composed of terpenoids, phenylpropanoids, aldehydes, and ketones—whose composition determines their biological activity [3,4]. These oils, obtained from leaves and stems through steam distillation, exhibit antimicrobial, antioxidant, anti-inflammatory, and insecticidal activities derived from the combined action of their secondary metabolites [5,6].

Several aromatic species cultivated between 2500 and 3500 m above sea level (a.s.l.), such as *Tagetes filifolia* (anise), *Aloysia citrodora* (lemon verbena), *Cymbopogon citratus* (lemongrass), *Eucalyptus globulus* (eucalyptus), *Piper aduncum* (matico), *Minthostachys mollis* (poleo), *Chamaemelum nobile* (chamomile), and *Rosmarinus officinalis* (rosemary), contain essential oils with specific chemical compositions dominated by monoterpenes, sesquiterpenes, and oxygenated compounds associated with their biological effects [7–9]. In addition, these species are traditionally used in folk medicine and gastronomy for their digestive, sedative, and anti-inflammatory properties, as well as for their ability to modify the aroma and flavor of foods [10–15].

The chemical characterization of these oils is essential to understand the relationship between their chemical composition and bioactive properties. A detailed analysis of the metabolites present in different species allows the identification of variations in chemical profiles that reflect genetic, environmental, and methodological differences, providing a framework for studying chemical diversity and its biological effects [16,17]. For chemical characterization, several complementary techniques have been employed, as recently reported in studies on essential oils, including confocal Raman spectroscopy [18], Ultra-high-performance liquid chromatography (UHPLC) [19], gas chromatography–mass spectrometry (GC–MS) [20], Fourier-transform infrared spectroscopy (FTIR) [21] and nuclear magnetic resonance (NMR) [22]. Confocal Raman spectroscopy is a rapid, non-destructive technique that provides direct spectral information relative to aromatic rings, enabling the identification of characteristic compounds such as terpenes, phenols, and others [23]. Likewise, GC–MS is used to identify and quantify volatile compounds, allowing the verification of results obtained through Raman spectroscopy [24–27]. The integration of both techniques therefore provides a more comprehensive view of the chemical composition, as Raman spectroscopy offers structural information that can be confirmed by GC–MS, which reveals the identity and relative abundance of the components [28,29]. This dual analytical approach enhances the robustness of chemical characterization by integrating information on compound identity, abundance, and structural features within essential oils. Moreover, it enables the establishment of correlations between chemical composition and the bioactive properties of the food matrix [29,30].

In this context, the present study aimed to characterize the chemical and structural composition of essential oils obtained from eight aromatic medicinal plants using Raman spectroscopy and GC–MS analyses. These techniques were complemented by multivariate analyses to group the predominant terpenoids according to the compounds shared among species, which are previously known for their medical, cosmetic, and industrial relevance.

2. Results and Discussion

2.1. Analysis of Raman Spectral Data

Raman spectroscopic analysis allowed the identification of the main vibrational bands associated with the predominant compounds present in the EO of *T. filifolia*, *A. citrodora*, *E. globulus*, *C. citratus*, *C. nobile*, *P. aduncum*, *M. mollis*, and *R. officinalis* (Figures 1 and S2). The observed signals, reported here for the first time, reflected vibrational modes characteristic of C–H, C=C, C–O, and C=O bonds, which are linked to aromatic, olefinic, and functional structures typical of terpenoids, phenylpropanoids, and oxygenated compounds [5,31,32].

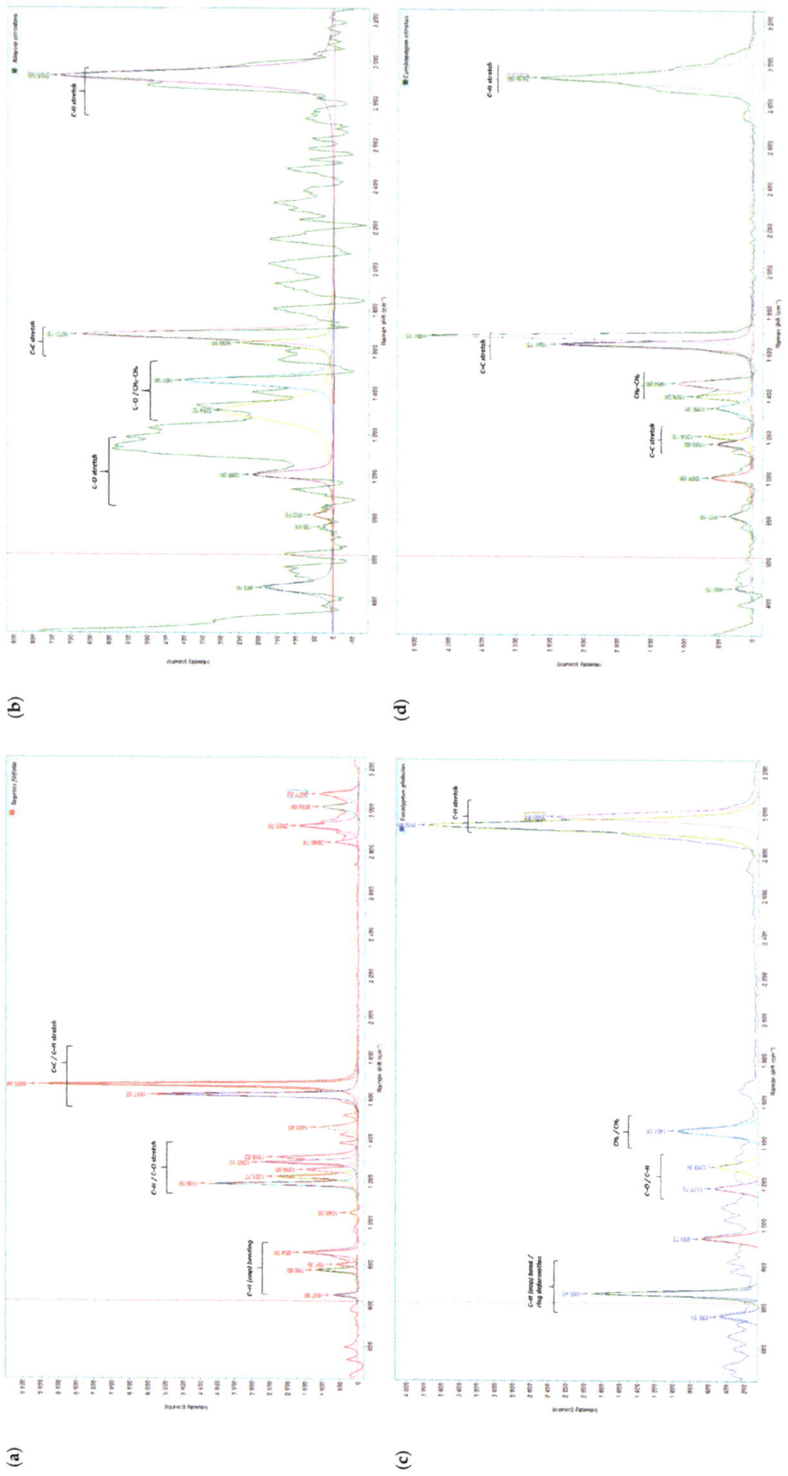

Figure 1. *Cont.*

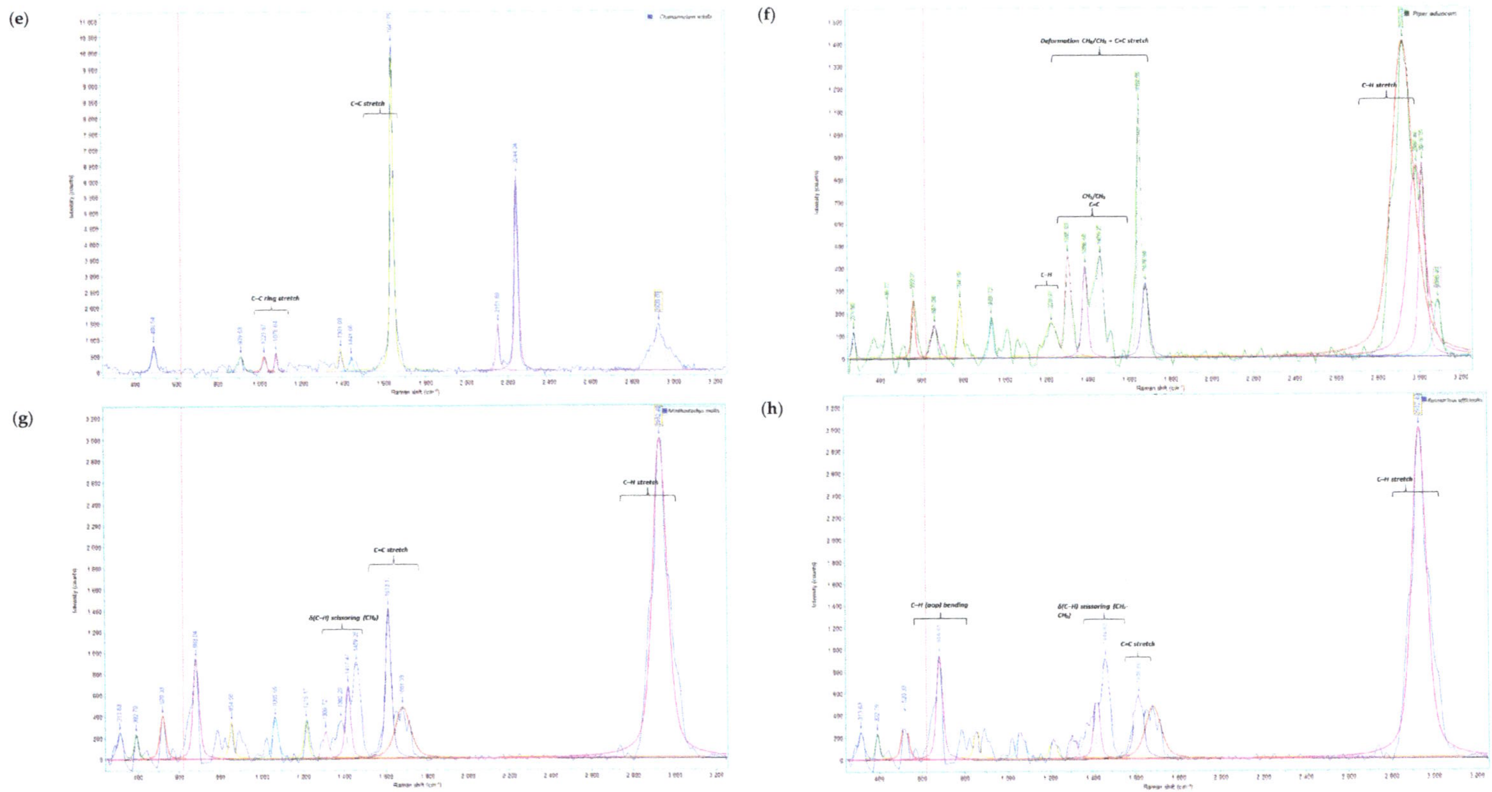

Figure 1. Representative Raman spectra of (**a**) *Tagetes filifolia*, (**b**) *Aloysia citrodora*, (**c**) *Eucalyptus globulus*, (**d**) *Cymbopogon citratus*, (**e**) *Chamaemelum nobile*, (**f**) *Piper aduncum*, (**g**) *Minthostachys mollis*, and (**h**) *Rosmarinus officinalis* in the spectral region of 200–3100 cm^{-1}. The average spectra for each essential oil are presented in Figure S2 of the Supplementary Material.

Raman bands identified in each essential oil are reported, aiming to associate the observed peaks with the functional groups and predominant metabolites of each species. In the essential oil of *T. filifolia*, vibrational modes were identified between 641 and 1656 cm^{-1}. Signals in the 647–854 cm^{-1} range corresponded to out-of-plane C–H bending in cyclic rings, primarily associated with monoterpenes [24,33]. Vibrations between 1186 and 1318.5 cm^{-1} were assigned to C–H and C–O stretching, typical of esters, ketones, and aromatic phenols, and are linked to the presence of oxygenated phenylpropanoids [34]. Finally, bands in the 1609–1670 cm^{-1} range corresponded to C=C and C–H stretching, reflecting oxygenated aromatic compounds [21,27,35].

The essential oil of *A. citrodora* exhibited bands between 242 and 2916 cm^{-1}. Signals below 750 cm^{-1} were associated with carbon skeleton deformations and out-of-plane C–H vibrations. C–O vibrations, typical of acyclic terpenic alcohols, were observed between 809 and 1154 cm^{-1}, confirming the presence of oxygenated compounds in the oil [36,37]. Between 1310 and 1460 cm^{-1}, C–O and CH_2/CH_3 deformations were observed, while bands at 1638–1680 cm^{-1} corresponded to C=C stretching of unsaturated compounds [34]. The high-intensity band at 2920–1930 cm^{-1} reflected C–H stretching vibrations associated with oxygenated monoterpenes, particularly alcohols or phenolic derivatives [36,37].

The essential oil of *E. globulus* showed peaks in the range of 550 to 2971 cm^{-1}. Bands between 550 and 815 cm^{-1} were attributed to ring bending and out-of-plane C–H deformations characteristic of monocyclic terpenoids. Between 1170 and 1318 cm^{-1}, C–H deformations of aromatic compounds were identified, whereas the band at 1457.05 cm^{-1} represented CH_2/CH_3 bending associated with aliphatic groups present in hydrocarbons and terpenoids. Finally, the signal at 2936.8 cm^{-1} corresponded to C-H stretching of unsaturated compounds such as alkenes and aromatic structures [23,33,38].

In the essential oil of *C. citratus*, bands between 1100 and 1340 cm^{-1} were related to symmetric CH_3 deformations and C–C stretching (terpenic compounds), while bands at 1632–1681 cm^{-1} corresponded to C=C stretching of unsaturated aldehydes (citral isomers) [39,40]. The signal at 2928 cm^{-1} reflected C–H stretching, indicating the presence of conjugated aldehydes typical of monoterpenes or oxygenated phenylpropanoids, contributing to the citrus aroma and antioxidant capacity [41,42]. These signals are consistent with profiles reported for *Cymbopogon* oils with a citral-dominant profile [43].

In the essential oil of *C. nobile*, the band at 477.3 cm^{-1} was assigned to carbon skeleton bending modes, characteristic of cyclic and terpenic structures. Signals at 1021.7 and 1078.84 cm^{-1} corresponded to C–C stretching of the ring, indicating the presence of aromatic and cyclic rings typical of oxygenated monoterpenes and sesquiterpenes [44]. The C=C stretching at 1641.75 cm^{-1} indicated the presence of conjugated double bonds typical of aromatic sesquiterpenes. Finally, the peak at 2928.09 cm^{-1} corresponded to C-H stretching, consistent with oxygenated terpenoids [45].

The essential oil of *P. aduncum* exhibited bands between 262 and 770 cm^{-1}, corresponding to carbon skeleton bending, and between 1294 and 1460 cm^{-1}, associated with olefinic CH_3 deformations. Between 1500 and 1600 cm^{-1}, bending vibrations of CH_2 and CH_3 groups ($\delta CH_2/CH_3$) and C=C stretching were observed, corresponding to cyclic and unsaturated monoterpenes as well as oxygenated phenylpropanoids [45]. High-frequency bands at 1650.6, 2923.7, and 2987.4 cm^{-1} were attributed to C=C and aliphatic C–H stretching vibrations, which are characteristic of monoterpenes and sesquiterpenes. These signals, indicate the presence of oxygenated and unsaturated compounds typically found in this essential oil [46].

The essential oil of *M. mollis* was characterized by bands ranging from 303 to 2921 cm^{-1}. Signals between 303 and 680 cm^{-1} were associated with ring deformations characteristic of α-pinene and limonene, while bands at 860 and 1065.65 cm^{-1} corresponded to out-of-plane

C–H deformations and ν(C–C) + δ(C–H) stretching [47]. The band at 1309.72 cm^{-1} was assigned to C–O stretching, typical of alcohols, and the band at 1459.25 cm^{-1} was attributed to CH_2 scissoring δ(C–H) [5]. Finally, signals at 2921 and 2940 cm^{-1} were assigned to aliphatic C–H stretching of cyclic monoterpenes, consistent with Raman profiles reported for essential oils of the genus *Mentha* [48].

The essential oil of *R. officinalis* exhibited peaks between 600 and 800 cm^{-1}, attributed to ring deformations and out-of-plane C–H modes characteristic of cyclic monoterpenes such as α- and β-pinene, as well as 1,8-cineole [49,50]. Additionally, the signal at 1294.6 cm^{-1} was related to =CH wagging typical of olefinic bonds, while the band at 1378.9 cm^{-1} corresponded to CH_2/CH_3 bending associated with aliphatic chains of monoterpenes. The band at 1444.6 cm^{-1} was assigned to δ(C–H), also linked to the aliphatic skeleton. The band at 1638.7 cm^{-1} was attributed to C=C stretching, and the band at 2932.5 cm^{-1} to C–H stretching (double bond), both indicative of unsaturated structures characteristic of monoterpenes and sesquiterpenes present in the essential oil [50–52].

2.2. Multivariate Analysis of Raman Spectra

Principal component analysis (PCA) applied to the Raman spectra allowed a clear differentiation of the analyzed essential oils, showing a distinct separation between species (Figure 2a). The first two principal components, PC1 and PC2, collectively explained 75.73% of the total variance (40.25% and 35.48%, respectively), demonstrating the model's adequate capacity to represent chemical differences between species.

The essential oils of *E. globulus* and *R. officinalis* formed closely related clusters, reflecting their structural similarity associated due to presence of oxygenated monoterpenes [23,50]. Similarly, *M. mollis* and *P. aduncum*, also clustered together, which is consistent with our results, as both share oxygenated functional groups (alcohols, ketones, and esters) and have a composition rich in monoterpenes and sesquiterpenes [53]. Another cluster, composed of *C. citratus* and *A. citrodora*, was identified, suggesting a high structural similarity between these oils, although distinct from the previously mentioned groups due to their greater separation in the projection plane. Finally, although *T. filifolia* was positioned close to *M. mollis* due to similarity in ester and alcohol content, the *M. mollis* oil exhibited a more dispersed distribution, likely associated with its complex composition of phenylpropanoids, unsaturated hydrocarbons, and terpenic rings [1,54], resulting in a more heterogeneous spectrum.

(a)
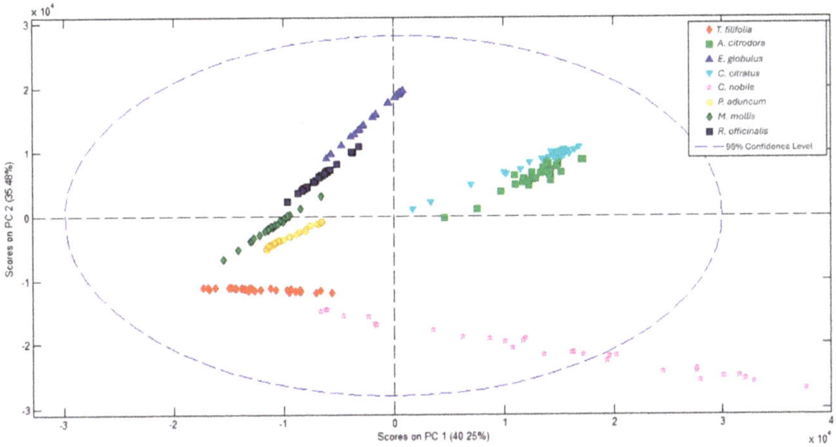

Figure 2. *Cont.*

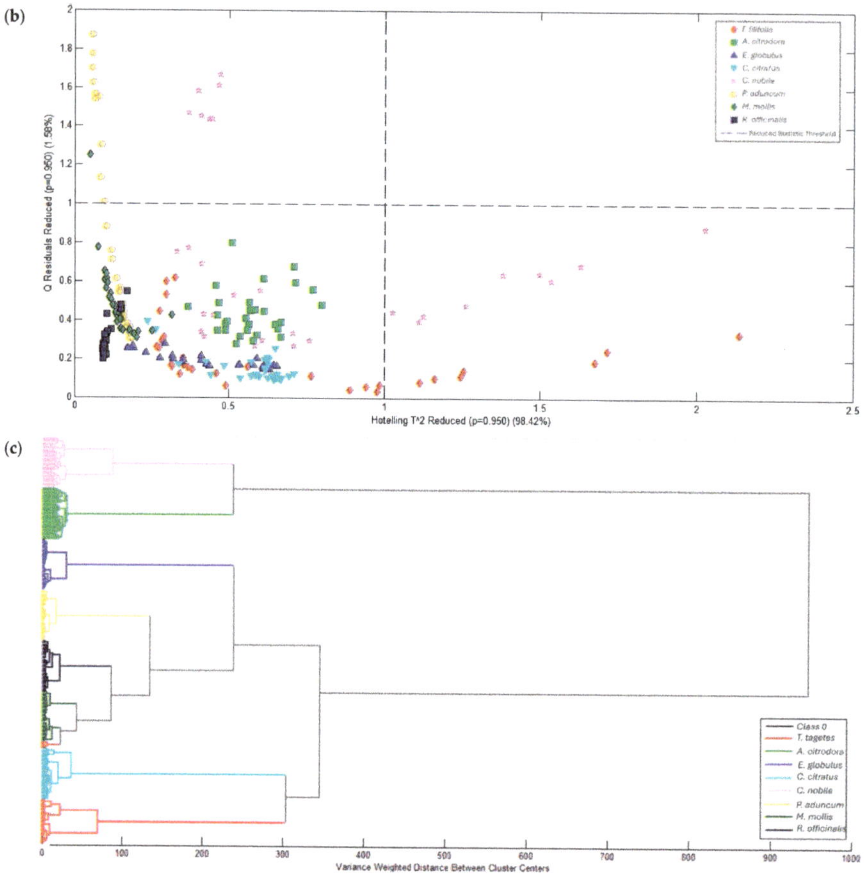

Figure 2. Principal Component Analysis (PCA) and Hierarchical Classification of Plant Samples: (**a**) PCA scores, (**b**) Q vs Hotelling's T^2 in the PCA model, and (**c**) dendrogram.

The Hotelling T^2 versus residual Q plot (Figure 2b) allowed verification of the PCA model validity (95% confidence limit). The observed distribution shows that most species are located within the optimal fit region (lower left quadrant), indicating that the model adequately describes the variability of their chemical characteristics. The dispersion of *C. nobile* and *T. filifolia* in the lower right quadrant highlights chemical compositions that distinguish them from the rest. The absence of outliers in the critical region confirms that there are no observations with high influence (elevated T^2) and high fitting error (high residual Q), supporting the stability and robustness of the PCA model.

Hierarchical cluster analysis (Figure 2c) complemented the PCA results by showing the similarity relationships between species through hierarchical clustering based on weighted variance distance. The dendrogram revealed the formation of two main groups. The first group consisted of *A. citrodora* and *C. nobile*, characterized by a predominance of oxygenated monoterpenes [55]. The second group included *E. globulus*, *P. aduncum*, *R. officinalis*, *M. mollis*, *T. filifolia*, and *C. citratus*, whose Raman signals were dominated by aromatic compounds and aliphatic C–H stretching [18,28,56].

This hierarchical organization was consistent with the chemical structure of the oils, as it clustered species with a high proportion of oxygenated monoterpenes together, while separating those dominated by aromatic or aliphatic compounds. Overall, the agreement between the results obtained from PCA, T^2–Q, and HCA confirms that Raman spectroscopy, combined with multivariate analysis, constitutes an effective tool for discriminating essential oils based on their molecular composition and botanical origin.

2.3. Gas Chromatography–Mass Spectrometry (GC–MS) Results Analysis

After characterizing the essential oils (EO) using Raman spectroscopy, which allowed the identification of their main functional groups and vibrational profiles, chemical analysis was conducted by gas chromatography coupled with mass spectrometry (GC–MS). After characterizing the essential oils (EO) using Raman spectroscopy, which allowed the identification of their main functional groups and vibrational profiles, chemical analysis was conducted by gas chromatography coupled with mass spectrometry (GC–MS). This technique enables the precise identification, and a relative quantification of volatile compounds present in the EO, complementing the information obtained from Raman spectroscopy by providing data on the specific molecular composition of chemical families.

It is important to note that the GC-MS data correspond to semi-quantitative relative peak areas (%), obtained after signal normalization. Therefore, the reported values represent the relative abundance of each compound in the oil and should not be interpreted as absolute concentrations or gravimetric percentages. This method allows for a reliable comparison of the relative chemical composition between essential oils, although it does not consider the response factors of each detector.

GC–MS analysis identified a total of 224 compounds (Table S1) belonging to nine chemical families: alcohols, aldehydes, esters, hydrocarbons, ketones, phenols, phenylpropanoids, terpenoids, and a residual group classified as "others," in the EO of *T. filifolia*, *A. citrodora*, *E. globulus*, *C. citratus*, *P. aduncum*, *M. mollis*, and *R. officinalis*. In most essential oils, terpenoids constituted the predominant fraction, representing over 70% of the total relative peak area (Figure 3), followed by lower proportions of alcohols, hydrocarbons, ketones, and esters. This pattern aligns with the findings of Fenghour & Bouabida [25], who report terpenoids as the main constituents of EO, with abundant compounds such as α-curcumene (7.84–16.91%) in *R. officinalis* [10], β-cyclocitral (21.46% relative peak area), germacrene D (11.7–62.9% relative peak area), and β-caryophyllene (17.34% relative peak area) in *P. aduncum* [57], as well as α-terpinyl acetate and citral (17% relative peak area) in *A. citrodora* [58].

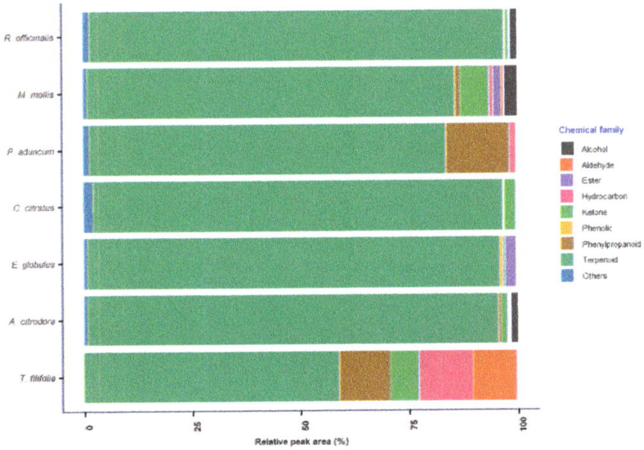

Figure 3. Distribution of chemical families identified in essential oils by GC–MS analysis.

The essential oil of *R. officinalis* exhibited notable relative peak areas of D-limonene (8.59%), linalool (2.91%), α-terpinene (2.51%), and β-caryophyllene (18.26%) (Table 1). These values are higher than those reported by Alvarado-García et al. [53], who identified 10.76% relative peak area of β-caryophyllene in *R. officinalis* samples through GC–MS analysis. In contrast, Malik & Upadhyay [59] recorded only 1.54% relative peak area of this compound in EO from *R. officinalis* originating from India. Similarly, the essential oil of *P. aduncum* presented high semiquantitative contributions based on GC–MS signal intensity of β-caryophyllene (11%), D-limonene (4.59%), and β-pinene (3.37%), slightly higher than those reported by Jedidi et al. [60] in samples collected in northwestern Tunisia, where a β-caryophyllene relative peak area of 9.49% was reported.

Table 1. Areas of relative peaks (%) of the main terpenoids identified in the essential oils of seven medicinal species.

Group	Compound	*R. officinalis*	*M. mollis*	*P. aduncum*	*C. citratus*	*E. globulus*	*A. citrodora*	*T. filifolia*
Terpenoid	cis,cis-Nepetalactone	0.00	30.16	0.00	0.00	0.00	0.00	0.00
	α-terpinene	2.51	0.00	1.35	0.00	0.13	0.00	0.00
	Citral	0.00	0.00	0.00	0.00	0.40	0.00	2.35
	Myrcene	0.00	0.39	0.00	0.00	8.49	0.00	0.58
	Sabinene	0.49	0.86	0.00	0.00	0.00	0.00	0.00
	β-caryophyllene	18.26	0.00	11.00	0.22	0.00	6.00	0.00
	Citronellol	0.00	0.00	0.00	0.00	0.00	10.92	0.00
	Linalool	2.91	9.63	0.33	2.33	0.90	0.59	1.61
	Terpinen-4-ol	0.00	0.00	2.03	0.00	3.54	0.15	0.00
	Thymol	0.00	2.04	0.00	0.00	0.00	0.00	0.00
	D-limonene	8.59	6.87	4.59	0.28	0.00	6.31	2.81
	Eucalyptol	0.00	0.29	0.00	0.00	0.00	1.42	0.35
	Geraniol	0.00	0.00	0.00	6.06	3.13	3.17	0.39
	β-pinene	0.00	0.00	3.37	0.00	8.21	0.00	0.91

The table shows the relative peak areas (%) of the main terpenoids detected in essential oils by GC–MS, derived from signal normalization. The values represent semi-quantitative data.

The observed variations in β-caryophyllene relative peak area can be attributed to both environmental factors and post-harvest and processing conditions. On one hand, parameters such as temperature, humidity, altitude, and soil composition in the cultivation areas directly influence the synthesis and accumulation of terpenoids [10]. On the other hand, technological aspects related to drying and extraction methods also modify the chemical profile of the oil. In this regard, Mohammed et al. [61] demonstrated that prolonged drying of *R. officinalis* reduces the semiquantitative contribution of volatile compounds, mainly monoterpenes such as D-limonene, linalool, and α-terpinene, while also altering the proportion of sesquiterpenes such as β-caryophyllene. Similarly, Zhang et al. [62] showed that the extraction method (hydrodistillation, steam distillation, or solvent extraction) significantly modifies the EO relative peak areas, whereas Zheljazkov et al. [63] reported that both the distillation time and the sample condition (dry or fresh) affect the yield and bioactivity of *R. officinalis* essential oil.

Additionally, the part of the plant used for essential oil extraction is also a determining factor. In *R. officinalis*, it has been observed that young leaves concentrate higher levels of monoterpenes, whereas structural tissues such as stems and bark present lower proportions of these metabolites [64]. Similarly, in species of the genus *Origanum*, EO extracted from leaves show higher concentrations of volatile aromatic compounds, such as carvacrol

(30.73%), thymol (18.81%), and β-caryophyllene (8.21%), compared to those obtained from stems, where the concentrations were lower (carvacrol 6.02%, thymol 3.46%, and β-caryophyllene, which was not detected) [65].

Likewise, in *P. aduncum*, the plant part also influences the semiquantitative profile of the essential oil. For instance, Valadares et al. [66] compared essential oils obtained from leaves and flowers, finding differences in the relative peak areas of β-caryophyllene (7.2% in flowers and 5.4% in leaves) and other minor compounds such as terpinen-4-ol and myrcene. Moreover, antifungal activity varied depending on the plant organ, as the oil obtained from flowers completely inhibited the growth of *Sclerotinia sclerotiorum* at doses above 30 μL, whereas the leaf essential oil required 50 μL to achieve 98.74% inhibition. These findings confirm that the distribution of monoterpenes and sesquiterpenes is closely associated with the structural characteristics of different plant organs, highlighting the importance of considering the plant part used in EO characterization.

The essential oil of *M. mollis* exhibited a unique composition, characterized by the dominant presence of cis,cis-nepetalactone at concentrations above 20%, a compound absent in the other evaluated EO in this study. The absence of β-caryophyllene, along with the predominance of this lactone, highlights the metabolic uniqueness of *M. mollis* oil compared to other members of the Lamiaceae family. This finding represents a significant chemical discovery, as this study constitutes the first report of cis,cis-nepetalactone in this species; previous investigations have not documented the presence of this lactone in *M. mollis* [1,54,67,68].The detection of this lactone in *M. mollis* essential oil differentiates the chemical composition of the plant, which is relevant from both a chemical and functional perspective. This finding suggests the existence of a population with distinct chemical characteristics, whose profile could be less toxic and potentially more bioactive than other *M. mollis* varieties dominated by pulegone [69,70].

Alongside nepetalactone, linalool (9.63%) and D-limonene (6.87%) were identified as relevant secondary components, with relative peak areas higher than those reported by Cravero Ponso et al. [1] for the same species (<5%). These oxygenated monoterpenes are known for their antibacterial, anti-inflammatory, anxiolytic, and contraceptive properties, reinforcing the bioactive potential of the essential oil. The differences in EO relative peak areas compared to previous studies are mainly attributed to natural variations in plant chemistry and the environmental conditions in which the plants grew, such as climate, soil type, and altitude. For example, van Baren et al. [54] reported marked variations in linalool relative peak areas (6–84%) among Argentine populations of *M. mollis*, while Quispe-Sanchez et al. [17] documented higher proportions of D-limonene (14.57%) and linalool (15.30%) in samples collected in the same Andean region as this study (Peru), suggesting that even under similar environmental conditions, significant differences can arise due to genetic, phenological, or methodological factors, including the collection period, which is associated with the plant's developmental stage and seasonal environmental conditions.

Comparatively, the composition observed in this study differs from the classical profiles of *M. mollis*, where pulegone, menthone, and 1,8-cineole are the characteristic components [71–74]. The analysis of the essential oil of *C. citratus* allowed the identification of linalool (2.33%) and geraniol (6.06%) relative peak areas. Although the overall profile of the species is characterized by the predominance of citrals (a mixture of geranial and neral), the detection of these terpenoids at proportions above 1% is relevant due to their contribution to both the aroma and the bioactive properties of the oil essential. For instance, studies indicate that the presence of geraniol in *C. citratus* can enhance the antibacterial activity of the essential oil [1,15,54]. Monoterpenic alcohols have been shown to disrupt bacterial cell membrane integrity, promoting ion leakage and disorganization of enzymatic systems, thereby increasing efficacy against Gram-positive bacteria (*Staphylococcus aureus* and *Bacillus subtilis*) and Gram-negative bacteria (*Escherichia coli, Pasteurella multocida*) [28,75,76]. The absence or low relative peak areas of other chemical compounds (β-caryophyllene, myrcene, citral) in the essential oil may favor the stability of active components such as geraniol and linalool, preserving their antibacterial activity and preventing potential interactions that could reduce efficacy [77].

The essential oil of white *E. globulus* exhibited a chemical profile dominated by myrcene (8.49%), β-pinene (8.21%), and terpinen-4-ol (3.54%), relative peak areas higher than those reported in other species of the genus. For instance, Belhachemi et al. [78] reported the presence of these terpenoids in red *E. globulus* essential oil, albeit at lower relative peak areas (<2%). Similarly, dos Santos et al. [79] reported low β-pinene levels in lemon-scented *E. globulus* (1.98%), values much lower than those observed in the present study. Additionally, Ait benlabchir et al. [23] detected myrcene (1.51%) and terpinen-4-ol (2.43% in directly distilled oil and up to 15.24% in purified fractions) in white *E. globulus* essential oils, demonstrating that although these compounds are usually present at low to moderate relative peak areas, their values can vary widely depending on the fraction analyzed or the processing method applied.

In Mediterranean regions, a comparative analysis of *E. globulus* essential oil showed that myrcene and β-pinene rarely exceed 2–3%, while terpinen-4-ol remains below 5% [26]. Likewise, studies on commercial *E. globulus* oils indicate that these monoterpenes are secondary components: myrcene (<2%), β-pinene (1–2%), and terpinen-4-ol (1%) [80]. In contrast, the relative peak areas of these compounds in this study are significantly higher.

The high proportion of β-pinene and myrcene observed in the study could be explained by the presence of specific chemotypes of *E. globulus*, such as those characterized by the combinations γ-terpinene + o-cymene + β-pinene or α-pinene + 1,8-cineole. These chemotypes correspond to populations that produce essential oils with differentiated chemical profiles, resulting from genetic or environmental variations specific to each group [81]. Additionally, it is possible that the metabolic pathways responsible for monoterpene formation favored the production of β-pinene and myrcene in this population over other compounds. In this regard, it has been demonstrated in similar species, such as tea tree (*Melaleuca alternifolia*), that the activity of genes associated with the methylerythritol phosphate (MEP) pathway, which generates the precursors necessary for monoterpene synthesis, directly influences both the quantity and type of monoterpenes produced, supporting this possible explanation [82].

The essential oils of *A. citrodora* and *T. filifolia*, both recognized for their high content of biologically active monoterpenes, exhibited differentiated chemical compositions. In the essential oil of *A. citrodora*, the predominant compounds were citronellol (10.92%), β-caryophyllene (6.0%), and D-limonene (6.31%), whereas in *T. filifolia*, citral (2.35%), linalool (1.61%), and D-limonene (2.81%) were the main components. In the case of *A. citrodora*, β-caryophyllene and D-limonene were present at similar relative peak areas; however, D-limonene showed notable differences between the two oils, with a higher proportion in *A. citrodora* compared to *T. filifolia*.

In contrast with our findings, Al-Maharik et al. [10] reported for the essential oil of *A. citrodora* the presence of β-caryophyllene (0.62%), citral (10.79%), limonene (0.21%), and D-limonene (19.03%), along with other minor compounds such as 1-terpineol (0.07%), 4,4-dimethyl-2-butenolide (0.17%), α-thujone (0.16%), and ethyl 3-(2-furyl)propanoate (0.09%). Although the qualitative composition (the same compounds) partially coincides with that of our study, the quantitative differences in relative peak areas could be explained by the effects of cultivation factors such as temperature and light intensity, which modulate the expression of genes encoding terpene synthases enzymes that determine both the biosynthetic pathway and the proportion in which precursors are converted into different monoterpenes [83,84].

For instance, it has been demonstrated that the expression of monoterpene biosynthetic genes can vary depending on the time of day, increasing around midday when light and temperature are higher, coinciding with peaks in the production of terpenoids such as D-limonene or linalool in lavender [23]. Similarly, studies in other species, including alpine plants [85] and sandalwood [86], have observed that abrupt changes in temperature or radiation levels affect volatile terpene emission, as lighter or less-oxygenated compounds can be lost or rearranged in response to thermal stress. Therefore, it is likely that in the analyzed population of *A. citrodora*, local conditions—such as high daytime temperatures or intense irradiance—favored the biosynthetic preference toward monoterpenes like D-limonene or citral, resulting in higher relative peak areas than those reported by Al-Maharik et al. [10].

Similarly, Rashid et al. [30] reported in the same species levels of β-caryophyllene (5.09%), D-limonene (18.80%), and linalool (0.38%). In this case, the relative peak area of β-caryophyllene was comparable to that observed in our study, suggesting that this compound exhibits low variation throughout the plant's vegetative cycle; however, this assumption remains to be verified. In contrast, the variations observed in D-limonene and linalool could be explained by differences in the availability of precursor compounds in the plant and by changes induced by temperature and solar radiation. Recent studies have demonstrated that environmental and physiological conditions directly modify the proportion of these compounds [87]. For example, in *Zanthoxylum bungeanum*, D-limonene content increases under higher temperature and solar radiation, whereas linalool decreases under the same conditions, indicating an inverse relationship between these two compounds [88,89].

Moreover, Dumitrescu et al. [90] analyzed the essential oil of *T. filifolia*, identifying linalool (1.90%) and D-limonene (10.01%) as the main components. Compared to our results, the relative peak area of linalool was similar, whereas D-limonene was considerably higher. Beyond extraction methodological differences, this variation can be attributed to several physiological factors, such as leaf characteristics and soil properties, which directly influence the relative peak area of the essential oil. In particular, the physiological maturity and position of leaves on the stem at harvest significantly affect the proportion of monoterpenes. In plants such as geranium, younger leaves exhibit higher biosynthetic activity and accumulate greater levels of alcohols like linalool, whereas older leaves tend to present a more hydrocarbon-rich profile, with predominance of limonene [91]. Additionally, soil properties, especially pH and mineral composition, modify the availability of essential nutrients and the efficiency of enzymes involved in monoterpene synthesis [92].

Unlike the other essential oils, the essential oil of *C. nobile* could not be characterized by GC–MS due to its low extraction yield and the high viscosity of the resulting concentrate (Table S2), which exhibited a pasty and dense consistency, that prevented adequate volatilization during chromatographic analysis. This behavior can be attributed to the predominance of higher molecular weight and less volatile compounds, such as oxygenated sesquiterpenes, high-molecular-weight esters, or traces of natural waxes [93,94], which hinder their detection and separation by gas chromatography.

2.4. Multivariate Analysis of GC-MS Data

Multivariate analysis using Principal Component Analysis (PCA) and Hierarchical Cluster Analysis (HCA) allowed the identification of chemical relationships among the essential oils evaluated and the determination of the functional groups responsible for their differentiation. Overall, the first two dimensions of the PCA (Figure 4) explained 72.7% of the total variance (PC1 = 53.1% and PC2 = 19.6%), demonstrating an adequate representation of the chemical variability of the samples.

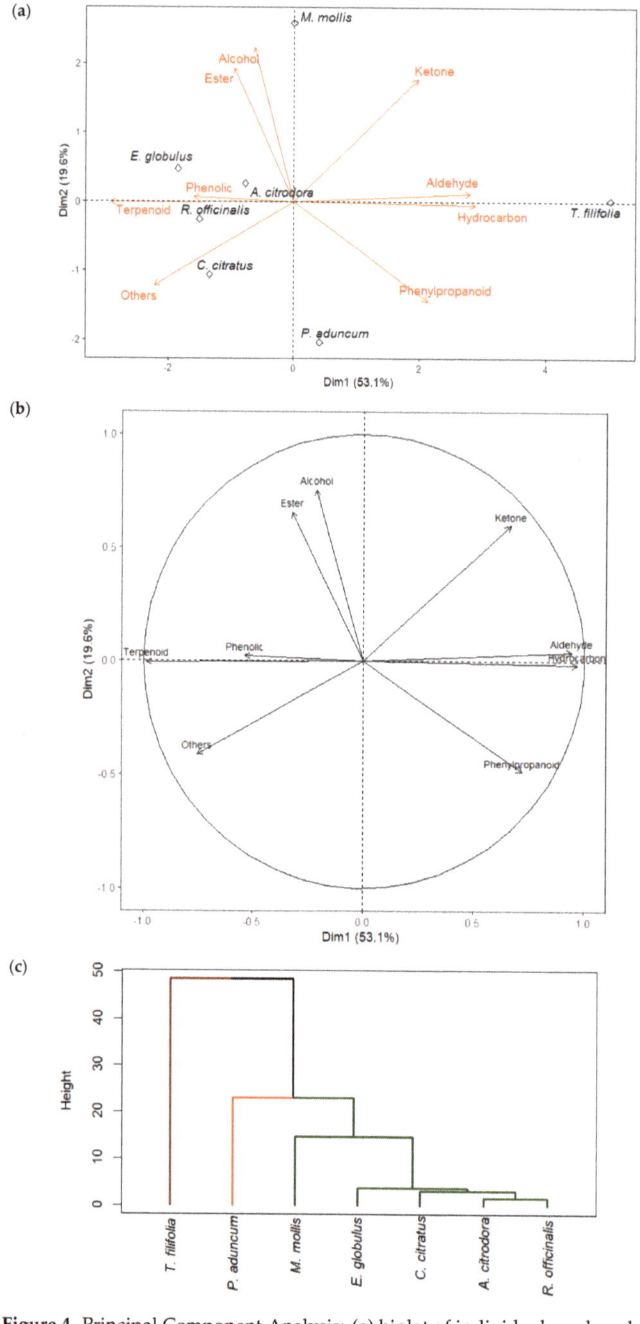

Figure 4. Principal Component Analysis: (**a**) biplot of individuals and explanatory variables, and (**b**) correlation circle, and Hierarchical Analysis: (**c**) dendrogram.

The essential oil of *T. filifolia* was projected at the far right, close to the vectors of aldehydes and hydrocarbons, indicating a composition rich in these compounds and a clear differentiation from the other oils. In contrast, *M. mollis* was located at the upper part of the plot, associated with the vectors of alcohols and esters, suggesting a composition dominated by more polar oxygenated metabolites [95,96]. The oils of *E. globulus* and *R. officinalis* clustered in the left region, linked to terpenoids and phenolic compounds, consistent with their characteristic content of oxygenated monoterpenes, in agreement with Čmiková et al. [97], who highlight 1,8-cineole as a main component. *P. aduncum* was projected in the lower right quadrant, defined by phenylpropanoids, while *C. citratus* was positioned toward the lower left quadrant, in the area associated with the "others" group, suggesting a distinctive chemical profile with minor or species-specific compounds.

These groupings observed in the PCA were confirmed by hierarchical cluster analysis (Figure 4c), which revealed three main levels of chemical similarity. The oil of *T. filifolia* exhibited a singular profile dominated by aldehydes and hydrocarbons, based on relative peak areas (%). Secondly, *P. aduncum* formed an independent group, consistent with its predominance of phenylpropanoids [98]. Finally, the cluster formed by *M. mollis*, *E. globulus*, *C. citratus*, *A. citrodora*, and *R. officinalis* showed the highest similarity, reflecting a more homogeneous composition based on terpenoids and alcohols, as reported in this study.

The multivariate analysis demonstrated that the differences in the distribution of essential oils are driven by the chemical nature of their dominant metabolites. Thus, PCA and cluster analysis not only confirmed the consistency among the reported chemical compositions but also facilitated the identification of characteristic patterns that distinguish each species based on its profile of terpenoids, phenylpropanoids, aldehydes, or alcohols.

During the spectroscopic analysis, additional bands within the 200–3100 cm^{-1} range were also identified, associated with families such as aldehydes, esters, alcohols, and organic acids. Complementarily, the GC–MS analysis allowed the detection of compounds from diverse families, including ketones and phenylpropanoids, in addition to the major terpenoids. However, these metabolites were not discussed in detail in the present study because their relative concentrations were considerably lower, and the main focus was on the characterization and interpretation of the major metabolites (primarily terpenoids), as they largely determine the chemical profile, bioactive properties, and variability among species.

It should be noted that the hierarchical classification obtained from Raman spectroscopy (Figure 2c) differs from the GC–MS data (Figure 4c). This is understandable, given that Raman detects vibrational characteristics related to functional groups and molecular structures, while GC–MS provides semi-quantitative information on specific volatile compounds [99]. Therefore, the classification patterns obtained with each technique should be interpreted in a complementary manner, rather than combined directly.

2.5. Limitations and Future Perspectives

The complementary analysis of Raman spectroscopy and gas chromatography–mass spectrometry (GC–MS) enabled the identification of the main metabolites in the EO of *P. anisum*, *A. citrodora*, *E. globulus*, *C. citratus*, *M. chamomilla*, *P. aduncum*, *M. mollis*, and *R. officinalis*. Terpenoids were detected that clearly differentiate each species. In *P. anisum*, citrals (2.35%) and D-limonene (2.81%) were observed. In *A. citrodora*, citronellol (10.92%) and β-caryophyllene (6%) were predominant, whereas *E. globulus* was characterized by myrcene (8.49%) and β-pinene (8.21%). *C. citratus* showed geraniol (6%), and *P. aduncum* was distinguished by β-caryophyllene (11%). *M. mollis* was notable for a high semiquantitative contribution based on GC–MS signal intensity of cis,cis-nepetalactone (30.16%), while *R. officinalis* exhibited a predominance of β-caryophyllene (18.26%).

The high content of these terpenoids suggests the anti-inflammatory, antioxidant, sedative, muscle-relaxant, antimicrobial, and antitumor properties attributed to these plants [78,100]. For example, β-caryophyllene has been described as an insecticidal agent, a bacterial biofilm inhibitor, and a modulator of apoptosis in cancer cells [101,102]. Additionally, da Costa Sobral et al. [103] demonstrated that the combination of β-caryophyllene with drugs such as pregabalin enhances seizure control, whereas Nunes et al. [104] reported that camu camu essential oil, rich in (E)-caryophyllene (14.02–34.31%), exhibits anticancer and anti-inflammatory potential by inhibiting tumor cells and reducing inflammatory processes in vitro.

The discovery of cis,cis-nepetalactone in the essential oil of *M. mollis* represents the first record of this lactone in the species. Its presence indicates the existence of populations with distinct chemical profiles, likely influenced by genetic or environmental factors affecting the production of bioactive compounds [77,105]. This finding opens new research avenues to evaluate how genetics and environmental conditions influence the production of cis,cis-nepetalactone and other bioactive compounds in *M. mollis*. Understanding these factors could allow the identification of previously unstudied varieties with higher semiquantitative contributions based on GC–MS signal intensity of functional metabolites and low toxicity risk. Additionally, these essential oils could have applications in natural cosmetics and aromatherapy, particularly due to their antimicrobial, anti-inflammatory properties and characteristic fresh aroma, further highlighting their functional and commercial potential [67,106].

Given that reports of nepetalactone in Lamiaceae family are scarce, it is recommended to confirm its structural identification through retention index comparison, co-injection with authentic standards, and the use of complementary techniques such as GC-MS/MS or NMR [107–109]. This approach will validate the authenticity of the compound and consolidate its role as a chemical marker distinguishing *M. mollis* in Peru.

Although the techniques employed enabled the characterization of the essential oils, Raman spectroscopy showed limitations in differentiating structural isomers or stereoisomers and could not detect compounds or chemical families present at very low concentrations. Likewise, GC–MS encountered difficulties with dense or low-volatility oils, such as that of *C. nobile*, restricting the complete identification of all constituents. Future studies are therefore recommended to optimize analytical methods for detecting minor or low-volatility compounds not captured by current techniques; to investigate how environmental conditions in different growing areas may influence the biosynthesis of volatile compounds—potentially leading to the formation of distinct ecotypes; and to evaluate the industrial applications of these essential oils in light of their unique chemical composition and bioactive properties.

3. Materials and Methods

3.1. Plant Material and Sampling

Eight species of aromatic medicinal plants were studied, as detailed in Table 2. The analyzed species were anise (*T. filifolia*), lemon verbena (*A. citrodora*), eucalyptus (*E. globulus*), lemongrass (*C. citratus*), matico (*P. aduncum*), poleo (*M. mollis*), and rosemary (*R. officinalis*) (Figure 5 and Figure S1). The taxonomic identification of the species was verified by the Kuelap Herbarium of the Universidad Nacional Toribio Rodríguez de Mendoza (UNTRM). Plant material was collected during the peak flowering stage, obtaining stems, leaves, and flowers depending on the species, between February and March 2024, in the Amazonas region, Peru. The species *T. filifolia*, *E. globulus*, *C. citratus*, *M. mollis*, *C. nobile*, *P. aduncum*, and *R. officinalis* were collected directly from the field, whereas *A. citrodora* was purchased fresh from the central market of Chachapoyas. For each species, 15 kg of plant material were collected, distributed in three replicates of 5 kg each. The collected leaves were transported in containers to the Engineering Laboratory of UNTRM for washing and subsequent air-drying at 21 °C, with an average drying time of two days, prior to distillation.

Table 2. Taxonomic and geographical characterization of aromatic medicinal plants.

Common Name	Scientific Name	Abbreviation	Botanical Family	Growth Habit	Geographical Origin	Latitude–Longitude
Anise	*Tagetes filifolia*	*T.filifolia*	Asteraceae	Herb	Chachapoyas-Levanto	−6.30625, −77.8987
Lemon verbena	*Aloysia citrodora*	*A. citrodora*	Verbenaceae		Chachapoyas-Leymebamba	−6.7023, −77.7966
Eucalyptus	*Eucalyptus globulus*	*E. globulus*	Myrtaceae	Shrub	Chachapoyas-Huancas	−6.17500, −77.7966
Lemongrass	*Cymbopogon citratus*	*C. citratus*	Poaceae	Herb	Utcubamba-Bagua Grande	−5.7329, −78.4279
Matico	*Piper aduncum*	*P. aduncum*	Piperaceae	Shrub	Chachapoyas-Chachapoyas	−6.2528, −77.9016
Chamomile	*chamaemelum nobile*	*C. nobile*	Asteraceae	Herb		−6.2523, −77.8325
Pennyroyal	*Minthostachys mollis*	*M. mollis*	Lamiaceae	Shrub		−6.1795, −77.8702
Rosemary	*Rosmarinus officinalis*	*R. officinalis*	Lamiaceae	Herb	Chachapoyas-Levanto	−6.3121, −77.9066

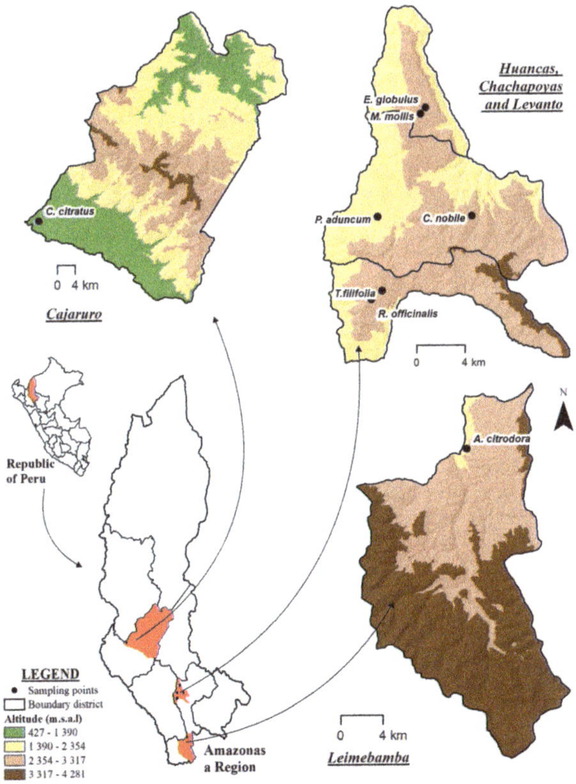

Figure 5. Growth location of aromatic plants.

3.2. Essential Oil Extraction (EO)

The extraction of essential oils (EO) from the eight aromatic medicinal plants was performed following the procedure described by Vasquez-Gomez et al. [110], with minor modifications. Steam distillation was carried out using a TECNAL TE-2761 apparatus (Piracicaba, Brazil) at a temperature of 130 °C. For each species, 15 kg of plant material were processed in three independent replicates, with a distillation time of 4 h per sample. The water–oil mixture obtained after distillation was cooled to 1 °C to promote phase separation (aqueous and oily layers). The essential oil phase was then carefully decanted and collected for each species. The EO were stored in amber glass vials with airtight caps at 4 °C until analysis.

The essential oil yield (Table S1) was calculated according to Barbosa et al. [57] using the following equation

$$Y\% = \frac{V}{W} * 100 \quad (1)$$

where Y is the essential oil yield (%, v/w), V is the volume of essential oil obtained (mL), and W is the weight of the plant material used (g).

3.3. Raman Spectral Data Acquisition

Raman spectra of the essential oils from the eight aromatic medicinal plants were acquired using a Horiba XploRA Plus confocal Raman microscope system (Horiba Scientific, Montpellier, France), equipped with an Olympus BX41 microscope (Olympus, Tokyo, Japan) and a 50× long working distance (LWD) objective. Excitation was performed with a 532 nm laser operating at 40 mW power, using a 1200 lines/mm diffraction grating. The Raman signal was detected with a charge-coupled device (CCD) detector, covering an extended spectral range from 100 to 5000 cm^{-1} with a spectral resolution of 3 cm^{-1} [111].

For each measurement, approximately 20 µL of essential oil was placed in an aluminum crucible. Spectra were recorded with an integration time of 5 s and 10 spectral accumulations, yielding a total of 30 spectra per sample. To minimize sample heterogeneity, three independent replicates were analyzed for each essential oil, resulting in a total of 90 spectra per species. Raman peaks were assigned according to previously published literature.

3.4. Chemical Characterization by Gas Chromatography–Mass Spectrometry (GC–MS)

The chemical composition of the essential oils (EO) was determined using gas chromatography coupled with mass spectrometry (GC–MS) (GC System 7890B, MSD 5977B, Agilent Technologies, Santa Clara, CA, USA) following Adams [112]. The EO were diluted (1 µL of essential oil + 49 µL of hexane), and 0.5 µL of the resulting solution was injected in splitless mode. Chromatographic separation was performed on a DB-5MS UI capillary column (60 m × 0.25 mm × 1.0 µm). Helium was used as the carrier gas at a constant flow rate of 1 mL/min. The injector, detector, transfer line, and ion source were maintained at 220, 150, 240, and 280 °C, respectively. The oven temperature was set at 60 °C, then increased at a rate of 3 °C/min up to 246 °C and held for 15 min. Mass spectra were recorded in scan mode over a mass range of 40–600 m/z. Detected compounds were identified by comparison with the National Institute of Standards and Technology Library (NIST Library 17). Data were deconvoluted using MassHunter Unknowns Analysis software (version 10.1 build 10.1.733.0, Agilent Technologies), considering a similarity index ≥ 80%. The linear retention index (LRI) was calculated to confirm compound identification by injecting a commercial n-alkane standard mixture containing (C8–C20, ~40 mg/L each, in hexane, Supelco, Sigma-Aldrich, St. Louis, MO, USA) under the same chromatographic conditions applied to the essential oil samples.

The data obtained by GC-MS were processed by integrating the areas of the peaks of the total ion chromatogram (TIC) corresponding to each compound. The areas were normalized with respect to the total signal and expressed as relative peak areas (%), which provided semi-quantitative information that was subsequently used in multivariate analyses (PCA and HCA).

3.5. Analysis of Raman and GC–MS Spectral Data

Raman spectral data from the essential oils of the eight aromatic medicinal plants were preprocessed following the protocols described by Castro-Alayo et al. [113]. The preprocessing included data smoothing, baseline correction (200–3100 cm^{-1}), vector normalization, and substrate removal using Solo+MIA 8.1 software (Eigenvector Research, Inc., Wenatchee, WA, USA).

Subsequently, a principal component analysis (PCA) was performed to reduce data dimensionality and visualize clustering patterns among the essential oils based on their chemical composition. Hotelling's T^2 and Q residual statistics were computed at a 95% confidence level to identify and remove anomalous spectra considered as outliers. Additionally, a hierarchical cluster analysis (HCA) was conducted to complement the PCA and confirm the formation of distinct groups among the essential oils. The combination of these multivariate methods enabled the identification of similarity patterns and validation of the obtained groupings.

For the multivariate processing of GC–MS data, the detected compounds were classified according to their chemical families, and PCA and HCA were again applied using RStudio software (version 4.3.3, Boston, MA, USA). This approach facilitated the identification of similarity patterns, visualization of clustering based on chemical composition, and verification of group consistency, thereby complementing the results obtained from Raman analysis.

4. Conclusions

The integration of Raman spectroscopic and chromatographic (GC–MS) analyses enabled a comprehensive characterization of essential oils obtained from eight aromatic medicinal species (*T. filifolia*, *A. citrodora*, *C. citratus*, *E. globulus*, *C. nobile*, *P. aduncum*, *M. mollis*, and *R. officinalis*). The results revealed that interspecific differences were primarily associated with the relative proportion of oxygenated compounds (alcohols, aldehydes, and esters) versus aromatic hydrocarbons and terpenoids. Raman spectra exhibited characteristic bands corresponding to C–H, C=C, C–O, and C=O stretching vibrations, reflecting the presence of aromatic and olefinic structures typical of mono- and sesquiterpenes. These findings were corroborated by GC–MS, which identified 224 compounds across nine chemical families, with terpenoids predominating (>70% in most species). Major metabolites included β-caryophyllene (18.26% and 11% in *R. officinalis* and *P. aduncum*, respectively), citronellol (10.92% in *A. citrodora*), citral (2.35% in *T. filifolia*), geraniol (6.06% in *C. citratus*), β-pinene and myrcene (8.21% and 8.49% in *E. globulus*), and cis,cis-nepetalactone (30.16% in *M. mollis*). The detection of cis,cis-nepetalactone in *M. mollis* constitutes a significant scientific contribution, representing the first report of this lactone in the species and opening new perspectives for research on its structure and biological potential. Furthermore, the observed correlation between vibrational profiles and GC–MS quantifications highlights the effectiveness of the complementary Raman and GC–MS approach as a rapid, reproducible, and non-destructive method for essential oil characterization.

Supplementary Materials: The following supporting information can be downloaded at: https://www.mdpi.com/article/10.3390/ijms262311254/s1.

Author Contributions: Conceptualization, M.G.-S., E.M.C.-A., J.L.M. and J.R.D.-V.; methodology, M.G.-S., L.T.-V., E.M.C.-A., M.G.V.-E. and A.C.C.; software, M.G.-S., K.R.-G., L.T.-V., A.C.C. and E.M.C.-A.; validation, M.G.-S., E.M.C.-A., J.L.M. and C.R.B.-Z.; formal analysis, M.G.-S., K.R.-G., L.T.-V., E.M.C.-A. and A.C.C.; investigation, M.G.-S., E.M.C.-A., J.R.D.-V. and J.L.M.; resources, M.G.-S., E.M.C.-A. and J.L.M.; data curation, M.G.-S., K.R.-G., M.G.V.-E. and C.R.B.-Z.; writing—original draft preparation, M.G.-S., K.R.-G., E.M.C.-A., I.S.C.-C. and C.R.B.-Z.; writing—review and editing, M.G.-S., K.R.-G., I.S.C.-C., J.R.D.-V., J.L.M. and C.R.B.-Z.; visualization, M.G.-S., K.R.-G., L.T.-V., M.G.V.-E., E.M.C.-A.; supervision, I.S.C.-C., E.M.C.-A., J.R.D.-V. and J.L.M.; project administration, M.G.-S., J.L.M. and C.R.B.-Z.; funding acquisition, M.G.-S., J.L.M. and C.R.B.-Z. All authors have read and agreed to the published version of the manuscript.

Funding: This research was funded by Consejo Nacional de Ciencia, Tecnología e Innovación (Concytec) hrough the PROCIENCIA Program in the framework of "Tesis de Pregrado y Postgrado en Ciencia, Tecnología e Innovación Tecnológica (E073-2023-01)", according to Grant N° PE501085393-2023-PROCIENCIA. The APC was funded by Vicerrectorado de Investigación–Universidad Nacional Toribio Rodríguez de Mendoza de Amazonas.

Institutional Review Board Statement: Not applicable.

Informed Consent Statement: Not applicable.

Data Availability Statement: The raw data supporting the conclusions of this article will be made available by the authors on request.

Acknowledgments: The authors thank the Laboratorio de Biotecnología Agroindustrial Laboratoriode Ingeniería, Laboratorio de Investigación en Ingeniería de los Alimentos y Poscosecha of the Universidad Nacional Toribio Rodríguez de Mendoza de Amazonas, and Instituto de Investigación para el Desarrollo Sustentable de Ceja de Selva (UNTRM, Perú).

Conflicts of Interest: The authors declare no conflicts of interest.

References

1. Cravero Ponso, C.; Juncos, N.S.; Di Francisco, G.; Olmedo, H.R. Antioxidant efficiency of *Minthostachys mollis* (Benth) Griseb. essential oil with low pulegone/MENT relation in combination with BHT. *Food Biosci.* **2025**, *68*, 106787. [CrossRef]
2. Wawoczny, A.; Gillner, D. The Most Potent Natural Pharmaceuticals, Cosmetics, and Food Ingredients Isolated from Plants with Deep Eutectic Solvents. *J. Agric. Food Chem.* **2023**, *71*, 10877–10900. [CrossRef] [PubMed]
3. Bolouri, P.; Salami, R.; Kouhi, S.; Kordi, M.; Asgari Lajayer, B.; Hadian, J.; Astatkie, T. Applications of Essential Oils and Plant Extracts in Different Industries. *Molecules* **2022**, *27*, 8999. [CrossRef]
4. Damtie, D. A Review Article on the Antimicrobial and Antioxidant Essential Oils of Aromatic Plants in Ethiopia. *J. Herb. Med.* **2024**, *48*, 100948. [CrossRef]
5. Aimad, A.; Sanae, R.; Anas, F.; Abdelfattah, E.M.; Bourhia, M.; Mohammad Salamatullah, A.; Alzahrani, A.; Alyahya, H.K.; Albadr, N.A.; Abdelkrim, A.; et al. Chemical Characterization and Antioxidant, Antimicrobial, and Insecticidal Properties of Essential Oil from *Mentha pulegium* L. *Evid. Based Complement. Alt. Med.* **2021**, *2021*, 1108133. [CrossRef]
6. Nabi, M.H.B.; Ahmed, M.M.; Mia, M.S.; Islam, S.; Zzaman, W. Essential Oils: Advances in Extraction Techniques, Chemical Composition, Bioactivities, and Emerging Applications. *Food Chem. Adv.* **2025**, *8*, 101048. [CrossRef]
7. Lawson, S.K.; Satyal, P.; Setzer, W.N. The Volatile Phytochemistry of Seven Native American Aromatic Medicinal Plants. *Plants* **2021**, *10*, 1061. [CrossRef] [PubMed]
8. Mirković, S.; Martinović, M.; Tadić, V.M.; Nešić, I.; Jovanović, A.S.; Žugić, A. Antimicrobial and Antioxidant Activity of Essential Oils from Selected *Pinus* Species from Bosnia and Herzegovina. *Antibiotics* **2025**, *14*, 677. [CrossRef]
9. Mukaila, Y.O.; Pfukwa, T.M.; Fawole, O.A. Essential Oils from South African Indigenous Plants: Extraction Techniques, Phytochemistry, Biological Activities and Applications. *S. Afr. J. Bot.* **2025**, *180*, 774–794. [CrossRef]

10. Al-Maharik, N.; Salama, Y.; Al-Hajj, N.; Jaradat, N.; Jobran, N.T.; Warad, I.; Hamdan, L.; Alrob, M.A.; Sawafta, A.; Hidmi, A. Chemical Composition, Anticancer, Antimicrobial Activity of *Aloysia citriodora* Palau Essential Oils from Four Different Locations in Palestine. *BMC Complement. Med. Ther.* **2024**, *24*, 94. [CrossRef]
11. Elangovan, S.; Mudgil, P. Antibacterial Properties of *Eucalyptus Globulus* Essential Oil against MRSA: A Systematic Review. *Antibiotics* **2023**, *12*, 474. [CrossRef]
12. Monzote, L.; Scull, R.; Cos, P.; Setzer, W.N. Essential Oil from *Piper aduncum*: Chemical Analysis, Antimicrobial Assessment, and Literature Review. *Medicines* **2017**, *4*, 49. [CrossRef]
13. Nieto, G.; Ros, G.; Castillo, J. Antioxidant and Antimicrobial Properties of Rosemary (*Rosmarinus officinalis* L.): A Review. *Medicines* **2018**, *5*, 98. [CrossRef]
14. Olivas-Méndez, P.; Chávez-Martínez, A.; Santellano-Estrada, E.; Guerrero Asorey, L.; Sánchez-Vega, R.; Rentería-Monterrubio, A.L.; Chávez-Flores, D.; Tirado-Gallegos, J.M.; Méndez-Zamora, G. Antioxidant and Antimicrobial Activity of Rosemary (*Rosmarinus officinalis*) and Garlic (*Allium sativum*) Essential Oils and Chipotle Pepper Oleoresin (*Capsicum annum*) on Beef Hamburgers. *Foods* **2022**, *11*, 2018. [CrossRef]
15. Valková, V.; Ďúranová, H.; Galovičová, L.; Borotová, P.; Vukovic, N.L.; Vukic, M.; Kačániová, M. *Cymbopogon citratus* Essential Oil: Its Application as an Antimicrobial Agent in Food Preservation. *Agronomy* **2022**, *12*, 155. [CrossRef]
16. El-Kased, R.F.; El-Kersh, D.M. GC–MS Profiling of Naturally Extracted Essential Oils: Antimicrobial and Beverage Preservative Actions. *Life* **2022**, *12*, 1587. [CrossRef] [PubMed]
17. Quispe-Sanchez, L.; Mestanza, M.; Oliva-Cruz, M.; Rimarachín, N.; Caetano, A.C.; Chuquizuta, T.; Goñas, M.; Gill, E.R.A.; Chavez, S.G. Oxidative Stability and Physicochemical Changes of Dark Chocolates with Essential Oils Addition. *Heliyon* **2023**, *9*, e18139. [CrossRef]
18. Huang, X.; Li, H.; Ruan, Y.; Li, Z.; Yang, H.; Xie, G.; Yang, Y.; Du, Q.; Ji, K.; Yang, M. An Integrated Approach Utilizing Raman Spectroscopy and Chemometrics for Authentication and Detection of Adulteration of Agarwood Essential Oils. *Front. Chem.* **2022**, *10*, 1036082. [CrossRef] [PubMed]
19. Elshibani, F.A.; Farouk, A.; Naili, E.E.; Mahmoud, E.; Othman, A.; Mohammed, H.A.; Elkazza, N.; Sharkasi, M.A.; Alzunaidy, N.A. GC-MS and HPLC Chemical Profile, Antioxidant, Anti-Acetylcholinesterase, and Anti-Diabetic Activities of Libyan Salvia Lanigera Herb Extract and Essential Oil. *Sci. Rep.* **2025**, *15*, 31853. [CrossRef] [PubMed]
20. Gamal El-Din, M.I.; Youssef, F.S.; Altyar, A.E.; Ashour, M.L. GC/MS Analyses of the Essential Oils Obtained from Different Jatropha Species, Their Discrimination Using Chemometric Analysis and Assessment of Their Antibacterial and Anti-Biofilm Activities. *Plants* **2022**, *11*, 1268. [CrossRef]
21. Sander, A.; Bival Štefan, M.; Radetić, A.; Petračić, A.; Kučić Grgić, D.; Cvetnić, M.; Parlov Vuković, J. Advanced Spectroscopic Characterization, Antioxidant and Antibacterial Activity Evaluation, and Trace Metal Analyses of Essential Oils from Star Anise, Nutmeg, Clove, Oregano, Bay Leaves, and Lemon Peel. *Appl. Sci.* **2024**, *14*, 11094. [CrossRef]
22. Zhao, J.; Wang, M.; Lee, J.; Khan, I.A. Characterization, differentiation, and adulteration detection of peppermint essential oil: An NMR approach. *J. Pharm. Biomed. Anal.* **2025**, *263*, 116941. [CrossRef]
23. Ait benlabchir, A.; Fikri-Benbrahim, K.; Moutawalli, A.; Alanazi, M.M.; Halmoune, A.; Benkhouili, F.Z.; Oubihi, A.; Kabra, A.; Hanoune, E.; Assila, H.; et al. GC–MS Characterization and Bioactivity Study of *Eucalyptus globulus* Labill. (Myrtaceae) Essential Oils and Their Fractions: Antibacterial and Antioxidant Properties and Molecular Docking Modeling. *Pharmaceuticals* **2024**, *17*, 1552. [CrossRef] [PubMed]
24. El-Kersh, D.M.; Mostafa, N.M.; Fayez, S.; Al-Warhi, T.; Abourehab, M.A.S.; Eldehna, W.M.; Salem, M.A. GC-MS Metabolites Profiling of Anethole-Rich Oils by Different Extraction Techniques: Antioxidant, Cytotoxicity and in-Silico Enzymes Inhibitory Insights. *J. Enzyme Inhib. Med. Chem.* **2022**, *37*, 1974–1986. [CrossRef]
25. Fenghour, H.; Bouabida, H. GC/MS Analysis, Antimicrobial, Fungicidal Activities and Toxicity of *Rosmarinus Officinalis* Essential Oil Plant from Algeria. *Pol. J. Environ. Stud.* **2025**, *34*, 2665–2672. [CrossRef] [PubMed]
26. Ieri, F.; Cecchi, L.; Giannini, E.; Clemente, C.; Romani, A. GC-MS and HS-SPME-GC×GC-TOFMS Determination of the Volatile Composition of Essential Oils and Hydrosols (By-Products) from Four *Eucalyptus* Species Cultivated in Tuscany. *Molecules* **2019**, *24*, 226. [CrossRef] [PubMed]
27. Qasem, A. GC-MS Analysis, Antimicrobial Activity, and Genotoxicity of *Pimpinella Anisum* Essential Oil: In Vitro, ADMET and Molecular Docking Investigations. *Phyton* **2025**, *94*, 809–824. [CrossRef]
28. Huang, W.; Wang, Y.; Tian, W.; Cui, X.; Tu, P.; Li, J.; Shi, S.; Liu, X. Biosynthesis Investigations of Terpenoid, Alkaloid, and Flavonoid Antimicrobial Agents Derived from Medicinal Plants. *Antibiotics* **2022**, *11*, 1380. [CrossRef]
29. Oliveira, M.F.L.D.; Pinheiro, J.F.; Freitas, A.S.; Mendonça, C.D.J.S.; Silva, G.S.D.; Sousa, E.R.D. Exploratory Analysis of Adulterant Detection in Essential Oils Using Infrared Spectroscopy and Chemometric Tools. *J. Essent. Oil Res.* **2025**, *37*, 357–367. [CrossRef]

30. Rashid, H.M.; Mahmod, A.I.; Afifi, F.U.; Talib, W.H. Antioxidant and Antiproliferation Activities of Lemon Verbena (*Aloysia citrodora*): An In Vitro and In Vivo Study. *Plants* **2022**, *11*, 785. [CrossRef]
31. Abd-Elhafeez, E.M.A.; Ramadan, B.R.; Abou-El-Hawa, S.H.M.; Rashwan, M.R.A. Chemical Composition and Antimicrobial Activity of Anise and Fennel Essential Oils. *Assiut J. Agric. Sci.* **2023**, *54*, 127–140. [CrossRef]
32. Siatis, N.G.; Kimbaris, A.C.; Pappas, C.S.; Tarantilis, P.A.; Daferera, D.J.; Polissiou, M.G. Rapid Method for Simultaneous Quantitative Determination of Four Major Essential Oil Components from Oregano (*Oreganum* Sp.) and Thyme (*Thymus* Sp.) Using FT-Raman Spectroscopy. *J. Agric. Food Chem.* **2005**, *53*, 202–206. [CrossRef]
33. Baranska, M.; Schulz, H.; Reitzenstein, S.; Uhlemann, U.; Strehle, M.A.; Krüger, H.; Quilitzsch, R.; Foley, W.; Popp, J. Vibrational Spectroscopic Studies to Acquire a Quality Control Method of Eucalyptus Essential Oils. *Biopolymers* **2005**, *78*, 237–248. [CrossRef]
34. Jentzsch, P.V.; Ramos, L.A.; Ciobotă, V. Handheld Raman Spectroscopy for the Distinction of Essential Oils Used in the Cosmetics Industry. *Cosmetics* **2015**, *2*, 162–176. [CrossRef]
35. Murphy, B.J.; Wilson, T.M.; Ziebarth, E.A.; Bowerbank, C.R.; Carlson, R.E. Authentication of Fennel, Star Anise, and Anise Essential Oils by Gas Chromatography (GC/MS) and Stable Isotope Ratio (GC/IRMS) Analyses. *Plants* **2024**, *13*, 214. [CrossRef] [PubMed]
36. Henry, M.S. Chemical Composition and Bioactivity of Essential Oils from Lippia Alba and Aloysia Citrodora. *Rev. Latinoam. Quím.* **2024**, *51*, 13–15. [CrossRef]
37. Sprea, R.M.; Fernandes, L.H.M.; Pires, T.C.S.P.; Calhelha, R.C.; Rodrigues, P.J.; Amaral, J.S. Volatile Compounds and Biological Activity of the Essential Oil of *Aloysia aitrodora* Paláu: Comparison of Hydrodistillation and Microwave-Assisted Hydrodistillation. *Molecules* **2023**, *28*, 4528. [CrossRef]
38. Shala, A.Y.; Gururani, M.A. Phytochemical Properties and Diverse Beneficial Roles of *Eucalyptus globulus* Labill.: A Review. *Horticulturae* **2021**, *7*, 450. [CrossRef]
39. Agnish, S.; Sharma, A.D.; Kaur, I. Nanoemulsions (O/W) Containing *Cymbopogon pendulus* Essential Oil: Development, Characterization, Stability Study, and Evaluation of in Vitro Anti-Bacterial, Anti-Inflammatory, Anti-Diabetic Activities. *BioNanoScience* **2022**, *12*, 540–554. [CrossRef]
40. Noorbakhsh, F.; Ghasemi, M.M.; Maghbool, M.; Sorouri, M.; Firooziyan, S.; Osanloo, M. Preparation, Characterization, and Antibacterial Evaluation of Nanoemulsions and Chitosan Nanoparticles Containing Lemongrass Essential Oil and Citral against *Staphylococcus aureus* and *Pseudomonas aeruginosa*. *BioNanoScience* **2025**, *15*, 210. [CrossRef]
41. Khasanah, L.U.; Ariviani, S.; Purwanto, E.; Praseptiangga, D. Chemical composition and citral content of essential oil of lemongrass (*Cymbopogon citratus* (DC.) Stapf) leaf waste prepared with various production methods. *J. Agric. Food Res.* **2025**, *19*, 101570. [CrossRef]
42. Silva, T.L.M.D.; Rosa, G.I.D.; Santos, M.A.L.D.; Graf, S.L.; Maia, B.H.L.D.N.S.; Beltrame, F.L.; Ferrari, P.C. Lemongrass Essential Oil (*Cymbopogon citratus* (DC) Stapf.) Seasonal Evaluation and Microencapsulation by Spray-Drying. *Braz. Arch. Biol. Technol.* **2023**, *66*, e23230016. [CrossRef]
43. Rösch, P.; Popp, J.; Kiefer, W. Raman and Surface Enhanced Raman Spectroscopic Investigation on *Lamiaceae* Plants. *J. Mol. Struct.* **1999**, *480–481*, 121–124. [CrossRef]
44. Schulz, H.; Baranska, M.; Belz, H.-H.; Rösch, P.; Strehle, M.A.; Popp, J. Chemotaxonomic Characterisation of Essential Oil Plants by Vibrational Spectroscopy Measurements. *Vib. Spectrosc.* **2004**, *35*, 81–86. [CrossRef]
45. Vargas Jentzsch, P.; Ciobotă, V. Raman Spectroscopy as an Analytical Tool for Analysis of Vegetable and Essential Oils. *Flavour Fragr. J.* **2014**, *29*, 287–295. [CrossRef]
46. Santos, F.A.D.; Iulianelli, G.C.V.; Tavares, M.I.B. The Use of Cellulose Nanofillers in Obtaining Polymer Nanocomposites: Properties, Processing, and Applications. *Mat. Sci. Appl.* **2016**, *7*, 257–294. [CrossRef]
47. Kampasakali, E.; Nakas, A.; Mertzanidis, D.; Kokkini, S.; Assimopoulou, A.N.; Christofilos, D. μ-Raman Determination of Essential Oils' Constituents from Distillates and Leaf Glands of Origanum Plants. *Molecules* **2023**, *28*, 1221. [CrossRef]
48. Vargas Jentzsch, P.; Sandoval Pauker, C.; Zárate Pozo, P.; Sinche Serra, M.; Jácome Camacho, G.; Rueda-Ayala, V.; Garrido, P.; Ramos Guerrero, L.; Ciobotă, V. Raman Spectroscopy in the Detection of Adulterated Essential Oils: The Case of Nonvolatile Adulterants. *J. Raman Spectrosc.* **2021**, *52*, 1055–1063. [CrossRef]
49. Barak, T.H.; Bölükbaş, E.; Bardakcı, H. Evaluation of Marketed Rosemary Essential Oils (*Rosmarinus officinalis* L.) in Terms of European Pharmacopoeia 10.0 Criteria. *Turk J. Pharm. Sci.* **2023**, *20*, 253–260. [CrossRef] [PubMed]
50. Micić, D.; Đurović, S.; Riabov, P.; Tomić, A.; Šovljanski, O.; Filip, S.; Tosti, T.; Dojčinović, B.; Božović, R.; Jovanović, D.; et al. Rosemary Essential Oils as a Promising Source of Bioactive Compounds: Chemical Composition, Thermal Properties, Biological Activity, and Gastronomical Perspectives. *Foods* **2021**, *10*, 2734. [CrossRef]
51. Omar, J.; Sarmiento, A.; Olivares, M.; Alonso, I.; Etxebarria, N. Quantitative Analysis of Essential Oils from Rosemary in Virgin Olive Oil Using Raman Spectroscopy and Chemometrics. *J. Raman Spectrosc.* **2012**, *43*, 1151–1156. [CrossRef]

52. Székely-Szentmiklósi, I.; Rédai, E.M.; Vlad, R.-A.; Szabó, Z.; Kovács, B.; Gergely, A.-L.; Albert, C.; Székely-Szentmiklósi, B.; Sipos, E. Microencapsulation of *Salvia officinalis* L. Essential Oil by Complex Coacervation Technology. *Stud. UBB Chem.* **2024**, *69*, 221–241. [CrossRef]
53. Alvarado-García, P.; Soto-Vásquez, M.; Jara-Aguilar, D.; Gavidia-Valencia, J.; Guzmán-Rodríguez, N.; Rodrigo-Villanueva, E.; Beltrán, I. Antidepressant, Anxiolytic, and Antioxidant Properties of *Piper Aduncum* Essential Oil from Northern Peru. *Pharmacogn. J.* **2024**, *16*, 1252–1258. [CrossRef]
54. van Baren, C.M.; Di Leo Lira, P.; Elechosa, M.A.; Molina, A.M.; Juárez, M.A.; Martínez, A.; Perelman, S.; Bandoni, A.L. New insights into the chemical biodiversity of *Minthostachys mollis* in Argentina. *Biochem. Syst. Ecol.* **2014**, *57*, 374–383. [CrossRef]
55. de Luna, A.V.; Fagundes, T.D.S.F.; Ramos, Y.J.; de Araújo, M.H.; Muzitano, M.F.; Calixto, S.D.; Simão, T.L.B.V.; de Queiroz, G.A.; Guimarães, E.F.; Marques, A.M.; et al. UHPLC-HRMS/MS Chemical Fingerprinting of the Bioactive Partition from Cultivated *Piper aduncum* L. *Molecules* **2024**, *29*, 1690. [CrossRef] [PubMed]
56. Strehle, K.R.; Rösch, P.; Berg, D.; Schulz, H.; Popp, J. Quality Control of Commercially Available Essential Oils by Means of Raman Spectroscopy. *J. Agric. Food Chem.* **2006**, *54*, 7020–7026. [CrossRef]
57. Barbosa, A.J.C.; Bezerra, J.J.L.; de Araujo, H.L.S.; da Luz, M.A.; França, J.F.; Barbosa, M.F.D.S.; da Silva, M.V.; de Menezes Torres, M.D.C. Chemical Composition, Antibacterial Activity, and Antibiotic-Potentiating Effect Mediated by the Essential Oil of the Leaves of *Croton urticifolius* Lam. (*Euphorbiaceae*). *Pharmacol. Res. Nat. Prod.* **2025**, *8*, 100369. [CrossRef]
58. Recio-Cázares, S.L.; López-Malo, A.; Ramírez-Corona, N.; Palou, E. Relationship Between the Chemical Composition and Transport Properties with the Antimicrobial Activity of Essential Oil from Leaves of Mexican Lippia (*Aloysia citriodora*) Extracted by Hydro-Distillation. *Biointerface Res. Appl. Chem.* **2023**, *13*, 560. [CrossRef]
59. Malik, P.; Upadhyay, P. GC-MS Chemical profile, Antioxidant Activity, and Sun Protection Factor of Essential Oil of Tea Tree (*Melaleuca alternifolia*) and Rosemary (*Rosmarinus officinalis* L.). *Orient. J. Chem.* **2022**, *38*, 1266–1275. [CrossRef]
60. Jedidi, S.; Aloui, F.; Selmi, H.; Jdaidi, N.; Ghribi, S.; Sebai, H. Phytochemical Profile and Antioxidant Properties of Essential Oils Isolated from the Aerial Parts of *Rosmarinus officinalis* L. Cultivated in Northwestern of Tunisia. *Mor. J. Agric. Sci.* **2022**, *3*, 229–233. [CrossRef]
61. Mohammed, H.A.; Al-Omar, M.S.; Mohammed, S.A.A.; Aly, M.S.A.; Alsuqub, A.N.A.; Khan, R.A. Drying Induced Impact on Composition and Oil Quality of Rosemary Herb, *Rosmarinus officinalis* Linn. *Molecules* **2020**, *25*, 2830. [CrossRef]
62. Zhang, G.; Ma, Z.; Piao, Y.; Li, N.; Bian, C.; Ren, C.; Gao, M.; Yue, W.; Guan, T. Revealing the Potential of Star Anise Essential Oil: Comparative Analysis and Optimization of Innovative Extraction Methods for Enhanced Yield, Aroma Characteristics, Chemical Composition, and Biological Activities. *Food Sci. Nutr.* **2024**, *12*, 9540–9554. [CrossRef]
63. Zheljazkov, V.D.; Astatkie, T.; Zhalnov, I.; Georgieva, T.D. Method for Attaining Rosemary Essential Oil with Differential Composition from Dried or Fresh Material. *J. Oleo Sci.* **2015**, *64*, 485–496. [CrossRef]
64. Bellumori, M.; Innocenti, M.; Congiu, F.; Cencetti, G.; Raio, A.; Menicucci, F.; Mulinacci, N.; Michelozzi, M. Within-Plant variation in *Rosmarinus officinalis* L. terpenes and phenols and their antimicrobial activity against the rosemary phytopathogens alternaria alternata and *Pseudomonas viridiflava*. *Molecules* **2021**, *26*, 3425. [CrossRef] [PubMed]
65. Han, F.; Ma, G.; Yang, M.; Yan, L.; Xiong, W.; Shu, J.; Zhao, Z.; Xu, H. Chemical Composition and Antioxidant Activities of Essential Oils from Different Parts of the Oregano. *J. Zhejiang Univ. Sci. B.* **2017**, *18*, 79–84. [CrossRef]
66. Valadares, A.C.F.; Alves, C.C.F.; Alves, J.M.; De Deus, I.P.B.; De Oliveira Filho, J.G.; Dos Santos, T.C.L.; Dias, H.J.; Crotti, A.E.M.; Miranda, M.L.D. Essential Oils from *Piper aduncum* Inflorescences and Leaves: Chemical Composition and Antifungal Activity against *Sclerotinia sclerotiorum*. *An. Acad. Bras. Ciênc.* **2018**, *90*, 2691–2699. [CrossRef]
67. Meza, E.T.V.; Vasquez-Kool, J.; Sánchez, N.I.C.; Vieira, A.; Rodrigues, R.A.F.; Sartoratto, A.; Granados, A.D.P.F.; Tello, C.L.M.; Ruiz, A.L.T.G. Chemical Composition and Anti-Proliferative Activity of Essential Oils from Some Medicinal Plants from Cachicadán, Región La Libertad, Perú. *Nat. Prod. Res.* **2024**, *38*, 2145–2150. [CrossRef] [PubMed]
68. Peralta-Canchis, L.P.; Schneid Kroning, I.; Zandoná, G.P.; Kleinubing, N.R.; Larre Oliveira, T.; Florentini, Á.M.; Valmor Rombaldi, C. Chemical composition of *Minthostachys setosa* (Briquet) and *Piper elongatum* (Vahl) essential oils, antistaphylococcal activity and effect on *Staphylococcus aureus* biofilm removal. *Biocatal. Agric. Biotech.* **2024**, *58*, 103170. [CrossRef]
69. Ghosh, A.; Zhu, E.V.; Wang, H.; Zurek, L.; Zhu, J.J. Antibacterial Activities of Nepetalactones Against Public Health-Related Pathogens. *Nat. Prod. Commun.* **2021**, *16*, 1934578X211004875. [CrossRef]
70. Voigt, V.; Franke, H.; Lachenmeier, D.W. Risk Assessment of Pulegone in Foods Based on Benchmark Dose–Response Modeling. *Foods* **2024**, *13*, 2906. [CrossRef]
71. Assaggaf, H.; El Hachlafi, N.; Elbouzidi, A.; Taibi, M.; Benkhaira, N.; El Kamari, F.; Alnasseri, S.M.; Laaboudi, W.; Bouyahya, A.; Ardianto, C.; et al. Unlocking the combined action of *Mentha pulegium* L. essential oil and Thym honey: In vitro pharmacological activities, molecular docking, and in vivo anti-inflammatory effect. *Heliyon* **2024**, *10*, e31922. [CrossRef]

72. Kokkini, S.; Hanlidou, E.; Karousou, R.; Lanaras, T. Variation of Pulegone Content in Pennyroyal (*Mentha pulegium* L.) Plants Growing Wild in Greece. *J. Essent. Oil Res.* **2002**, *14*, 224–227. [CrossRef]
73. Luís, Â.; Domingues, F. Screening of the Potential Bioactivities of Pennyroyal (*Mentha pulegium* L.) Essential Oil. *Antibiotics* **2021**, *10*, 1266. [CrossRef] [PubMed]
74. Mollaei, S.; Ebadi, M.; Hazrati, S.; Habibi, B.; Gholami, F.; Sourestani, M.M. Essential oil variation and antioxidant capacity of *Mentha pulegium* populations and their relation to ecological factors. *Biochem. Syst. Ecol.* **2020**, *91*, 104084. [CrossRef]
75. Fajdek-Bieda, A.; Pawlińska, J.; Wróblewska, A.; Łuś, A. Evaluation of the Antimicrobial Activity of Geraniol and Selected Geraniol Transformation Products against Gram-Positive Bacteria. *Molecules* **2024**, *29*, 950. [CrossRef]
76. Khwaza, V.; Aderibigbe, B.A. Antibacterial Activity of Selected Essential Oil Components and Their Derivatives: A Review. *Antibiotics* **2025**, *14*, 68. [CrossRef]
77. Sahal, G.; Woerdenbag, H.J.; Hinrichs, W.L.J.; Visser, A.; Tepper, P.G.; Quax, W.J.; Van Der Mei, H.C.; Bilkay, I.S. Antifungal and biofilm inhibitory effect of *Cymbopogon citratus* (lemongrass) essential oil on biofilm forming by Candida tropicalis isolates; an in vitro study. *J. Ethnopharmacol.* **2020**, *246*, 112188. [CrossRef]
78. Belhachemi, A.; Maatoug, M.; Canela-Garayoa, R. GC-MS and GC-FID analyses of the essential oil of *Eucalyptus camaldulensis* grown under greenhouses differentiated by the LDPE cover-films. *Ind. Crops Prod.* **2022**, *178*, 114606. [CrossRef]
79. dos Santos, J.F.; Possate, M.E.D.; Almeida, M.B.D.; Melo, E.M.D.; Furtado, R.A.; Santos, L.D.S.M.; Honório, I.C.G.; de Almeida-Junior, S. Anticoagulant activity of Eucalyptus essential oils: An in vitro approach and a bioinformatics-based pharmacokinetic-pharmacodynamic analysis. *J. Pharmacol. Toxicol. Methods* **2025**, *135*, 108381. [CrossRef]
80. Luís, Â.; Duarte, A.P.; Pereira, L.; Domingues, F. Chemical Profiling and Evaluation of Antioxidant and Anti-Microbial Properties of Selected Commercial Essential Oils: A Comparative Study. *Medicines* **2017**, *4*, 36. [CrossRef]
81. Barbosa, L.C.A.; Filomeno, C.A.; Teixeira, R.R. Chemical Variability and Biological Activities of *Eucalyptus* spp. Essential Oils. *Molecules* **2016**, *21*, 1671. [CrossRef] [PubMed]
82. Webb, H.; Lanfear, R.; Hamill, J.; Foley, W.J.; Külheim, C. The Yield of Essential Oils in Melaleuca alternifolia (*Myrtaceae*) Is Regulated through Transcript Abundance of Genes in the MEP Pathway. *PLoS ONE* **2013**, *8*, e60631. [CrossRef] [PubMed]
83. Friedel, M.; Frotscher, J.; Nitsch, M.; Hofmann, M.; Bogs, J.; Stoll, M.; Dietrich, H. Light Promotes Expression of Monoterpene and Flavonol Metabolic Genes and Enhances Flavour of Winegrape Berries (*Vitis vinifera* L. Cv. Riesling). *Aust. J. Grape Wine Res.* **2016**, *22*, 409–421. [CrossRef]
84. Seira, E.; Poulaki, S.; Hassiotis, C.; Poulios, S.; Vlachonasios, K.E. Gene Expression of Monoterpene Synthases Is Affected Rhythmically during the Day in *Lavandula angustifolia* Flowers. *Physiologia* **2023**, *3*, 433–441. [CrossRef]
85. Dobhal, P.; Kant Purohit, V.; Chandra, S.; Rawat, S. Climate-induced changes in essential oil production and terpene composition in alpine aromatic plants. *Plant Stress* **2024**, *12*, 100445. [CrossRef]
86. Zhang, X.; Niu, M.; Teixeira da Silva, J.A.; Zhang, Y.; Yuan, Y.; Jia, Y.; Xiao, Y.; Li, Y.; Fang, L.; Zeng, S.; et al. Identification and Functional Characterization of Three New Terpene Synthase Genes Involved in Chemical Defense and Abiotic Stresses in *Santalum album*. *BMC Plant Biol.* **2019**, *19*, 115. [CrossRef]
87. Malik, T.G.; Sahu, L.K.; Gupta, M.; Mir, B.A.; Gajbhiye, T.; Dubey, R.; Clavijo McCormick, A.; Pandey, S.K. Environmental Factors Affecting Monoterpene Emissions from Terrestrial Vegetation. *Plants* **2023**, *12*, 3146. [CrossRef]
88. Wu, Y.; Zhuo, Z.; Qian, Q.; Xu, D. Chemotaxonomic Variation of Volatile Components in Zanthoxylum Bungeanum Peel and Effects of Climate on Volatile Components. *BMC Plant Biol.* **2024**, *24*, 793. [CrossRef]
89. Zhang, T.; Zheng, Y.; Fu, C.; Yang, H.; Liu, X.; Qiu, F.; Wang, X.; Wang, Z. Chemical Variation and Environmental Influence on Essential Oil of *Cinnamomum camphora*. *Molecules* **2023**, *28*, 973. [CrossRef]
90. Dumitrescu, E.; Muselin, F.; Tîrziu, E.; Folescu, M.; Dumitrescu, C.S.; Orboi, D.M.; Cristina, R.T. *Pimpinella anisum* L. Essential Oil a Valuable Antibacterial and Antifungal Alternative. *Plants* **2023**, *12*, 2428. [CrossRef]
91. Eiasu, B.K.; Dyafta, V.; Araya, H.T. Effect of Leaf Age on Essential Oil Yield and Composition in Rose-Scented Geranium. *Hort. Sci.* **2022**, *57*, 1524–1528. [CrossRef]
92. Oğuz, I.K.; Kaplan, M. El efecto de la altitud y las propiedades del suelo sobre los componentes de aceites esenciales en la salvia turca (*Salvia fruticosa* Mill.). *Bol. Latinoam. Caribe Plantas Med. Aromát.* **2023**, *22*, 204–213. [CrossRef]
93. Benković-Lačić, T.; Orehovec, I.; Mirosavljević, K.; Benković, R.; Ćavar Zeljković, S.; Štefelová, N.; Tarkowski, P.; Salopek-Sondi, B. Effect of Drying Methods on Chemical Profile of Chamomile (*Matricaria chamomilla* L.) Flowers. *Sustainability* **2023**, *15*, 15373. [CrossRef]
94. Slavik, B.; Roehrer, S.; Loos, H.M.; Minceva, M.; Buettner, A. Isolation of Sesquiterpenoids from *Matricaria chamomilla* by Means of Solvent Assisted Flavor Evaporation and Centrifugal Partition Chromatography. *Anal. Bioanal. Chem.* **2021**, *413*, 4387–4396. [CrossRef]

95. Rojas-Molina, J.O.; Pino, J.A.; Cevallos-Carvajal, E.R.; Zambrano-Ochoa, Z.E.; Vaca-Castro, C.E.; Molina-Borja, F.A.; Mena-Herrera, K.R. Aceite esencial de hojas de *Minthostachys mollis* [HBK] Griseb. del Ecuador: Extracción, composición química, capacidad antioxidante y actividad antimicrobiana. *Bol. Latinoam. Caribe Plantas Med. Aromát.* **2024**, *23*, 437–447. [CrossRef]
96. Ruiz-Durán, J.; Torres, R.; Stashenko, E.E.; Ortiz, C. Antifungal and Antibiofilm Activity of Colombian Essential Oils against Different *Candida* Strains. *Antibiotics* **2023**, *12*, 668. [CrossRef]
97. Čmiková, N.; Galovičová, L.; Schwarzová, M.; Vukic, M.D.; Vukovic, N.L.; Kowalczewski, P.Ł.; Bakay, L.; Kluz, M.I.; Puchalski, C.; Kačániová, M. Chemical Composition and Biological Activities of *Eucalyptus globulus* Essential Oil. *Plants* **2023**, *12*, 1076. [CrossRef] [PubMed]
98. Morocho, V.; Chamba, A.; Pozo, P.; Montalván, M.; Suárez, A.I. Chemical Characterization and Enantioselective Analysis of *Tagetes filifolia* Lag. Essential Oil and Crude Extract. *Plants* **2024**, *13*, 1921. [CrossRef]
99. Cebi, N.; Arici, M.; Sagdic, O. The Famous Turkish Rose Essential Oil: Characterization and Authenticity Monitoring by FTIR, Raman and GC–MS Techniques Combined with Chemometrics. *Food Chem.* **2021**, *354*, 129495. [CrossRef]
100. Khaiper, M.; Poonia, P.K.; Dhanda, S.K.; Beniwal, R.; Verma, P.; Nasir, M. Seasonal variation in chemical composition and bioactivity of *Eucalyptus tereticornis* leaf essential oil. *Biochem. Syst. Ecol.* **2025**, *121*, 104988. [CrossRef]
101. Niu, Y.; Xu, L.; Qiao, M.; Wang, Y. The Anti-Depression Effect and Mechanism of *Harmonious rosemary* Essential Oil and Its Application in Microcapsules. *Materials Today Bio* **2025**, *31*, 101546. [CrossRef]
102. Raffo, A.; Baiamonte, I.; De Benedetti, L.; Lupotto, E.; Marchioni, I.; Nardo, N.; Cervelli, C. Exploring volatile aroma and non-volatile bioactive compounds diversity in wild populations of rosemary (*Salvia rosmarinus* Schleid.). *Food Chem.* **2023**, *404*, 134532. [CrossRef] [PubMed]
103. da Costa Sobral, K.G.; Neuberger, B.; Mello, F.K.; Mallmann, M.P.; Sampaio, T.B.; Oliveira, M.S. Anticonvulsant Activity of β-Caryophyllene in Association with Pregabalin in a Seizure Model in Rats. *Epilepsy Res.* **2022**, *179*, 106842. [CrossRef]
104. Nunes, J.D.A.; Teixeira, L.L.; Nascimento, W.M.O.; Lopes, D.C.F.; da Silva, J.K.R.; Pinto, L.C.; Santos, P.V.L.; Figueiredo, P.L.B. Seasonal variation on chemical composition and in vitro cytotoxic and anti-inflammatory activities of *Myrciaria dubia* (Kunth) McVaugh essential oil from Amazon. *Biochem. Syst. Ecol.* **2025**, *121*, 105001. [CrossRef]
105. Sánchez-Tito, M.A.; Cartagena-Cutipa, R.; Flores-Valencia, E.; Collantes Díaz, I. Chemical composition and antimicrobial activity of essential oil from *Minthostachys mollis* against oral pathogens. *Rev. Cuba Estomatol.* **2025**, *58*, e3647.
106. Juncos, N.S.; Cravero, C.F.; Grosso, N.R.; Olmedo, R.H. Pulegone/MENT ratio as an indicator of antioxidant activity for the selection of industrial cultivars of *peperina* with high antioxidant potential. *Ind. Crops Prod.* **2024**, *216*, 118770. [CrossRef]
107. Aničić, N.; Gašić, U.; Lu, F.; Ćirić, A.; Ivanov, M.; Jevtić, B.; Dimitrijević, M.; Anđelković, B.; Skorić, M.; Nestorović Živković, J.; et al. Antimicrobial and Immunomodulating Activities of Two Endemic Nepeta Species and Their Major Iridoids Isolated from Natural Sources. *Pharmaceuticals* **2021**, *14*, 414. [CrossRef]
108. Batume, C.; Mulongo, I.M.; Ludlow, R.; Ssebaale, J.; Randerson, P.; Pickett, J.A.; Mukisa, I.M.; Scofield, S. Evaluating Repellence Properties of Catnip Essential Oil against the Mosquito Species Aedes Aegypti Using a Y-Tube Olfactometer. *Sci. Rep.* **2024**, *14*, 2269. [CrossRef] [PubMed]
109. Srivastava, A.; Gupta, S.; Singh, S.; Verma, R.S.; Srivastava, R.K.; Gupta, A.K.; Lal, R.K. Genetic Variability and Elite Line Selection for High Essential Oil and Nepetalactone Content in Catmint (*Nepeta cataria* L.). *Am. J. Plant Sci.* **2021**, *12*, 1135–1154. [CrossRef]
110. Vasquez-Gomez, K.L.; Mori-Mestanza, D.; Caetano, A.C.; Idrogo-Vasquez, G.; Culqui-Arce, C.; Auquiñivin-Silva, E.A.; Castro-Alayo, E.M.; Cruz-Lacerna, R.; Perez-Ramos, H.A.; Balcázar-Zumaeta, C.R.; et al. Exploring Chemical Properties of Essential Oils from Citrus Peels Using Green Solvent. *Heliyon* **2024**, *10*, e40088. [CrossRef]
111. Balcázar-Zumaeta, C.R.; Maicelo-Quintana, J.L.; Salón-Llanos, G.; Barrena, M.; Muñoz-Astecker, L.D.; Cayo-Colca, I.S.; Torrejón-Valqui, L.; Castro-Alayo, E.M. A Novel Technique Using Confocal Raman Spectroscopy Coupled with PLS-DA to Identify the Types of Sugar in Three Tropical Fruits. *App. Sci.* **2024**, *14*, 8476. [CrossRef]
112. Adams, R.P. *Identification of Essential Oil Components by Gas Chromatography/Mass Spectrometry*, 4.1 ed.; Allured Business Media: Carol Stream, IL, USA, 2017.
113. Castro-Alayo, E.M.; Torrejón-Valqui, L.; Cayo-Colca, I.; Cárdenas-Toro, F.P. Evaluation of the Miscibility of Novel Cocoa Butter Equivalents by Raman Mapping and Multivariate Curve Resolution–Alternating Least Squares. *Foods* **2021**, *10*, 3101. [CrossRef] [PubMed]

International Journal of
Molecular Sciences

MDPI

Article

Bioactive Properties of Selected European *Phellinus* Species: A Comprehensive Study

Grzegorz Świderski [1,*], Monika Kalinowska [1], Ewa Zapora [2], Marek Wołkowycki [2], Marcin Stocki [2], Ewa Ciszkowicz [3], Aleksandra Bocian [3], Marcin Jaromin [3], Mirosław Tyrka [3], Katarzyna Lecka-Szlachta [3], Elżbieta Wołejko [1], Urszula Wydro [1], Małgorzata Pawłowska [4], Paweł Golianek [4], Małgorzata Zawadzka [1], Qëndrim Ramshaj [5], Carolina Elena Girometta [6] and Mitko Karadelev [7]

1 Department of Chemistry, Biology and Biotechnology, Bialystok University of Technology, Wiejska 45E, 15351 Białystok, Poland; m.kalinowska@pb.edu.pl (M.K.); e.wolejko@pb.edu.pl (E.W.); u.wydro@pb.edu.pl (U.W.); malgorzata.zawadzka@pb.edu.pl (M.Z.)

2 Department of Silviculture and Forest Utilization, Bialystok University of Technology, Wiejska 45E, 15351 Bialystok, Poland; e.zapora@pb.edu.pl (E.Z.); m.wolkowycki@pb.edu.pl (M.W.); m.stocki@pb.edu.pl (M.S.)

3 Department of Biotechnology and Bioinformatics, Faculty of Chemistry, Rzeszow University of Technology, Powstańców Warszawy 6, 35959 Rzeszow, Poland; eciszkow@prz.edu.pl (E.C.); bocian@prz.edu.pl (A.B.); mjaromin@prz.edu.pl (M.J.); mtyrka@prz.edu.pl (M.T.); szlachta@prz.edu.pl (K.L.-S.)

4 Faculty of Environmental Engineering, Lublin University of Technology, Nadbystrzycka 40B, 20618 Lublin, Poland; m.pawlowska@pollub.pl (M.P.); p.golianek@pollub.pl (P.G.)

5 Department of Biology, Faculty of Mathematics and Natural Sciences, University of Prishtina, St. "Eqrem Çabej", 10 000 Prishtina, Kosovo; qendrim.ramshaj@uni-pr.edu

6 Department of Earth and Environmental Sciences, University of Pavia, Via Sant'Epifanio 14, 27100 Pavia, Italy; carolinaelena.girometta@unipv.it

7 Institute of Biology, Faculty of Natural Sciences and Mathematics, St. Cyril and Methodius University, Arhimedova 3, 1000 Skopje, North Macedonia; mitkok@pmf.ukim.mk

* Correspondence: g.swiderski@pb.edu.pl

Abstract: This study conducted a multi-directional evaluation of the chemical potential and biological properties of selected European fungal species of the genus *Phellinus*. We investigated
30 samples belonging to 22 *Phellinus* species. Fruiting bodies were collected, among other specimens, in the Białowieża Forest (Poland); Village Kozhle (North Macedonia); Estremadura, Sesimbra, and Lagoa de Albufeira (Portugal); Zlatari close to Prishtina (Kosovo); and Spoleto and the Bosco Siro Negri State Nature Reserve (Italy). Morphological identification of the collected fungi was carried out, and genetic tests were performed to confirm the identity of the collected specimens. Methanol extracts for biological activity tests were prepared. Screening of antimicrobial activity of 30 methanolic extracts was performed on strains of bacteria (*Klebsiella pneumoniae*, *Pseudomonas aeruginosa*, *Staphylococcus aureus*, *Staphylococcus epidermidis*, and *Kocuria rhizophila*) and fungi (*Candida albicans*). Antioxidant activity tests (DPPH and ABTS) were also performed. The three most biologically active fungi species were selected (*Phellinus igniarius*, *Fomitiporia robusta*, and *Porodaedalea pini*) for further research. The chemical composition of the extracts was determined using GC-MS analysis. Thermal decomposition studies and spectroscopic analysis of the dry fruiting bodies were performed. The extracts were tested for their antimicrobial activity against antibiotic-resistant bacteria. Cytotoxic activity was also tested.

Keywords: *Phellinus*; medical fungi; cytotoxic study; antimicrobial study

1. Introduction

The genus *Phellinus sensu lato* is a significant group within the family Hymenochaetaceae Imazeki & Toki, comprising approximately 220 species globally [1,2]. These fungi are predominantly lignicolous, meaning they grow on wood, and are known for their roles in forest ecosystems as both decomposers and pathogens [3]. *Phellinus* species are found in diverse habitats, including temperate and tropical forests, where they contribute to the decomposition of wood and nutrient cycling [4]. In Europe, several species of *Phellinus* are recognized for their ecological and medicinal importance, with some species having been used in traditional medicine for centuries [5].

Taxonomically, *Phellinus sensu lato* is a complex genus with a rich diversity of species, characterized by their perennial fruiting bodies and tough, woody consistency [4]. The genus has been subject to extensive taxonomic revisions, with recent studies employing molecular phylogenetic analyses to better understand the relationships within the group [6]. These analyses have led to the reclassification of some species and the identification of new ones, highlighting the genus's diversity and evolutionary history [7].

1.1. Chemical Composition

The genus *Phellinus*, a group of medicinal fungi, is known for its diverse chemical composition—particularly its secondary metabolites. These fungi produce a wide array of compounds, including polysaccharides, flavones, coumarins, terpenes, steroids, and styrylpyranones, which have been extensively studied for their potential health benefits [1]. Among these, polysaccharides are the most prominent bioactive macromolecules, exhibiting various health-promoting effects such as immunomodulatory, anticancer, and antioxidant activities [8,9]. Specific species such as *Phellinus igniarius* (L.) Quél., *Porodaedalea pini* (Brot.) Murrill [former *Phellinus pini* (Brot.) A. Ames], *Phellinus tuberculosus* (Baumg.) Niemelä [syn. *P. pomaceus* (Pers.) Maire], and *Fomitiporia robusta* (P. Karst.) Fiasson & Niemelä [former *P. robustus* (P. Karst.) Bourdot & Galzin] have been found to contain polysaccharides, phenolic compounds, and terpenoids that contribute to their biological activities, including anticancer and antiviral effects [5]. *Sanghuangporus baumii* (Pilát) L.W. Zhou & Y.C. Dai [former *Phellinus baumii* Pilát] has yielded unique sesquiterpenes and alkyl-benzene derivatives, such as phellinbaumins A and B, which have shown moderate anti-inflammatory activity [10]. *Fuscoporia gilva* (Schwein.) T. Wagner & M. Fisch. [former *Phellinus gilvus* (Schwein.) Pat.] contains significant amounts of carbohydrates, proteins, and phenols, which contribute to its antioxidant potential [11]. *Tropicoporus linteus* (Berk. & M.A. Curtis) L.W. Zhou & Y.C. Dai [former *Tropicoporus linteus* (Berk. & M.A. Curtis) Teng] is notable for its acidic proteo-heteroglycan—a complex molecule composed of polysaccharides and proteins—which has been isolated and characterized for its unique structural properties [12]. Additionally, *P. igniarius* has been found to produce pyrano [4,3-c] isochromen-4-one derivatives, such as phelligridins H, I, and J, which possess novel carbon skeletons and exhibit cytotoxic activity against cancer cell lines [13]. These findings highlight the chemical diversity within the genus *Phellinus* and underscore the potential of these fungi as sources of novel bioactive compounds.

1.2. Bioactivity

Anticancer activity

Phellinus sensu lato species, particularly *Tropicoporus linteus* and *Phellinus nigricans*, (Fr.) P. Karst. (the latter being currently embedded into *P. igniarius* according to both Mycobank and Index Fungorum) have been extensively studied for their anticancer properties, with several in vivo studies highlighting their potential. In an in vivo study, polysaccharides extracted from *PT. linteus* significantly reduced tumor growth in a mouse model by modulating the Wnt/β-catenin signaling pathway, which is crucial for cell proliferation and survival [14]. Additionally, *P. linteus* extracts have been found to selectively induce apoptosis in cancer cells while exhibiting low toxicity towards normal cells, suggesting a potential for targeted cancer therapy [15]. Proteoglycans isolated from the mycelium of *P. nigricans* have demonstrated antitumor effects in vivo against Sarcoma 180 in mice. These effects were primarily attributed to the immunomodulatory properties of the proteoglycans, which enhanced the immune response rather than directly killing the tumor cells. The study by the authors of [16] reported increased spleen and thymus weights and elevated levels of tumor necrosis factor-alpha (TNF-α), indicating an activated immune system capable of combating tumor growth. Moreover, hispolon—a compound found in *Tropicoporus linteus*—has been identified as a potent anticancer agent. In vivo studies have shown that hispolon can induce apoptosis in cancer cells by disrupting mitochondrial membrane potential and activating caspase pathways, which are essential for programmed cell death [17]. This compound has also been isolated from other *Phellinus* species, including *P. igniarius*, *Sanghuangporus. lonicerinus*, (Bondartsev) Sheng H. Wu, L.W. Zhou & Y.C. Dai [fromer *P. lonicerinus* (Bondartsev) Bondartsev & Singer], *Fulvifomes. merrillii*, (Murrill) Baltazar & Gibertoni [former *P. merrillii* (Murrill) Ryvarden], and *Inonotus hispidus*,(Bull.) P. Karst. and has been shown to modulate several signaling pathways involved in cancer progression, such as cell cycle arrest and inhibition of angiogenesis and metastasis [18].

Immunomodulatory activity

Selected *Phellinus* species also demonstrate notable immunomodulatory effects. Polysaccharides have been shown to enhance both hormonal and cell-mediated immune functions, which can help in managing various immune-related conditions [2,9,19], *Tropicoporus linteus* polysaccharides have demonstrated significant anti-inflammatory effects by inhibiting pathways such as MAPK and NF-kappaB, which are involved in the expression of inflammatory cytokines. This inhibition helps to reduce inflammation in various models, including those induced by lipopolysaccharides (LPSs) [20–23]. Additionally, *Tropicoporus linteus* has been shown to modulate immune responses by increasing the production of cytokines such as IL-2, IL-12, and IFN-gamma, which are crucial for immune function [20–23]. *Sanghuangporus baumii* also exhibits anti-inflammatory properties, particularly in conditions such as inflammatory bowel disease. Its polysaccharides can decrease the expression of pro-inflammatory genes and increase anti-inflammatory markers, contributing to tissue repair and immune modulation [24,25]. This species has been noted for its potential in enhancing immune responses, especially in immunosuppressed conditions, by improving lymphocyte proliferation and natural killer cell activity [24].

Antioxidant activity

The genus Phellinus is known for its diverse species that exhibit significant antioxidant activities, which have been the focus of various studies. *Porodaedalea pini* and *Phellinus sp.* collected in Foshan have demonstrated potent antioxidant activities, with IC50 values for DPPH and hydroxyl radicals comparable to ascorbic acid, indicating their strong radical scavenging capabilities [26]. Similarly, *Fuscoporia torulosa* (Pers.) T. Wagner & M. Fisch. [former *Phellinus torulosus* (Pers.) Bourdot & Galzin] and *P. igniarius* have been used to fortify yogurt, enhancing its antioxidant profile significantly, with *P. torulosus* showing superior DPPH inhibition and lipid peroxidation neutralizing activities [27]. *Sanghuangporus baumii* extracts, which are rich in phenolics, have shown high antioxidant and free radical scavenging activities, particularly in methanol and hot water extracts, exhibiting 80–90% inhibition rates in various assays [28]. The ethanolic extracts of *S. baumii* have been effective in protecting PC12 cells from oxidative damage, with strong correlations between flavonoid content and antioxidant capacity [29]. *Phellinus rimosus* (Berk.) Pilát has also been noted for its significant antioxidant and antihepatotoxic activities, with its ethyl acetate extract showing potent free radical scavenging and liver protective effects [30]. *Tropicoporus linteus*, extracted with ethanol, has demonstrated strong antioxidant activities, comparable to vitamin C, and has shown potential antitumor effects due to its anti-angiogenic properties [31]. In studies conducted in Northeast Thailand, *Fuscoporia torulosa* and *Porodaedalea pini* exhibited potent radical scavenging activities, with high total phenolic content correlating with their antioxidant efficacy [32]. Additionally, *Phellinus rimosus* has shown excellent antioxidant and antimutagenic activities, particularly in ethanol extracts, suggesting its potential in disease prevention and treatment [33].

Antiviral activity

Among the species studied, *T. linteus* and *P. igniarius* have shown promising antiviral effects. *Tropicoporus linteus*, for instance, has been investigated for its potential against dengue virus type-2, where compounds such as phelignan A and dimeric ferulates exhibited significant inhibitory activity [34]. Additionally, *Tropicoporus linteus* has demonstrated potential in inhibiting SARS-CoV-2, with key metabolites such as hispidin and hypholomine B reducing viral entry by affecting angiotensin-converting enzyme 2 gene expression [35]. *Phellinus igniarius* has also been highlighted for its antiviral properties. A new sesquiterpenoid isolated from this species showed significant activity against the H5N1 influenza virus, suggesting its potential as a neuraminidase inhibitor [14]. Furthermore, polysaccharides and flavonoids derived from *Phellinus igniarius* have been shown to enhance immune responses in immunocompromised mice, which could indirectly contribute to antiviral defense [19].

Antidiabetic effects

Phellinus spp. have shown promising antidiabetic effects in various in vivo studies. *Tropicoporus linteus* polysaccharides (PLPs) have been demonstrated to ameliorate insulin resistance in mice fed a high-fat, high-fructose diet. The administration of PLP significantly reduced fasting blood glucose levels and improved glucose intolerance. This effect was attributed to the modification of hepatic phospholipid metabolism and the enhancement of insulin signaling transduction, as well as the stimulation of gut bacteria capable of synthesizing vitamin B12, which plays a role in metabolic processes [36]. Similarly, *S. baumii* exopolysaccharides (EPSs) have exhibited substantial hypoglycemic effects in streptozotocin-induced diabetic rats. The EPS treatment led to a significant reduction in plasma glucose levels and improved liver function, as indicated by decreased activities of liver enzymes such as alanine aminotransferase and aspartate aminotransferase. These findings suggest that *S. baumii* EPS could be beneficial for managing diabetes mellitus [37]. *Phellinus igniarius* has also been studied for its antidiabetic properties. A polyphenol-rich extract from this species was shown to lower fasting blood glucose levels and improve glucose tolerance in KK-Ay mice with spontaneous type 2 diabetes. The extract enhanced the expression of GLUT4 and activated the AMPK pathway; these processes are crucial for glucose uptake and metabolism [38]. Furthermore, *S. baumii* extract (PBE) demonstrated hypoglycemic effects in diabetic mice by reducing fasting blood glucose levels and improving insulin sensitivity. The extract modulated the gut microbiota, promoting beneficial bacteria and inhibiting harmful ones, which may contribute to its antidiabetic effects [39].

Phellinus species are valuable medicinal fungi with a wide array of health-promoting properties, including antitumor, immunomodulatory, anti-inflammatory, antioxidant, and antidiabetic effects. These properties are primarily attributed to their rich composition of bioactive compounds, making them a significant focus of research for developing functional foods and pharmaceuticals.

The aim of this study was a multi-directional evaluation of the chemical potential and biological properties of selected European fungal species of the genus *Phellinus*.

As part of the project, 30 specimens of *Phellinus* mushrooms were collected. Morphological identification of the collected fungi was carried out, and genetic tests were performed to confirm the identity of the collected specimens. Methanol extracts for biological activity tests were prepared. Screening of the antimicrobial activity of 30 methanolic extracts was performed on strains of bacteria (*Klebsiella pneumoniae, Pseudomonas aeruginosa, Staphylococcus aureus, Staphylococcus epidermidis*, and *Kocuria rhizophila*) and fungi (*Candida albicans*). Antioxidant activity tests (DPPH and ABTS) were also performed.

Three species of fungi with the highest activity were selected for further research (*Phellinus igniarius, Fomitiporia robusta*, and *Porodaedalea pini*) The chemical composition of the extracts was determined using GC-MS analysis. Thermal decomposition studies and spectroscopic analysis of the dry fruiting bodies were performed. The extracts were tested for their antimicrobial activity against antibiotic-resistant bacteria. Cytotoxic activity was also tested.

2. Results and Discussion

2.1. Antimicrobial Activity—Preliminary Studies

The antimicrobial activity of 30 samples of methanolic extracts of fruiting bodies of the genus *Phellinus* was tested. The antimicrobial properties of the tested fungal extracts are summarized in Table 1. Based on the obtained results, it was observed that the highest antimicrobial activity was shown by *Fomitiporia robusta* (mainly on strains *C. albicans*—MIC = 25 mg/mL, *S. aureus*—MIC = 25 mg/mL, and *K. rhizophila*—MIC = 25 mg/mL); in turn, *Porodaedalea pini* inhibited the growth of strains *C. albicans* (MIC = 12.5 mg/mL), *S. aureus* (MIC = 25 mg/mL), *K. rhizophila* (MIC = 25 mg/mL), and *S. epidermidis* (MIC = 25 mg/mL). Extract from *Phellinus igniarius* was the most effective against *S. aureus* (MIC = 6.3 mg/mL), *S. epidermidis* (MIC = 25 mg/mL), and *C. albicans* (MIC = 25 mg/mL).

Table 1. Antimicrobial properties (MIC) of examined *Phellinus* spp. methanolic extracts.

Sample No.	Species Name	*C. albicans* [mg/mL]	*P. aeruginosa* [mg/mL]	*S. aureus* [mg/mL]	*S. epidermidis* [mg/mL]	*K. pneumoniae* [mg/mL]	*K. rhizophila* [mg/mL]
16.D	*Porodaedalea chrysoloma*	>50	>50	50	>50	50	>50
17.D	*Porodaedalea pini*	>50	50	25	50	50	50
18.D	*Phellinus tremulae*	>50	50	12,5	50	50	50
19.D	*Phellinus laevigatus*	>50	>50	50	>50	>50	>50
21.D	*Phellinus tuberculosus*	>50	>50	25	>50	50	>50
22.D	*Fomitiporia robusta*	>50	50	25	25	50	25
25.D	*Phellinopsis conchata*	25	>50	25	>50	50	50
38.D	*Phellinus igniarius*	>50	>50	25	>50	>50	50
118.D	*Phellinus alni*	25	>50	25	>50	>50	50
160.D	*Fuscoporia ferruginosa*	50	50	25	50	50	25
254.D	*Phellinus alni*	25	>50	25	>50	>50	>50
255.D	*Phellinus alni*	50	>50	>50	50	>50	50
256.D	*Phellinidium ferrugineofuscum*	25	>50	50	>50	>50	50
257.D	*Fomitiporia punctata*	>50	50	50	50	50	50
258.D	*Fomitiporia punctata*	>50	50	>50	50	25	25
261.D	*Phellinus nigricans*	>50	>50	25	50	>50	>50
262.D	*Phellinus populicola*	>50	50	12.5	>50	50	50
266.D	*Fomitiporia robusta*	>50	>50	25	50	50	25
267.D	*Phellinus nigricans*	>50	>50	12.5	50	50	50
268.D	*Fomitiporia hartigii*	>50	50	25	25	50	25

Table 1. *Cont.*

Sample No.	Species Name	*C. albicans* [mg/mL]	*P. aeruginosa* [mg/mL]	*S. aureus* [mg/mL]	*S. epidermidis* [mg/mL]	*K. pneumoniae* [mg/mL]	*K. rhizophila* [mg/mL]
279.D	*Phellinus tuberculosus*	50	>50	25	>50	>50	>50
280.D	*Fomitiporia robusta*	25	50	25	50	50	25
298.D	*Porodaedalea pini*	12.5	50	25	25	50	25
300.D	*Phellinus nigricans*	25	>50	25	50	>50	>50
300.Da	*Phellinus nigricans*	>50	>50	6.3	50	>50	50
301.D	*Phellinus igniarius*	25	>50	6.3	25	50	50
319.D	*Fuscoporia torulosa*	50	>50	50	50	50	>50
320.D	*Fuscoporia torulosa*	25	>50	>50	50	50	50
322.D	*Fomitiporia mediterranea*	>50	50	50	50	50	50
329.D	*Porodaedalea chrysoloma*	50	>50	12.5	50	50	50

2.2. Elemental Analysis

The highest amounts of macroelements (P, K, and Ca) were found in *Fomitiporia robusta* extract (2930.88 mg/kg, 3218.03 mg/kg, and 324.89 mg/kg of extract weight) (Table 2). The potassium content in the extracts from *P. pini* and *P. igniarius* was at a similar level, but was significantly lower than in the extract from *F. robusta*. The lowest phosphorus content was recorded in the extract from *P. pini* (1224.77 mg/kg); in the extract from *P. igniarius*, the content of this element was almost twice as high. The calcium content was also the lowest in *P. pini* and amounted to 128.02%, which is almost 2.5 times less than in the extract from *F. robusta*. The iron content in the extracts from *P. pini* and *P. igniarius* was at a similar level of about 5 mg/kg of extract weight. The iron content in the *F. robusta* extract was twice as high. The *F. robusta* extract contained the most zinc, while the *P. pini* extract contained the most copper. Interestingly, the *F. robusta* and *P. pini* extracts did not contain magnesium while, in the *Phellinus igniarius* extract, the content of this element was over 1 g/kg of extract weight. The highest silicon and sulfur content was noted in the *F. robusta* extract, and the lowest in *P. igniarius*. The tested extracts also contained arsenic of about 1–2 mg per kg of extracts. The presence of bromine and chlorine was also noted in the tested extracts. The latter element was observed in significant amounts in the *P. pini* and *P. igniarius* extracts—over 3 g/kg of extract weight—while in the *F. robusta* extract, the chlorine content was over 25 times lower. Small amounts of nickel were also observed in the extracts of *P. pini* and *P. igniarius*.

Table 2. Elemental composition of examined *Phellinus* spp. extracts.

Element	Content of Elements [mg/kg of Extract]					
	Fomitiporia robusta	* SD	*Porodaedalea pini*	* SD	*Phellinus igniarius*	* SD
Cu	17.50	0.75	23.18	0.83	6.01	0.22
Fe	10.84	0.69	5.35	0.22	5.04	0.88
Mn	10.86	0.79	1.69	0.10	-	-
Ni	-	-	0.41	0.14	0.57	0.14
Zn	18.52	0.27	16.15	0.79	6.01	0.22
Ti	5.76	0.28	4.92	0.27	8.09	1.02
Rb	12.81	0.55	11.47	0.70	18.27	0.72
K	3218.03	0.50	2349.73	5.62	2287.27	2.91
Ca	324.89	1.44	128.02	2.91	146.06	1.70
Mg	-	-	-	-	1051.27	1.90
P	2930.88	1.87	1224.77	2.40	2117.81	0.94
S	710.19	0.64	581.22	0.94	252.63	1.13
Cl	139.01	0.35	3194.66	1.00	3240.80	0.62
Si	202.47	0.89	88.55	1.90	70.70	2.40
As	1.78	0.48	1.41	0.09	0.99	0.11
Br	3.80	0.23	21.54	2.22	4.18	0.38

* SD: standard deviation.

2.3. Thermal Analysis

Thermal analysis of dried fruiting bodies of *Fomitiporia robusta* (Figure 1A), *Porodaedalea pini* (Figure 1B), and *Phellinus igniarius* (Figure 1C) was performed. Dried and powdered fruiting bodies were heated in the temperature range of 30–660 °C in a nitrogen atmosphere. The first stage of thermal decomposition of the sample is evaporation of the remaining water. This process occurs in the temperature range of 50–200 °C. A rather mild decrease in mass is observed on the TG curves (Figure 1A–C) in this temperature range.

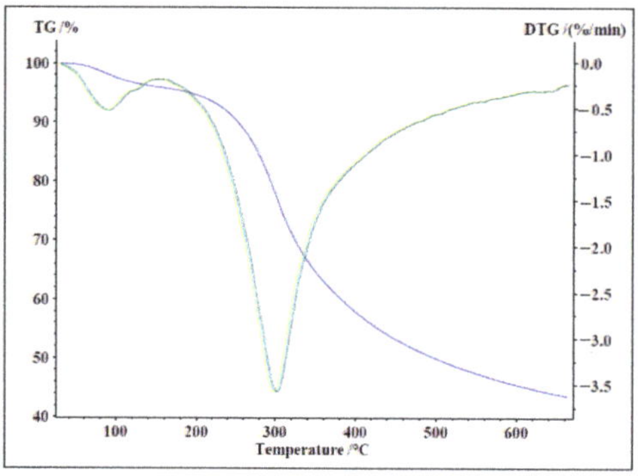

A

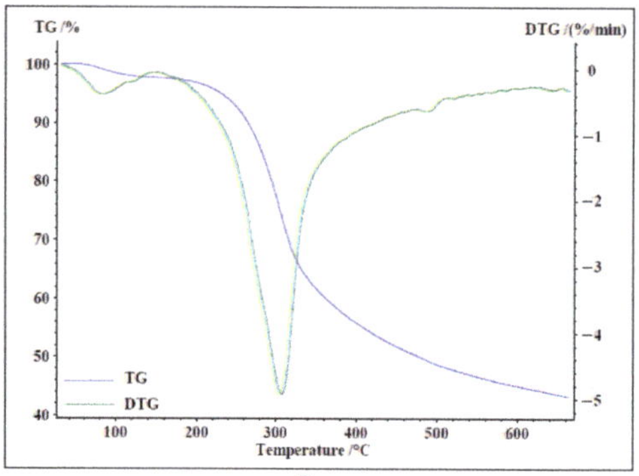

B

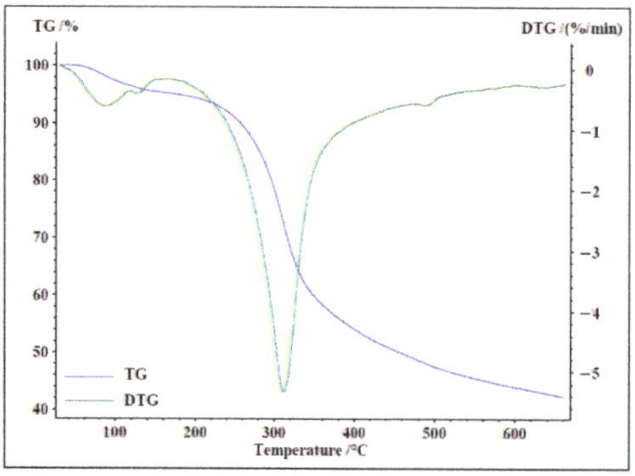

C

Figure 1. TG and DTG curves of dried fungal samples ((**A**)—*F. robusta*, (**B**)—*P. pini*, and (**C**)—*P. igniarius*).

Based on the mass loss, it can be stated that the dried sample of *Fomitiporia robusta* contained about 4.1% water, *Porodaedalea pini* contained 2.2% water, and the water content in *Phellinus igniarius* was 4.8%. The peak on the DTG curve reached a minimum at 93.0 °C (*F. robusta*), 86.7 °C (*P. pini*), and 90.3 °C (*P. igniarius*).

Further heating of the samples led to the decomposition of organic substances contained in the samples—mainly sugars, carboxylic acids, and sterols. The maximum peak on the DTG decomposition curve of the *F. robusta* sample was 302.7.0 °C, while for the *P. pini* sample it was 307.9 °C, and for the *P. igniarius* sample it was 313.0 °C. The thermal decomposition process was carried out to a temperature of 660 °C. At a temperature of 640 °C, the thermal decomposition curves began to flatten out. The masses of the sample residues after thermal decomposition were 43.57% (Figure 1A), 43.22% (Figure 1B), and 42.30% (Figure 1C), respectively. The residue from the decomposition process of the tested dried fruiting body samples is the residue of inorganic carbon and mineral compounds contained in the tested samples. The thermal decomposition curves for all three species of the *Phellinus* genus have a similar course, which indicates a similar composition of the matter constituting the fruiting bodies of the tested fungi.

2.4. FTIR Spectroscopy

Infrared spectroscopy (FTIR) was used to determine the main groups of organic compounds in the tested samples. Infrared spectra of dried fruiting bodies, *Phellinus igniarius, Fomitiporia robusta,* and *Porodaedalea pini,* and spectra of their extracts were recorded using the multireflection ATR technique in the spectral range of 2000–600 cm^{-1}. The spectra are presented in Figure 2, while the position of the bands and their assignment are provided in Table 3.

In the FTIR spectra of the dried fruiting body samples and their extracts, characteristic bands of high intensity are present, which are a result of the stretching vibrations of the carbonyl group, νC=O. In the spectra of dried fruiting bodies, these bands are present in the spectral range of 1647–1645 cm^{-1}, while in the spectra of fungal extracts, the bands are present in the range of 1735–1675 cm^{-1}. The shift of these bands in the spectra of extracts is caused by the presence of carboxylic acid dimers in the dried fruiting body samples. Other characteristic bands present in the spectra are related to the stretching vibrations, νC-O. These bands are located in the wavenumber range of 1055–1024 cm^{-1}. The presence of the νC=O and νC-O bands indicates the presence of compounds from the carboxylic acid group in the tested fruiting bodies and their extracts. Additionally, the spectra show bands associated with the vibrations of C-H and C-C bonds in aliphatic and aromatic systems. This indicates that the tested samples contain aromatic carboxylic acids and aliphatic carboxylic acids, including fatty acids.

Carbonylic group (νC-O) bands may also indicate the presence of carbohydrates, especially since the recorded spectra additionally show vibration bands originating from CH_2OH groups. These include the δCH_2OH bending vibration bands located in the wavenumber range of 1301–1242 cm^{-1}. Other bands present in the spectra of dried fruiting bodies and extracts associated with the presence of carbohydrates are the δCOH, δCCH, and δOCH bending vibration bands located at wavenumbers 946–779 cm^{-1}. In the spectra of the tested products, many bands originating from vibrations of the aromatic system—related to the presence of aromatic compounds—are also observed. These include bands originating from deformation vibrations δCHring in the wavenumber range of 1168–1088 cm^{-1}, bands related to stretching vibrations νCC (1556–1514 cm^{-1}), as well as bands related to deformation of the aromatic ring or de-fringing (697–624 cm^{-1}).

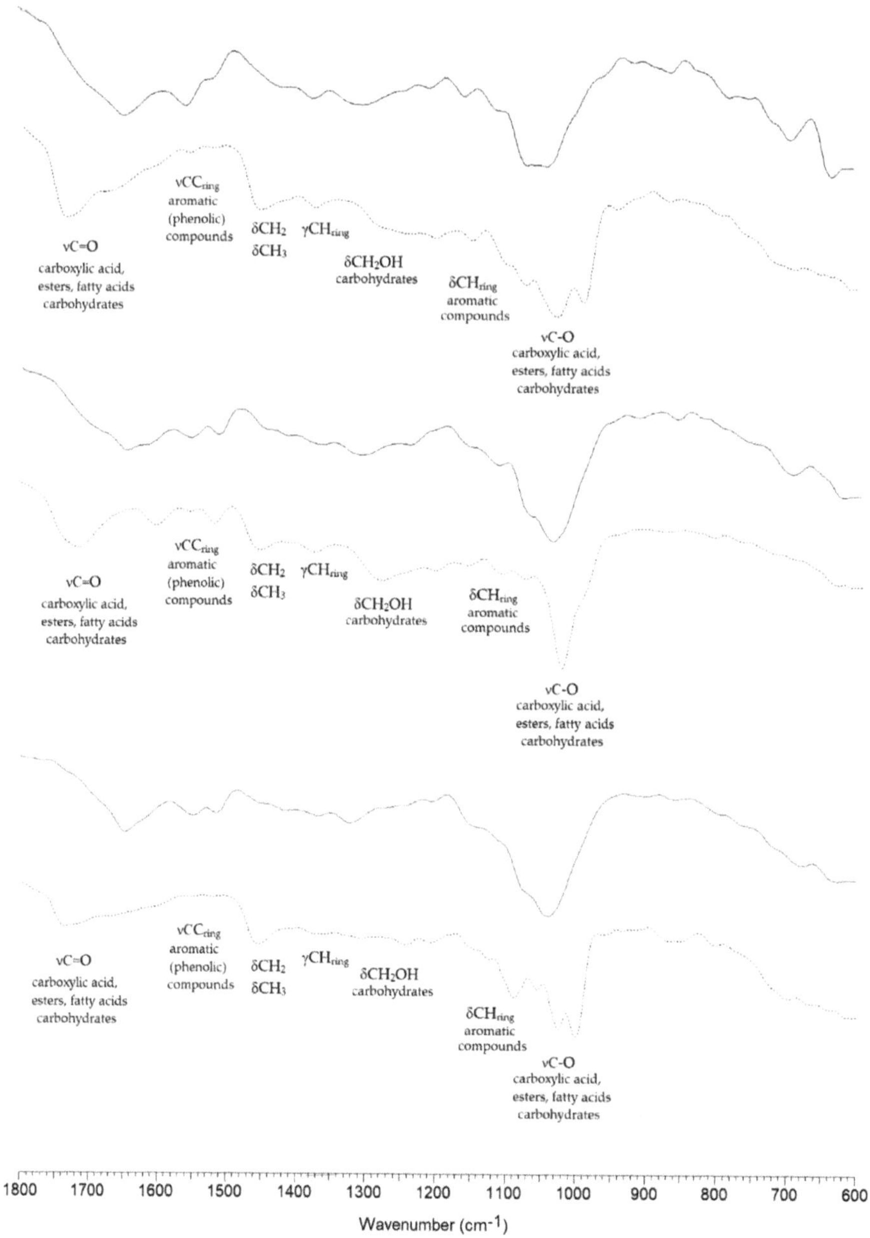

Figure 2. FTIR ATR spectra of *F. robusta* (black line), *P. pini* (blue line), and *P. igniarius* (green line). Continuous line—spectra of dried fruiting body; dashed line-spectra of methanolic extract.

Table 3. Wavenumbers (cm^{-1}), intensities, and assignments of bands occurring in the FTIR spectra of dried fruiting body samples and extracts of *F. robusta*, *P. pini*, and *P. igniarius*.

Fomitiporia robusta		*Porodaedalea pini*		*Phellinus igniarius*		Type of Vibration	Groups of Chemical Compounds
Dried Fruiting Body	Extract	Dried Fruiting Body	Extract	Dried Fruiting Body	Extract		
1645 s	1733 s 1675 m	1647 s	1711 vs	1647 s	1735 s	$\nu C = O$	Carboxylic acid, esters, fatty acids, and carbohydrates
1556 m	1554 m	1555 m	1552 m	1549 m	1553 w	νCC_{ring}	Aromatic/phenolic compounds
1515 m	1520 m	1514 m	1517 m	1515 m	1519 w	νCC_{ring}	Aromatic/phenolic compounds
-	1454 m	-	1451 s	-	1453 m	δCH_2, δCH_3	Aromatic and aliphatic compounds
-	-	1415 m	1410 m	1416 m	-	δCH_2, δCH_3, δOH	Aromatic and aliphatic compounds, and carbohydrates
1372 m	1374 m	1365 m	1374 m	1366 m	1374 m	γCH_{ring}	Aromatic/phenolic compounds
1301 m	-	1310 w	1280 m		1306 m	δCH_2OH	Carbohydrates
		1242 w		1245 m	1243 m	δCH_2OH	Carbohydrates
1206 w	1202 m	-	1202 m	1205 m	1205 m	νCO, νCC, δCOH	Carbohydrates
1168 m	1151 m	-	1155 m	-		δCH_{ring}	Aromatic/phenolic compounds
	1113 m	1116 m	1107 m	-	1123 m	δCH_{ring}	Aromatic/phenolic compounds
-	1075 m	-	1075 m	-	1088 s	δCH_{ring}	Aromatic/phenolic compounds
1054 vs, 1037 vs	1029 vs	1038 s	1024 s	1041 s	1055 s	$\nu C\text{-}O$	Carboxylic acids, fatty acids, and carbohydrates
-	993 m	-	-	-	1000 s		
913 w	946 w	915 w	-	899 m	-	δCOH, δCCH, δOCH	Carbohydrates
862 m	871 m	860 m	874 m	858 m	865 m	δCOH, δCCH, δOCH	Carbohydrates
779 m	779 m	-	809 m	-	800 m	δCOH, δCCH, δOCH	Carbohydrates
691 m	691 s	697 s	-	-	695 m	De-fring	Aromatic/phenolic compounds
632 s	657 s	624 m	633 m	627 s	635 s m	De-fring	Aromatic/phenolic compounds

ν—stretching, δ—deforming in plane, γ—deforming out of plane, ring—vibrations of atoms of an aromatic ring, and intensities: vs—very strong, s—strong, m—medium, w—weak.

2.5. GC/MS Analysis

The results of GC/MS analysis showed that the main components of the extracts of *Phellinus igniarius, Fomitiporia robusta,* and *Porodaedalea pini* are carbohydrates. The contents of monosaccharides and disaccharides constituting the polysaccharide fraction are 95.36% (*F. robusta*), 94.87% (*P. igniarus*), and 90.14% (*P. pini*) (Figure 3). The structural characteristics of polysaccharides from samples of *Phellinus* spp. are generally defined by their average molecular weight, monosaccharide composition, and chemical structure [40,41]. The main disaccharide building block of the polysaccharides of the tested samples is trehalose, the content of which is about 58% in *F. robusta* and *P. pini*, and as much as 68% in *P. igniarius*. High contents of polyhydric alcohols were also noted, i.e., the content of mannitol was about 19% in *F. robusta* and *P. pini* and 13% in *P. igniarius*; the content of ribitol was almost 15% in the extract from *F. robusta*, 11% in *P. igniarius*, and 4% in *P. pini*. In the extract of *P. igniarius*, small amounts of sugars such as arabinitol, β-fructofuranose, α-glucopyranose, β-glucopyranose, myo-inositol, xylitol, and galactitol (with a total content of less than 1%) were also noted. In the extract of *F. robusta*, sugars such as arabinitol, α-glucopyranose, β-glucopyranose, myo-inositol, deoxy-inositol, galactitol, and sucrose were determined. In the extract of *P. pini*, the presence of the same sugars as in the extract of *F. robusta* was noted, as well as additional sugars such as pinitol (1.83%), threitol (0.74%), D-ribose (0.23%), fructose (0.23%), β-L-fucopyranose, and smaller amounts of β-mannopyranose and α-galactopyranose.

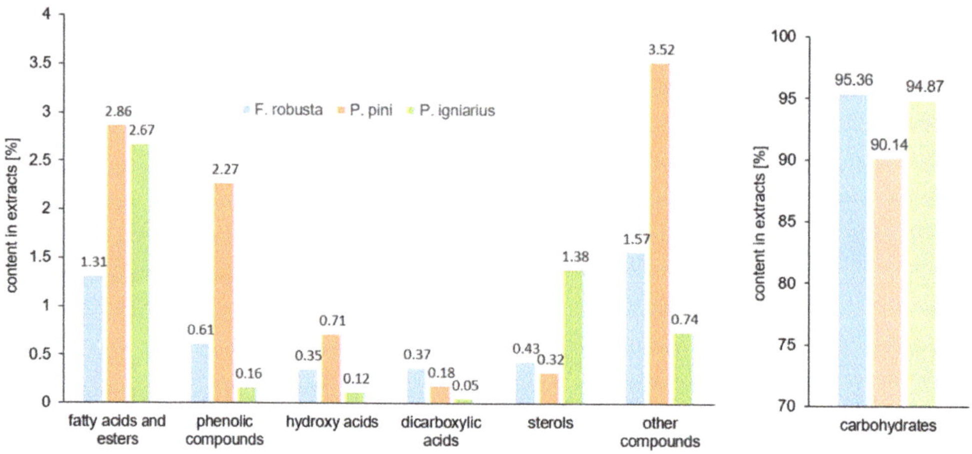

Figure 3. Main groups of compounds in the tested extracts of *F. robusta, P. pini,* and *P. igniarius*.

The content of fatty acids and their esters in the extracts of *Phellinus igniarius, Fomitiporia robusta,* and *Porodaedalea pini* was 2.67%, 1.31%, and 2.86%, respectively. The main acids identified in the extracts tested were palmitic acid (0.03%, 0.10%, and 0.59%, respectively), linoleic acid (0.20%, 0.35%, and 0.89%), and esters methyl palmitate (0.24%, 0.16%, and 0.25%) and methyl linoleate (1.07%, 0.47%, and 0.26%). In the extracts of *Phellinus igniarius* and *Porodaedalea pini*, the content of oleic acid (0.11% and 0.16%) was determined.

The highest content of phenolic acids was recorded in the extract from *P. pini* (2.27%), including 3-hydroxybenzoic acid (0.77%), protocatechuic acid (0.94%), caffeic acid (0.30%), and vanillic acid (0.10%), as well as small amounts of protocatechuic acid esters. These compounds are characterized by high antioxidant potential [42]. The *F. robusta* extract contained 0.61% of phenolic compounds, including 0.37% of 3-hydroxybenzoic acid, 0.12% of protocatechuic acid, and 0.05% of syringic acid. The lowest content of phenolic compounds was recorded in the extract from *P. igniarius* (0.16%), including 0.08% of 3-hydroxybenzoic acid, 0.05% of protocatechuic acid, and 0.04% of caffeic acid.

The extracts tested also contained compounds from the group of aliphatic hydroxy acids. The *P. igniarius* extract contained 0.12% of compounds from this group (lactic acid and malic acid), the *F. robusta* extract contained 0.35% (lactic acid, glycolic acid, malic acid, and citric acid), while the *P. pini* extract contained 0.71% (lactic acid, glycolic acid, glyceric acid, and malic acid).

The dicarboxylic acids present in the extracts tested were oxalic acid (only in *F. robusta*); succinic acid, which was present in all extracts in amounts below 0.15%; and fumaric acid, which was present in the amount of 0.27% in the *F. robusta* extract and 0.03% in the *P. igniarius* extract.

The sterol content in the *P. igniarius* extract was 1.38% (including ergosterol—0.9%, 3-hydroxyergosta-7,22-diene—0.27%, and 3-hydroxyergost-7-ene—0.12%). In the *F. robusta* extract, 0.43% sterols were noted (including 0.2% ergosterol, 0.14% 3-hydroxyergosta-7,22-diene, and 0.1% 3-hydroxyergost-7-ene). The *P. pini* extract contained 0.2% ergosterol and 0.11% 3-hydroxyergost-7-ene.

The extracts tested also contained small amounts of ethylene glycol (below 0.2%), 1,2-propanediol (below 0.15%), and pyroglutamic acid (below 0.2%). The glycerol content in *P. igniarius* was 0.51%, in *F. robusta* 0.19%, and in the extract from *P. pini* 1.42%. The extract from *P. pini* contained N-acetylglucosamine (0.59%), which is a component of biopolymers that build fungal cell walls.

2.6. Antioxidant Study

The antioxidant activity of methanolic extracts of *Fomitiporia robusta*, *Porodoedalea pini*, and *Phellinus igniarius* determined through tests with DPPH and ABTS radicals is presented in Figure 4. The concentration of extracts for which the level of radical inhibition was measured was 54 mg/L. The highest antioxidant activity in tests with ABTS and DPPH radicals was demonstrated by the extract from *P. pini*. Over 50% level scavenging of the ABTS radical and 48% level scavenging of the DPPH radical were noted. This extract was characterized by the highest content of phenolic compounds, which are mainly responsible for antioxidant activity. The lowest content of phenolic compounds was noted in the extract of *P. igniarius*; however, this extract did not show the lowest antioxidant potential among the extracts tested. A high content of sterols was found in this extract, which may also demonstrate antioxidant potential [43,44]. The level of ABTS radical neutralization by the *P. igniarius* extract was 45%, while that of the DPPH radical was 38%. The lowest antioxidant activity was demonstrated by the *F. robusta* extract, with 23% and 21.7% of radical removal, respectively, in the DPPH and ABTS tests. This extract is characterized by a low content of phenolic compounds, as well as a low content of sterols. For comparison, gallic acid at a concentration of 0.85 mg/L removes almost 100% of radicals in the ABTS test, and about 50% of radicals in the DPPH test.

Figure 5 presents the results of the fat oxidation inhibition test. The tested extracts at a concentration of 54 mg/L of *P. pini* and *P. igniarius* extracts caused nearly 50% inhibition of fat oxidation in the 5-day test. The *F. robusta* extract showed a slightly lower inhibition of oxidation fats (approximately 45%). The high sterol content in the *Phellinus igniarius* extract has a major impact on its antioxidant properties against fats.

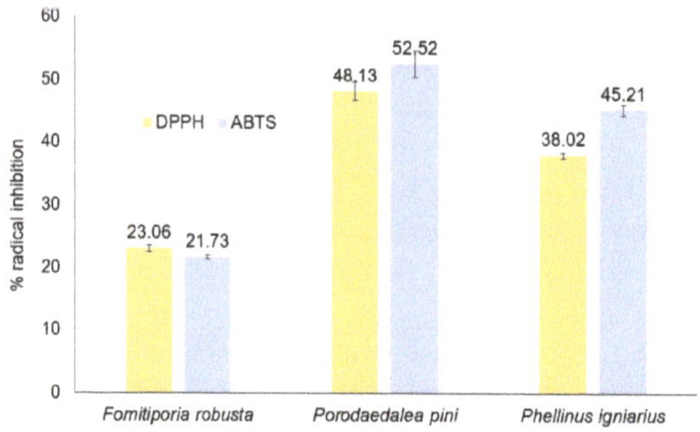

Figure 4. Antioxidant activity of *Phellinus* spp. extracts.

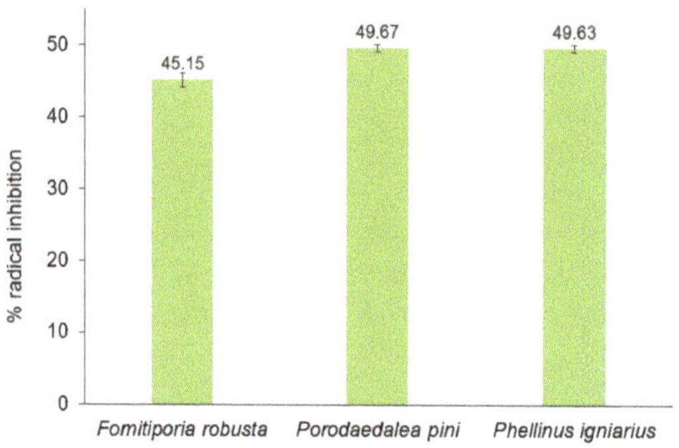

Figure 5. Lipid peroxidation inhibition after 5 days of reaction by *Phellinus* sample extracts.

2.7. Antimicrobial Study

To evaluate the therapeutic potential of tested fungal extracts, antimicrobial activity was assessed by determining the minimum inhibitory concentration (MIC), minimum bactericidal concentration (MBC), and minimum biofilm inhibitory concentration (MBIC). The MIC—the lowest concentration inhibiting bacterial growth—and the MBC—the concentration achieving bactericidal effects—were determined as described previously [45]. The MBIC is a concentration of tested agents that inhibits the formation of bacterial biofilm, which creates a protective barrier for bacteria, rendering them significantly more resistant to antibiotics [46]. Fungal extracts that inhibit biofilm formation can restore bacterial susceptibility to antibiotics, enhancing treatment efficacy. *Porodoedalea pini* (against *S. aureus* and *S. epidermidis*) and *P. igniarius* (against *S. epidermidis*) showed the highest anti-biofilm properties with MBIC values between 0.78 and 1.56 mg/mL against *Staphylococcus epidermidis* and <0.20 mg/mL against *S. aureus* clinical strains. *S. aureus* 4459 and 3907 strains showed increased resistance to the tested antibiotic, ERT, with MIC $\geq$ 0.50 mg/mL. Interestingly, in contrast to the other extracts tested, only the *P. pini* extract exhibited bactericidal activity against these two bacterial strains, as evidenced by its minimum bactericidal concentration (MBC) of 6.25 mg/mL. Furthermore, this extract demonstrated significant antibacterial activity with a minimum inhibitory concentration (MIC) of 1.56 mg/mL and a minimum biofilm inhibitory concentration (MBIC) below 0.20 mg/mL (Table 4, Figure 6). Planktonic bacteria—which are individual, free-floating cells—are more susceptible to antibacterial agents than biofilm bacteria, which form complex communities encased in a self-produced extracellular polymeric substance matrix. This matrix makes biofilm bacteria significantly more resistant, even up to 1000-fold [47]. *Phellinus iginarius* extract showed the lowest activity in inhibition of *P. aeruginosa* and *S. aureus* biofilm formation, which can be seen by comparing the MBIC results for all tested extracts. The lowest concentrations that inhibited biofilm formation of *S. aureus* and *S. epidermidis* were observed after treatment with *P. pini* extract, while *F. robusta* extract was the most effective against strains of *P. aeruginosa* (Figure 6).

Table 4. Antibacterial properties (MIC, MBC, and MBIC) of *Fomitiporia robusta*, *Porodoedalea pini*, and *Phellinus igniarius* methanolic extracts against clinical bacterial strains.

Methanolic Extract	Concentration [mg/mL]	Clinical Strains								
		Pseudomonas aeruginosa			*Staphylococcus aureus*			*Staphylococcus epidermidis*		
		1900	**1954**	**1955**	**3907**	**4069**	**4459**	**188**	**707**	**2542**
F. robusta	MIC	3.13	3.13	6.25	6.25	6.25	6.25	>6.25	>6.25	6.25
	MBC	BS [a]	BS [a]	BS [a]	BS [a]	BS [a]	BS [a]	BS [a]	BS [a]	BS [a]
	MBIC	3.13	3.13	6.25	6.25	3.13	6.25	- [c]	- [b]	6.25
P. pini	MIC	6.25	6.25	6.25	1.56	0.78	1.56	0.78	3.13	>12.5
	MBC	BS [a]	BS [a]	BS [a]	6.25	BS [a]	6.25	BS [a]	BS [a]	BS [a]
	MBIC	6.25	6.25	12.50	<0.20	<0.20	<0.20	0.78	- [b]	- [c]
P. igniarius	MIC	6.25	6.25	6.25	6.25	6.25	6.25	0.78	3.13	<0.78
	MBC	BS [a]	BS [a]	BS [a]	BS [a]	BS [a]	BS [a]	BS [a]	BS [a]	BS [a]
	MBIC	6.25	12.50	12.50	6.25	6.25	12.50	1.56	- [b]	<0.78
ERT	MIC	0.06	0.06	0.06	0.50	2×10^{-4}	>0.50	2×10^{-4}	2×10^{-3}	2×10^{-4}
	MBC	BS [a]	BS [a]	BS [a]	BS [a]	BS [a]	>0.50	4×10^{-3}	2×10^{-2}	4×10^{-3}
	MBIC	0.06	0.06	0.06	0.50	2×10^{-4}	- [c]	2×10^{-4}	- [b]	2×10^{-4}

[a] Bacteriostatic (BS) properties of compounds for which the MIC value was determined. [b] This bacterial strain does not form biofilm. [c] No inhibition of biofilm formation.

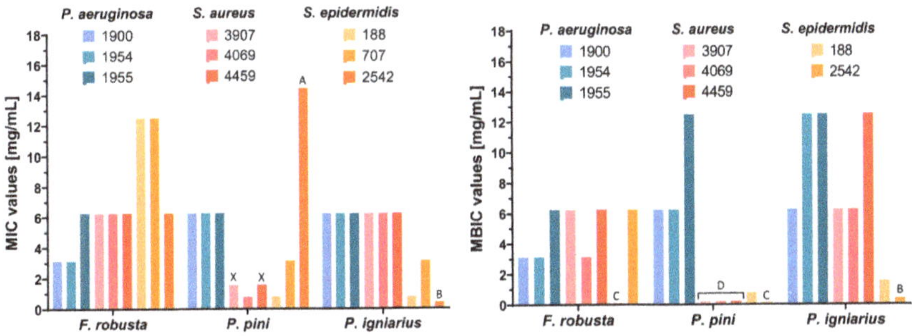

Figure 6. Antibacterial properties (MIC and MBIC) of *Fomitiporia robusta*, *Porodoedalea pini*, and *Phellinus igniarius* methanolic extracts against clinical bacterial strains; A—MIC > 12.5 mg/mL; B—MIC < 0.78 mg/mL; C—no inhibition of biofilm formation; D—MIC < 0.20 mg/mL; X—only for *S. aureus* 3907 and 4459 strains the MBC value was obtained (6.25 mg/mL).

2.8. Cytotoxicity

The cytotoxicity of the examined fungal extracts was evaluated using the Neutral Red (NR) assay, which measures cell viability based on the uptake of NR dye by functional lysosomes [48]. Only *F. robusta* extract showed a lack of cytotoxicity against human keratinocytes at 100 and 400 μg/mL; however, the highest concentration (800 μg/mL) of this extract caused an approximately 70% reduction in cell viability. The other two extracts (*P. pini* and *P. igniarius*) showed varying cytotoxicity at all three concentrations, with the *P.*

igniarius extract causing approximately 9, 43, and 69% decreases in HaCaT cell viability at concentrations of 100, 400, and 800, respectively (Figure 7).

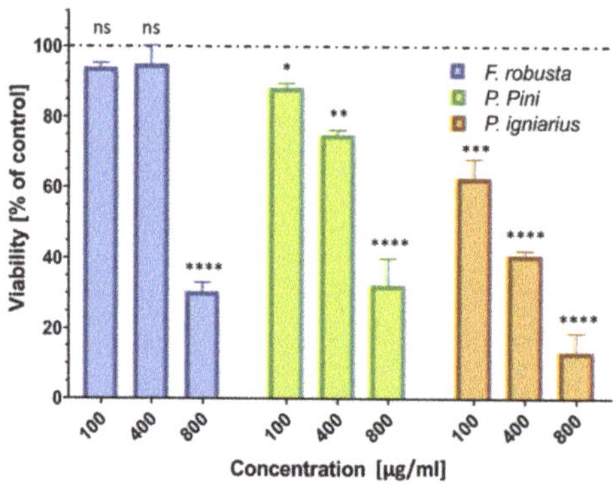

Figure 7. HaCaT cell viability after 24-h treatment with *Fomitiporia robusta*, *Porodoedalea pini*, and *Phellinus* igniarius methanolic extracts in the following concentrations: 100, 400, and 800 µg/mL. Viability was evaluated with the use of the NR assay. Statistical differences between the untreated control and cells treated with extracts were evaluated with the use of one-way ANOVA: * $p < 0.05$, ** $p < 0.005$, *** $p < 0.002$, and **** $p < 0.0001$.

2.9. Reactive Oxygen Species

Reactive oxygen species (ROS) generation is a critical mediator of cellular stress and oxidative damage that is implicated in diverse pathologies [49]. To investigate the effects of fungal extracts on intracellular ROS levels, the CellROX Green assay, a fluorogenic probe quantifying ROS through oxidation-induced fluorescence, was utilized. Only incubation with P. igniarius extract showed a statistically significant pro-oxidative effect against HaCaT cells (Figure 8).

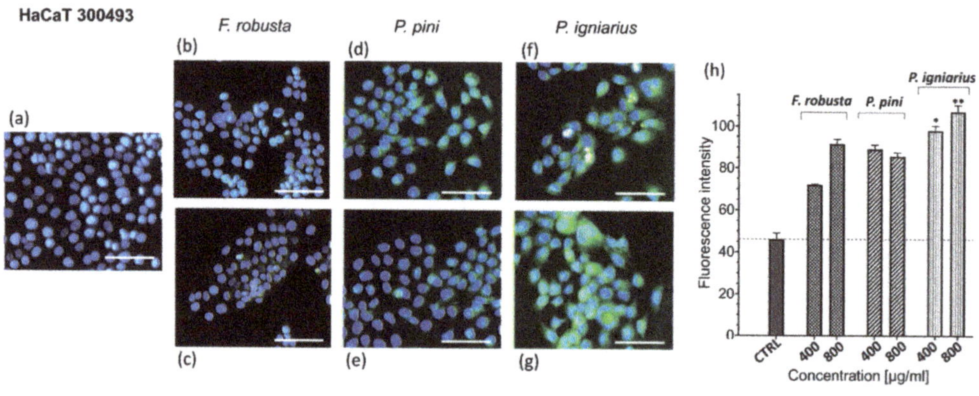

Figure 8. Fomitiporia robusta, Porodoedalea pini, and *Phellinus* igniarius extracts affect ROS production in HaCaT cells. Representative fluorescent images (40 ×) of CellROX ®Green Reagent-labeled cells, treated for 30 min: (**a**) untreated cells; 400 µg/mL (**b**) and 800 µg/mL (**c**) extracts of *F. robusta;* 400 µg/mL (**d**) and 800 µg/mL (**e**) extracts of *P. pini;* and 400 µg/mL (**f**) and 800 µg/mL (**g**) extracts of *Phellinus igniarius;* scale bar = 100 µm. Bar graph (**h**) summarizing the effect on ROS production after treatment with different extracts at concentrations of 400 and 800 µg/mL. Data are represented as mean ± SD; the presented results are not statistically significant, unless otherwise stated (* $p < 0.05$; ** $p < 0.01$).

3. Materials and Methods

3.1. Collection of Fungal Samples in the Field

This study investigated 30 samples belonging to 22 species of the genus *Phellinus sensu lato* (Table 5). Fruiting bodies were collected, among other specimens, in the Białowieża Forest (Poland); Village Kozhle (North Macedonia); Estremadura, Sesimbra, and Lagoa de Albufeira (Portugal); Zlatari close to Prishtina (Kosovo); and Spoleto and the Bosco Siro Negri State Nature Reserve (Italy). Each fruiting body found in the field was collected together with the substrate. Prior to collection, a photograph was taken under natural conditions, documenting the general habit and features of the fruiting body that may be lost during drying. The specimens were collected in paper envelopes on which the place and date of collection; the in situ fruiting body photo numbers; the type of substrate and host; and the habitat data were recorded. Additionally, in the case of saprobes growing on dead wood, the species was identified and information on the stage of decomposition of the substrate was recorded.

3.2. Morphological Identification

Slides were prepared under an OPTA-TECH MI6 binocular (OPTA-TECH) magnifier at 5–25x magnification. Thin sections of the fruiting body were obtained with a microtome or razor blade. If necessary, a preparation was made from the relevant part of the fruiting body—i.e., tube and context—and, for some species, also from the cap cover. The tube preparation was prepared in such a way as to make the following characteristics visible: the arrangement of the filaments in the trama; the appearance of the bristles in the hymenium and context, characteristic of the genus *Phellinus;* the appearance of the filament tips on the blades of the inner tube septa; and the appearance and elements of the hymenium at the bottom of the tubes. An OPTATECH LAB-40 light microscope with variable phase contrast and a NIKON ECLIPSE Ni with Nomarski contrast were used to observe microscopic features.

Preparations from fresh specimens were made in water and from dry specimens in 3–10% KOH. All available features were observed in these media. To improve observational comfort and obtain a more contrasting image, the slides were stained with Congo Red. This dye stains the cell walls and, less intensely, the cell contents. Given that microscopic features in the genus *Phellinus* s.l. have a dark brown color, lactophenol was used to lighten them. The amyloid and dextrorotatory properties of the microstructural elements were checked in Melzer's reagent and in 0.1% Cotton Blue in 60% lactic acid to determine the degree of cyanophilicity of the spore walls and filaments.

The primary sources of information for the taxonomic analysis of fungi of the genus *Phellinus* s.l. were the following keys and monographic studies: Domanski et al. (1967, 1973) [50,51]; Ryvarden and Gilbertson (1993, 1994) [52,53]; Larsen and Cobb-Poulle (1990) [54]; and Sell (2008) [55]. The nomenclature was adopted from MycoBank(available online: http://www.mycobank.org (accessed on 26 March 2025)). The correctness of the spelling used and the author of the name at the genus level was confirmed with the latest edition of the *Dictionary of the Fungi* (Kirk et al. 2008) [56]. Acronyms of herbaria follow the Index Herbariorum. Documentary material (dry fungal specimens) was deposited in the herbarium of the Institute of Forest Sciences, Bialystok University of Technology (BLS).

Table 5. Characteristics of evidentiary specimens of *Phellinus* spp. samples.

Sample No.	Evidence Specimen Number	Species Name	Substrate/Host (Species Name)	Collection Site	Collected and Identified
16.D	BLS-M-05505	***Porodaedalea chrysoloma*** *(Fries) Fiasson & Niemelä [syn. Phellinus chrysoloma (Fr.) Donk]*	*Picea abies*	Poland, Białowieża Forest	Leg. et det. Marek Wołkowycki
17.D	BLS-M-05498	***Porodaedalea pini*** *(Brot.) Murrill [syn. Phellinus pini (Brot.) A. Ames]*	*Pinus sylvestris*	Poland, Białowieża Forest	Leg. et det. Marek Wołkowycki
18.D	BLS-M-05506	***Phellinus tremulae*** *(Bondartsev) Bondartsev & P.N. Borissov*	*Populus tremula*	Poland, Białowieża Forest	Leg. et det. Marek Wołkowycki
19.D	BLS-M-05512	***Phellinus laevigatus*** *(Fries) Bourdot & Galzin*	*Betula pendula*	Poland, Białowieża Forest	Leg. et det. Marek Wołkowycki
21.D	BLS-M-05493	***Phellinus tuberculosus*** *(Baumgarten) Niemelä [syn. Phellinus pomaceus (Pers.) Maire]*	*Prunus domestica*	Poland, Białowieża Forest	Leg. et det. Marek Wołkowycki
22.D	BLS-M-05497	***Fomitiporia robusta*** *(P. Karst.) Fiasson & Niemelä [syn. Phellinus robustus (P. Karst.) Bourdot & Galzin]*	*Quercus robur*	Poland, Białowieża Forest	Leg. et det. Marek Wołkowycki
25.D	BLS-M-02596	***Phellinopsis conchata*** *(Persoon) Dai [syn. Phellinus conchatus (Pers.: Fr.) Quél]*	*Salix caprea*	Poland, Białowieża Forest	Leg. et det. Marek Wołkowycki
38.D	BLS-M-01027	***Phellinus igniarius*** *(L.) Quél.*	*Salix fragilis*	Poland, Sokolskie Hill, Losiniany	Leg. et det. Marek Wołkowycki
118.D	BLS-M-05609	***Phellinus alni*** *(Bondartsev) Parmasto*	*Alnus glutinosa*	Poland, Białowieża Forest	Leg. et det. Marek Wołkowycki
160.D	BLS-M-05495	***Fuscoporia ferruginosa*** *(Schrad ex J.F. Gmelin) Murill [syn. Phellinus ferruginosus (Schrad.: Fr.) Pat.]*	*Fraxinus excelsior*	Poland, Białowieża Forest	Leg. et det. Marek Wołkowycki
254.D	BLS-M-05492	***Phellinus alni*** *(Bondartsev) Parmasto*	*Malus domestica*	Poland, Białowieża Forest	Leg. et det. Marek Wołkowycki
255.D	BLS-M-05490	***Phellinus alni*** *(Bondartsev) Parmasto*	*Carpinus betulus*	Poland, Białowieża Forest	Leg. et det. Marek Wołkowycki
256.D	BLS-M-05448	***Phellinidium ferrugineofuscum*** *(P. Karst) Fiasson & Niemelä [syn. Phellinus ferrugineofuscus (P. Karst.) Bourdot]*	*Picea abies*	Poland, Białowieża Forest	Leg. et det. Marek Wołkowycki

Table 5. *Cont.*

Sample No.	Evidence Specimen Number	Species Name	Substrate/Host (Species Name)	Collection Site	Collected and Identified
257.D	BLS-M-05494	***Fomitiporia punctata*** *(Fries ex P. Karsten) Murill [syn. Phellinus punctatus (P.Karst.) Pilát]*	*Corylus avellana*	Poland, Białowieża Forest	Leg. et det. Marek Wołkowycki
258.D	BLS-M-05499	***Fomitiporia punctata*** *(Fries ex P. Karsten) Murill [syn. Phellinus punctatus (P.Karst.) Pilát]*	*Sorbus aucuparia*	Poland, Białowieża Forest	Leg. et det. Marek Wołkowycki
261.D	BLS-M-05509	***Phellinus nigricans*** *(Fries) P. Karsten*	*Betula pendula*	Poland, Białowieża Forest	Leg. et det. Marek Wołkowycki
262.D	BLS-M-05507	***Phellinus populicola*** *Niemelä*	*Populus tremula*	Poland, Białowieża Forest	Leg. et det. Marek Wołkowycki
266.D	BLS-M-02977	***Fomitiporia robusta*** *(P. Karst.) Fiasson & Niemelä [syn. Phellinus robustus (P. Karst.) Bourdot & Galzin]*	*Robinia pseudoacacia*	Poland, Białowieża Forest	Leg. et det. Marek Wołkowycki
267.D	BLS-M-04016	***Phellinus nigricans*** *(Fries) P. Karsten*	*Betula pubescens*	Poland, Białowieża Forest	Leg. et det. Marek Wołkowycki
268.D	BLS-M-05496	***Fomitiporia hartigii*** *(Allesch. & Schnabl) Fiasson & Niemelä [syn. Phellinus hartigii (Allesch. & Schnabl) Pat]*	*Abies alba*	Poland, Beskid Niski	Leg. et det. Anna Hreczka rev. Marek Wołkowycki
279.D	BLS-M-05599	***Phellinus tuberculosus*** *(Baumgarten) Niemelä [syn. Phellinus pomaceus (Pers.) Maire]*	*Cerasus avium*	Poland, Kuraszewo, wild fruit orchard	Leg. Konrad Wilamowski det. Marek Wołkowycki
280.D	BLS-M-05610	***Fomitiporia robusta*** *(P. Karst.) Fiasson & Niemelä [syn. Phellinus robustus (P. Karst.) Bourdot & Galzin]*	*Quercus rubra*	Poland, Białowieża Forest	Leg. et det. Marek Wołkowycki
298.D	BLS-M-10167	***Porodaedalea pini*** *(Brot.) Murrill (syn. Porodaedalea pini (Brot.) A. Ames)*	*Pinus pinea*	Portugal, Estremadura, Sesimbra, Lagoa de Albufeira	Leg. et det. Mitko Karadelev
300.D	BLS-M-10094	***Phellinus nigricans*** *(Fries) P. Karsten*	*Fagus sylvatica*	Poland, Białowieża Forest	Leg. Konrad Wilamowski det. Marek Wołkowycki
300.Da	BLS-M-10164	***Phellinus nigricans*** *(Fries) P. Karsten*	*Fagus sylvatica*	Poland, Baltic Coast, "Zagorska Struga" Valley	Leg. et det. Mirosław Wantoch-Rekowski rev. Marek Wołkowycki

Table 5. *Cont.*

Sample No.	Evidence Specimen Number	Species Name	Substrate/Host (Species Name)	Collection Site	Collected and Identified
301.D	BLS-M-10093	***Phellinus igniarius*** *(L.) Quél.*	*Sorbus intermedia*	Poland, Romnicka Forest	Leg. Konrad Wilamowski det. Marek Wołkowycki
319.D	BLS-M-10412	***Fuscoporia torulosa*** *(Pers.) T. Wagner & M. Fisch [syn. Phellinus torulosus (Pers.) Bourdot & Galzin]*	*Quercus ilex*	Italy, Perugia, Sacro Bosco di Monteluco, Spoleto	Leg. et det. Carolina Girometta
320.D	BLS-M-10413	***Fuscoporia torulosa*** *(Pers.) T. Wagner & M. Fisch [syn. Phellinus torulosus (Pers.) Bourdot & Galzin]*	*Robinia pseudoacacia*	Italy, Pavia, "Bosco Siro Negri" State Nature Reserve	Leg. et det. Carolina Girometta
322.D	BLS-M-10414	***Fomitiporia mediterranea*** *M. Fisch. [syn. Phellinus punctatus (P. Karst.) Pilát]*	*Robinia pseudoacacia*	Italy, Pavia, "Bosco Siro Negri" State Nature Reserve	Leg. et det. Carolina Girometta
329.D	BLS-M-05505	***Porodaedalea chrysoloma*** *(Fr.) [syn. Porodaedalea abietis] [syn. Phellinus chrysoloma (Fr.) Donk]*	*Picea abies*	Poland, Białowieża Forest	Leg. et det. Marek Wołkowycki

3.3. Genetic Identification

In order to confirm genetic identity, DNA was isolated from the fungal samples with the Plant & Fungi DNA Purification Kit (EURx, Gdańsk, Poland). The obtained preparations were characterized spectrophotometrically (Varioskan LUX, Thermo Fisher Scientific, Waltham, MA, USA) and fluorometrically (Qubit 2.0, Thermo Fisher Scientific, US). Whole genome sequencing was performed using the DNA library prepared with the ITD xGen DNA EZ kit on an Illumina NovaSeq S4 system. Paired reads (2 × 150 bp) were cleared and then subjected to de novo assembly using MEGAHIT v. 1.2.9 and CLC Genomics Workbench v. 12.0.3. The resulting contigs were searched to identify reference sequences, including the small subunit ribosomal RNA gene fragment, the internal transcribed spacer 1, the 5.8 S ribosomal RNA gene, the internal transcribed spacer 2, and the partial large subunit ribosomal RNA gene used in taxonomic analyses. Sequences were deposited at NCBI (PV125346.1 and PV186754.1). The fungal material was dried.

3.4. Preparation of Extracts

The fruiting bodies were cleaned and separated from the remaining substrate and then dried at a temperature not exceeding 40 °C. The fruiting bodies were crushed in a laboratory grinder. The samples were then macerated in methanol (99.8%; POCH Avantor Performance Materials, Poland) at a ratio of 1:6 [g/mL]. The maceration period was no less than 2 months, and the samples were shaken systematically.

The finished tinctures were filtered through 80 g/m^2 filter paper, and the solvent was evaporated in two steps. In the first step, most of the solvent was removed in a rotary evaporator Rotavapor® R-100 (Büchi, Switzerland) at 46 °C, rotation speed 3–4 units, and under the reduced pressure from 300 to 100 mbar. In the second step, the extract was collected from the walls of the round-bottomed extraction flask together with a small amount of methanol and placed in the glass extraction cells of a parallel evaporation system, MultivaporTM P-12 (Büchi), where it was incubated at 46 °C and under reduced pressure from 400 to 100 mbar. Each solid-state extract was stored in the dark at 10 °C in the Fungi Extract Bank® collection (available online: https://fungiextractbank.com/en (accessed on 1 April 2025)).

For fungal samples where the residue in the flask did not dissolve in methanol, the flask was rinsed with distilled water, then frozen at −25 °C, and lyophilized at −40 °C at a pressure of 0.04 mbar; these samples were stored for other tests. The methanolic extracts obtained had a plastic consistency, high viscosity, and colors from brown to dark brown. Dry methanolic extracts were used for further studies.

3.5. Antimicrobial Activity—Preliminary Studies

Preliminary screening tests of the antimicrobial activity of 30 samples of fungi were performed. Antimicrobial assays of fungal extracts were conducted as part of the project. For each extract, the MIC (minimum concentration of a substance that inhibits microbial growth) was determined using the microdilution method in Mueller–Hinton Broth II (MH II) medium using serial two-fold dilutions of the fungal extract, according to the procedure described by Balouiri et al. and Rašeta et al. [57,58]. The MIC value was expressed as the lowest concentration of fungal extract (in mg/mL) that completely inhibited the growth of the microorganism, as identified based on the turbidity of the culture after 24 h. To confirm the MIC value, cultures were performed on plates of three dilutions that showed the lowest or no turbidity. This study was performed on bacterial strains *Klebsiella pneumoniae* (ATCC-13883), *Pseudomonas aeruginosa* (ATCC-27853), *Staphylococcus aureus* (ATCC-25923), *Staphylococcus epidermidis*, and *Kocuria rhizophila* (ATCC-9341), and fungi *Candida albicans*

(ATCC-1023), which were obtained from the ATCC collection. Based on the MIC results, three extracts of *Phellinus* species were selected for further studies: *Fomitiporia robusta*, *Phellinus igniarius*, and *Porodaedalea pini*.

3.6. Elemental Analysis

The determination of selected elements in fungal extracts (*Phellinus igniarius*, *Fomitiporia robusta*, and *Porodaedalea pini* (Figure 9)) was performed using X-ray spectrofluorimetry. An amount of 900 μL of the aqueous extract solution was collected in an Eppendorf tube, and 100 μL of internal standard (Ga at 10 mg/L) was added to obtain a final concentration of 1 mg/L. An amount of 10 μL of the prepared solution was applied to measuring slides and dried at 50 °C. The prepared samples were analyzed using a TXRF (Total X-Ray Reflection Fluorescence) S2 PICOFOX apparatus, Bruker, Berlin, Germany.

Figure 9. Fruiting bodies: *Phellinus pini* syn., *Porodaedalea pini.*, *Phellinus robustus* syn. *Fomitiporia robusta*, and *Phellinus igniarius* (Photography: *M. Wołkowycki* and *M. Karadelev*).

3.7. Thermal Analysis

Thermal analysis of dried fruiting body samples of *Phellinus igniarius*, *Fomitiporia robusta*, and *Porodaedalea pini* was performed on a Libra thermal analyzer by TG 209 F1 (Netzsch company, Selb, Germany). Dried material weighing 10 ± 0.1 g was heated in ceramic crucibles in a nitrogen atmosphere (15 mL/min) with a temperature increase of 10 °C/min. The thermal decomposition process was carried out in the temperature range from 30 to 660 °C. The TG and DTG curves were recorded.

3.8. Gas Chromatography–Mass Spectrometry (GC-MS) Analysis

The fungus extract chemical composition was analyzed using an Agilent 7890A gas chromatograph coupled with an Agilent 5975C mass spectrometer. The methanol extract from fungus (10 mg) was diluted in 1 mL of pyridine, and 0.1 mL N,O-bis (trimethylsilyl)-trifluoroacetamide (BSTFA) was added. The sample was heated for 30 min at 60 °C. After silylation, 1 μL of the sample was entered into a GC/MS injector. The injector was set to a 1:10 split, and the temperature was 300 °C. The separation of compounds was conducted on a silica column HP-5MS (30 m × 0.25 mm × 0.25 μm), and the rate of helium flow was 1 mL/min. The starting column temperature was 50 °C, increased to 325 °C at 3 °C/min, and the final temperature was held for 10 min. The quadrupole and ion source temperatures were 150 °C and 230 °C, respectively. The energy of ionization was 70 eV, and detection was performed in a range of 41–800 units in a full scan mode. After peak integration, the percentage of every compound in the total ion current (% TIC) was calculated. Both mass spectra and retention indices were used for the identification of chemical compounds. The Isidorov (2020) [59], NIST (2020) [60], and Wiley (2020) [61] libraries of mass spectra were used for analysis. The experimental retention indices ($RI_{exp.}$) were calculated in relation to the retention times of C8-C40 n-alkanes. The $RI_{exp.}$ values were compared with the literature retention indices ($RI_{lit.}$).

3.9. Spectroscopy

The FTIR spectra of both extracts and dried fruiting bodies of 3 selected fungal samples, *Phellinus igniarius, Fomitiporia robusta,* and *Porodaedalea pini,* were obtained using the ATR multireflection technique. The spectra were recorded in the range of 2000–600 cm^{-1} using an Alfa spectrometer (Bruker, Billerica, MA, USA).

3.10. Antioxidant Properties and Lipid Peroxidation Analysis

The antioxidant properties were determined using DPPH and ABTS radical assays.

DPPH radical test

The determination of antioxidant activity via the DPPH radical assay was conducted following the method described in [62]. Briefly, 1 mL of the extract was mixed with 2 mL of a DPPH radical solution. A control sample was prepared simultaneously, in which 1 mL of solvent (methanol) was used instead of the extract. The samples were incubated for 1 h. After incubation, absorbance was measured at 516 nm using a UV–VIS spectrophotometer (NANOCOLOR® VIS, MACHEREY-NAGEL). Each sample was analyzed in triplicate. The free radical scavenging activity of DPPH was calculated based on absorbance measurements, with the average percentage of inhibition determined using the following equation:

$$\%I = \frac{A_c^{516} - A_s^{516}}{A_c^{516}} \cdot 100\% \quad (1)$$

$\%I$—percentage of inhibition; $A_c{}^{516}$—absorbance of the control sample at 516 nm; and $A_s{}^{516}$—absorbance of the sample containing the extract at 516 nm.

ABTS radical test

The antioxidant activity via the ABTS radical assay was determined based on the study by the authors of [63]. ABTS was prepared at a concentration of 7 mM and mixed with 2.45 mM of potassium persulfate ($K_2S_2O_8$) in a 1:1 ratio, followed by incubation in the dark at 20 °C for approximately 12 h to generate the $ABTS^+$ radical. For the assay, 1.5 mL of the working $ABTS^+$ solution was diluted with 30 mL of methanol. Then, 1 mL of the extract was mixed with 1 mL of the $ABTS^+$ solution. A control sample was prepared simultaneously, in which the extract was replaced with methanol. The samples were incubated for 7 min at 20 °C. After incubation, absorbance was measured at 734 nm using a UV–VIS spectrophotometer. Each sample was analyzed in triplicate. The average percentage of inhibition of the $ABTS^+$ cation radical was calculated using the following equation:

$$\%I = \frac{A_c^{734} - A_s^{734}}{A_c^{734}} \cdot 100\% \quad (2)$$

$\%I$—percentage of inhibition; $A_c{}^{734}$—absorbance of the control sample at 734 nm; $A_s{}^{734}$—absorbance of the sample containing the extract at 734 nm.

Lipid peroxidation

To evaluate the ability of extracts to inhibit lipid peroxidation, following the method described in [64], a linoleic acid emulsion was prepared by mixing 0.312 mL of linoleic acid, 0.256 mL of Tween 20, and 0.05 M of phosphate buffer in a 50 mL volumetric flask, with the final volume adjusted using the buffer. Then, 1 mL of each extract, prepared at a concentration of 45 mg/25 mL, was added to 1.5 mL of the emulsion. Each extract was tested in five replicates, and control samples were prepared in parallel by replacing the extract with 1 mL of methanol. The test tubes were sealed and incubated at 40 °C. Once the target temperature was reached, 0.1 mL of the reaction mixture was sampled from each tube and mixed with 4.7 mL of 75% ethanol and 0.05 mL of 30% ammonium thiocyanate. After a 3-min reaction period, 0.05 mL of 0.02 M $FeCl_2$ solution in 3.5% HCl was added. Absorbance was measured at 500 nm using a UV–VIS spectrophotometer, with 75% ethanol used as a reference. Measurements were performed every 24 h for 5 days, with 0.1 mL volumes withdrawn from the same emulsions at each time point. The percentage inhibition of linoleic acid peroxidation (%I) was calculated using the following formula:

$$\%I = \frac{A_c^{500} - A_s^{500}}{A_c^{500}} \cdot 100\% \quad (3)$$

%I—percentage of peroxidation inhibition; A_c^{500}—absorbance of the control sample at 500 nm; and A_s^{500}—absorbance of the sample containing the extract at 500 nm.

3.11. Antibacterial Activity

Bacterial Strains and Culture Conditions

Clinical bacterial isolates, specifically *Pseudomonas aeruginosa* (1900, 1954, and 1955), *Staphylococcus aureus* (3907, 4069, and 4459), and *Staphylococcus epidermidis* (188, 707, and 2542), were obtained from the Department of Medical Laboratory Diagnostics, Provincial Specialist Hospital in Rzeszow, and deposited at the Department of Biotechnology and Bioinformatics, Faculty of Chemistry, Rzeszow University of Technology. All bacterial strains were cultured at 37 °C in a New Brunswick Innova 40 Shaker (Eppendorf AG, Hamburg, Germany) until reaching a turbidity equivalent to 0.5 McFarland standard (approximately 10^8 colony-forming units per mL [CFU/mL]). The obtained cultures were then diluted to a final concentration of 10^5 CFU/mL. All bacterial cultures were prepared under aseptic conditions within an ESCO Airstream Laminar Flow Cabinet.

Antimicrobial Activity Assays

Minimum inhibitory concentrations (MICs) were determined using a micro-broth dilution method in Mueller–Hinton Broth (MHB), as previously described [65]. Serial two-fold dilutions of extracts (0.39–25 mg/mL) were prepared in MHB. Following 24-h incubation at 37 °C, the MIC was defined as the lowest concentration inhibiting visible growth, confirmed by OD_{600} measurement (BIO-RAD microplate reader). Minimum bactericidal concentrations (MBCs) were determined by subculturing MIC-, 2xMIC-, and 4xMIC-treated cultures onto Mueller–Hinton Agar (MHA) for 24 h at 37 °C. The MBC was defined as a $\geq$ 99.9% reduction in CFU compared with the untreated controls. Erythromycin (ERT) susceptibility was assessed using the microdilution method (0.03–500 µg/mL).

Anti-Biofilm Activity

The ability of extracts to inhibit biofilm formation was assessed using the MTT assay. Bacterial cultures, conditions, and density of the culture, as well as the 96-well plate preparation with extract dilutions, were the same as for the MIC analysis (see above in Section 2.5 Antimicrobial Activity—Preliminary Studies). After 24-h incubation, non-adherent cells were removed, and adherent cells were stained with 0.5% MTT in PBS for 2 h. If bacteria in biofilm are alive and metabolically active, their oxidoreductase enzymes convert the yellow MTT dye into insoluble purple formazan crystals. These crystals accumulate within the cells or on the biofilm surface. To quantify the formazan, a solubilization solution (DMSO) was added to dissolve the crystals, forming a homogeneous purple solution [65]. Subsequently, the absorbance was measured at 600 nm (Varioskan™ LUX multimode microplate reader, Thermo Scientific, Waltham, MA, USA). The minimum biofilm inhibitory concentration (MBIC) was defined as the lowest concentration inhibiting biofilm formation compared with media-only controls. The experiments were conducted in triplicate.

3.12. Cytotoxic Study

Cytotoxicity Assessment

Human immortalized epidermal cells (HaCaT 300493, Cytion, Eppelheim, Germany) were cultured in DMEM supplemented with 10% FBS and 1% penicillin–streptomycin, at 37 °C with 5% CO_2. Cell viability was determined via trypan blue exclusion using an Automatic Cell Counter TC20™ (Hercules, CA, USA) (Cytotoxicity was evaluated using the Neutral Red (NR) assay [48]. The cells were seeded in 96-well plates and exposed to extracts (100, 400, and 800 μg/mL in DMSO, final DMSO $\leq$1%) for 24 h. NR solution (0.033%) was added for 2 h; afterwards, plates were washed with PBS, and then destaining solution (50% ethanol, 49% water, and 1% acetic acid) was added. Absorbance was measured at 540 nm (Varioskan™ LUX multimode microplate reader, ThermoFisher Scientific, Waltham, MA, USA). Cell viability was expressed as a percentage of the untreated controls.

Reactive Oxygen Species (ROS) Quantification

HaCaT cells were treated with *Fomitiporia robusta, Porodaedalea pini*, and *Phellinus igniarius* extracts (400 and 800 μg/mL) for 2 h. Non-treated cells were used as a negative control. ROS production was assessed using the CellROX™ Green Oxidative Stress kit (ThermoFisher Scientific, Waltham, MA, USA). The cells were incubated with 5 μM CellROX Green for 30 min, fixed with 4% paraformaldehyde, and counterstained with Hoechst 33342. Fluorescence imaging was performed using the cellSens Software 4.3 (Olympus, Center Valley, PA, USA). CellROX fluorescence intensity was quantified using ImageJ 1.54 g [66].

3.13. Statistical Analysis

The final results obtained from triplicate experiments are presented as means ± SD. GraphPad Prism version 8.0.1 (GraphPad Software, Boston, MA, USA, available online: http://www.graphpad.com (accessed on 20 April 2025)) was used to compare the differences between groups. Statistical significance was indicated for p-values < 0.05.

4. Conclusions

Antimicrobial activity studies of 30 samples of methanol extracts (22 *Phellinus* species) of fungi from the genus *Phellinus* allowed for the identification of extracts with the highest activity. The identified extracts were subjected to further detailed studies. The antimicrobial activity of fungi from the *Porodaedalea pini*, *Fomitiporia robusta*, and *Phellinus igniarius* species was tested on strains of multidrug-resistant bacteria. It was observed that the habitat and origin of a given species have a significant influence on biological activity (related to the content of biologically active compounds). Studies of the content of active compounds in the analyzed extracts showed that each of the three selected extracts was characterized by a high content of sugars (over 90% of the extract mass). *Porodaedalea pini* extract was characterized by the highest content of phenolic compounds among the three samples analyzed in detail, which determined the highest antioxidant activity of the extract of this mushroom in antioxidant tests.

Antimicrobial tests performed on antibiotic-resistant bacteria (*Pseudomonas aeruginosa, Staphylococcus aureus*, and *Staphylococcus epidermidis*) showed that *P. pini* extract was characterized by the highest activity in neutralizing these microorganisms among the three extracts tested. Cytotoxicity tests of the three selected extracts from fungi of the *Phellinus* species against human keranocytes showed that only the *F. robusta* extract did not show a cytotoxic effect, while the remaining two extracts (*P. pini* and *P. igniarius*) were biocidal against keranocyte cell lines.

The effects of the three mushroom extracts on the level of reactive oxygen species (ROS) generation—which are critical mediators of cellular stress and oxidative damage associated with various pathologies—were also studied. The studies were performed on the human keratinocyte HaCaT cell line. It was observed that only the *P. igniarius* extract showed a statistically significant pro-oxidant effect against HaCaT cells.

The conducted experiments provided promising results that are expected to form the basis of further studies on the biological activity of *Phellinus sensu lato* fungi. It is anticipated that the use of other methods for extracting biologically active compounds from *Phellinus* fungi will allow for obtaining extracts with high biological activity, which may result in the application of these extracts as compounds with high antimicrobial, antioxidant, and/or anticancer potential.

Author Contributions: Conceptualization, G.Ś., M.K. (Monika Kalinowska), E.Z., E.C. and M.P.; methodology, G.Ś., M.P., E.C., E.Z., M.K. (Monika Kalinowska), E.W., U.W. and M.Z.; validation, G.Ś., M.P., E.C., E.Z. and M.K. (Monika Kalinowska); formal analysis, G.Ś., M.P., E.C., E.Z., M.K. (Monika Kalinowska), A.B., M.J., M.T., P.G., M.S., M.W. and M.Z.; investigation, G.Ś., M.P., E.C., E.Z., M.K. (Monika Kalinowska), A.B., M.J., M.T., P.G., K.L.-S., M.S. and M.K. (Mitko Karadelev); resources, G.Ś., M.P., E.C., E.Z. and M.K. (Monika Kalinowska); A.B., Q.R., C.E.G. and M.K. (Mitko Karadelev); data curation, M.P., G.Ś. and E.C.; writing—original draft preparation, G.Ś., M.P., E.C., E.Z., M.W., M.T., U.W. and E.W.; writing—review and editing, G.Ś., M.P., E.C., E.Z. and M.K. (Monika Kalinowska); visualization, G.Ś., M.P., E.C., M.W. and E.Z.; supervision, G.Ś.; project administration, G.Ś. All authors have read and agreed to the published version of the manuscript.

Funding: This study received funding from the commissioned task titled "VIA CARPATIA Universities of Technology Network named after the President of the Republic of Poland Lech Kaczyński" under the special purpose grant from the Minister of Science, contract no. MEiN/2022/DPI/2577/VC/WB-IIŚ/2/2023, as part of action "ISKRA—building inter-university research teams".

Institutional Review Board Statement: Not applicable.

Informed Consent Statement: Not applicable.

Data Availability Statement: The data presented in this study are available upon request from the corresponding author.

Acknowledgments: The authors thank the Bosco Siro Negri State Nature Reserve (University of Pavia) for authorization to collect samples and for sharing specimens. The authors thank Konrad Wilamowski and Slavica Tofilovska for collecting the samples for the study.

Conflicts of Interest: The authors declare no conflicts of interest.

References

1. He, P.; Zhang, Y.; Li, N. The phytochemistry and pharmacology of medicinal fungi of the genus: *Phellinus*: A review. *Food Funct.* **2021**, *12*, 1856–1881. [CrossRef]
2. Zhang, H.; Jiang, F.; Li, L.; Liu, X.; Yan, J.-K. Recent advances in the bioactive polysaccharides and other key components from *Phellinus* spp. and their pharmacological effects: A review. *Int. J. Biol. Macromol.* **2022**, *222*, 3108–3128. [CrossRef] [PubMed]
3. Pošta, A.; Matočec, N.; Kušan, I.; Tkalčec, Z.; Mešić, A. The Lignicolous Genus Entonaema: Its Phylogenetic–Taxonomic Position within Hypoxylaceae (*Xylariales*, Fungi) and an Overview of Its Species, Biogeography, and Ecology. *Forests* **2023**, *14*, 1764. [CrossRef]
4. Nankone, S.; Sanon, E.; Sawadogo, B.R.; Dabire, K.; Guissou, M.-L.K. Ecology and morphological characterization of the genus *Phellinus* sensu-lato (*Basidiomycetes*, *Hymenochaetaceae*) in Burkina Faso. *Afr. J. Plant Sci.* **2020**, *14*, 451–460.
5. Sułkowska-Ziaja, K.; Balik, M.; Muszyńska, B. Selected Species of the Genus *Phellinus*–Chemical Composition, Biological Activity, and Medicinal Applications. *Chem. Biodivers.* **2021**, *18*, e2100609. [CrossRef] [PubMed]

6. Salvador-Montoya, C.A.; Elias, S.G.; Popoff, O.F.; Robledo, G.L.; Urcelay, C.; Góes-Neto, A.; Martínez, S.; Drechsler-Santos, E.R. Neotropical Studies on Hymenochaetaceae: Unveiling the Diversity and Endemicity of Phellinotus. *J. Fungi* **2022**, *8*, 216. [CrossRef]
7. Zhou, L.W.; Vlasák, J.; Dai, Y.C. Taxonomy and phylogeny of Phellinidium (*Hymenochaetales, Basidiomycota*): A redefinition and the segregation of Coniferiporia gen. nov. for forest pathogens. *Fungal Biol.* **2016**, *120*, 988–1001. [CrossRef]
8. Luan, F.; Peng, X.; Zhao, G.; Zeng, J.; Zou, J.; Rao, Z.; Liu, Y.; Zhang, X.; Ma, H.; Zeng, N. Structural diversity and bioactivity of polysaccharides from medicinal mushroom *Phellinus* spp.: A review. *Food Chem.* **2022**, *397*, 133731. [CrossRef]
9. Yan, J.K.; Pei, J.J.; Ma, H.L.; Wang, Z.-B.; Liu, Y.-S. Advances in antitumor polysaccharides from *phellinus* sensu lato: Production, isolation, structure, antitumor activity, and mechanisms. *Crit. Rev. Food Sci. Nutr.* **2017**, *57*, 1256–1269. [CrossRef]
10. Zhao, Y.Q.; Dong, X.Y.; Ning, J.L.; Jin, Y.-X.; Wang, G.-K.; Feng, T. Chemical constituents from the fungus *Phellinus baumii* and their anti-inflammatory activity. *Phytochem. Lett.* **2023**, *54*, 18–22. [CrossRef]
11. Azeem, U. Antioxidant efficacy of a wood degrading fungus *Phellinus gilvus* (Schwein.) Pat., from Indian fungal flora. *J. Pharmacogn. Phytochem.* **2018**, *7*, 1017–1021.
12. Kim, G.Y.; Park, H.S.; Nam, B.H.; Lee, S.-J.; Lee, J.-D. Purification and characterization of acidic proteo-heteroglycan from the fruiting body of *Phellinus linteus* (Berk. & M.A. Curtis) Teng. *Bioresour. Technol.* **2003**, *89*, 81–87. [CrossRef] [PubMed]
13. Wang, Y.; Shang, X.Y.; Wang, S.J.; Mo, S.-Y.; Li, S.; Yang, Y.-C.; Ye, F.; Shi, J.-G.; He, L. Structures, biogenesis, and biological activities of pyrano [4,3-c] isochromen-4-one derivatives from the fungus *Phellinus igniarius*. *J. Nat. Prod.* **2007**, *70*, 296–299. [CrossRef]
14. Song, A.R.; Sun, X.L.; Kong, C.; Zhao, C.; Qin, D.; Huang, F.; Yang, S. Discovery of a new sesquiterpenoid from *Phellinus ignarius* with antiviral activity against influenza virus. *Arch. Virol.* **2014**, *159*, 753–760. [CrossRef]
15. Gao, M.; Huang, Y.; Hu, C.; Hu, J.; Wang, Y.; Chen, Y.; Huang, Y.; Song, G.; Song, Z.; Wang, Z. Selective Anticancer Effect of *Phellinus Linteus* on Epidermoid Cell Lines Studied by Atomic Force Microscopy: Anticancer Activity on A431 Cancer Cells and Low Toxicity on HaCat Normal Cells. *IEEE Nanotechnol. Mag.* **2021**, *15*, 4–16. [CrossRef]
16. Li, X.; Jiao, L.L.; Zhang, X.; Tian, W.-M.; Chen, S.; Zhang, L.-P. Anti-tumor and immunomodulating activities of proteoglycans from mycelium of *Phellinus nigricans* and culture medium. *Int. Immunopharmacol.* **2008**, *8*, 909–915. [CrossRef]
17. Chen, W.; He, F.Y.; Li, Y.Q. The apoptosis effect of hispolon from *Phellinus linteus* (Berkeley & Curtis) Teng on human epidermoid KB cells. *J. Ethnopharmacol.* **2006**, *105*, 280–285. [CrossRef]
18. Islam, M.T.; Ali, E.S.; Khan, I.N.; Shaw, S.; Uddin, S.J.; Rouf, R.; Dev, S.; Saravi, S.S.; Das, N.; Tripathi, S.; et al. Anticancer Perspectives on the Fungal-Derived Polyphenolic Hispolon. *Anti-Cancer Agents Med. Chem.* **2020**, *20*, 1636–1647. [CrossRef]
19. Zhu, X.; Guo, R.; Su, X.; Shang, K.; Tan, C.; Ma, J.; Zhang, Y.; Lin, D.; Ma, Y.; Zhou, M.; et al. Immune-enhancing activity of polysaccharides and flavonoids derived from *Phellinus igniarius* YASH1. *Front. Pharmacol.* **2023**, *14*, 1124607. [CrossRef] [PubMed]
20. Kim, H.G.; Yoon, D.H.; Lee, W.H.; Han, S.K.; Shrestha, B.; Kim, C.H.; Lim, M.H.; Chang, W.; Lim, S.; Choi, S.; et al. *Phellinus linteus* inhibits inflammatory mediators by suppressing redox-based NF-κB and MAPKs activation in lipopolysaccharide-induced RAW 264.7 macrophage. *J. Ethnopharmacol.* **2007**, *114*, 307–315. [CrossRef] [PubMed]
21. Xie, Z.; Wang, Y.; Huang, J.; Qian, N.; Shen, G.; Chen, L. Anti-inflammatory activity of polysaccharides from *Phellinus linteus* by regulating the NF-κB translocation in LPS-stimulated RAW264.7 macrophages. *Int. J. Biol. Macromol.* **2019**, *129*, 61–67. [CrossRef]
22. Shin, M.R.; Lee, J.H.; Lee, J.A.; Kim, M.J.; Park, H.-J.; Park, B.W.; Seo, S.B.; Roh, S.-S. Immunomodulatory and anti-inflammatory effects of *Phellinus linteus* mycelium. *BMC Complement. Med. Ther.* **2021**, *21*, 269. [CrossRef] [PubMed]
23. Hu, T.; Lin, Q.; Guo, T.; Yang, T.; Zhou, W.; Deng, X.; Yan, J.-K.; Luo, Y.; Ju, M.; Luo, F. Polysaccharide isolated from *Phellinus linteus* mycelia exerts anti-inflammatory effects via MAPK and PPAR signaling pathways. *Carbohydr. Polym.* **2018**, *200*, 487–497. [CrossRef] [PubMed]
24. Yoo, J.H.; Lee, Y.S.; Ku, S.K.; Lee, H.J. *Phellinus baumii* enhances the immune response in cyclophosphamide-induced immunosuppressed mice. *Nutr. Res.* **2020**, *75*, 15–31. [CrossRef]
25. Sun, Y.; Huo, J.; Zhong, S.; Zhu, J.; Li, Y.; Li, X. Chemical structure and anti-inflammatory activity of a branched polysaccharide isolated from *Phellinus baumii*. *Carbohydr. Polym.* **2021**, *268*, 118214. [CrossRef]
26. Zhu, F.; Lu, W.; Feng, W.; Song, Z.; Wang, C.; Chen, X. Preliminary Investigation on the Chemical Constituents and Antioxidant Activities of Two *Phellinus* Mushrooms Collected in Foshan. *Int. J. Org. Chem.* **2017**, *7*, 25–33. [CrossRef]
27. Dimitrova-Shumkovska, J.; Kosharkoska-Spasovska, F.; Krstanoski, L.; Karadelev, M. Antioxidant properties of fortified yogurt with medicinal mushrooms from *Phellinus* species. *J. Food Biochem.* **2022**, *46*, e14364. [CrossRef]
28. Shon, M.Y.; Kim, T.H.; Sung, N.J. Antioxidants and free radical scavenging activity of *Phellinus baumii* (*Phellinus* of *Hymenochaetaceae*) extracts. *Food Chem.* **2003**, *82*, 593–597. [CrossRef]

29. Yang, Y.; Wang, Q.; Liu, Y.; Wang, W.; Feng, N.; Wu, D. Antioxidant activities of extracts from the genus *Phellinus* species. In Proceedings of the 7th International Conference on Mushroom Biology and Mushroom Products (ICMBMP7), Arcachon, France, 4–7 October 2011.
30. Ajith, T.A.; Janardhanan, K.K. Antioxidant and antihepatotoxic activities of *Phellinus rimosus* (Berk) Pilat. *J. Ethnopharmacol.* **2002**, *81*, 387–391. [CrossRef]
31. Song, Y.S.; Kim, S.H.; Sa, J.H.; Jin, C.; Lim, C.-J.; Park, E.-H. Anti-angiogenic, antioxidant and xanthine oxidase inhibition activities of the mushroom *Phellinus linteus*. *J. Ethnopharmacol.* **2003**, *88*, 113–116. [CrossRef] [PubMed]
32. Seephonkai, P.; Samchai, S.; Thongsom, A.; Sunaart, S.; Kiemsanmuang, B.; Chakuton, K. DPPH radical scavenging activity and total phenolics of *Phellinus* mushroom extracts collected from northeast of Thailand. *Chin. J. Nat. Med.* **2011**, *9*, 441–445. [CrossRef]
33. Laovachirasuwan, P.; Judprakob, C.; Sinaphet, B.; Phadungkit, M. In vitro antioxidant and antimutagenic activities of different solvent extracts of *Phellinus* spp. *Int. Food Res. J.* **2016**, *23*, 2608.
34. Kim, Y.H.; Choi, D.C.; Ki, D.W.; Seo, W.-G.; Song, J.-H.; Ko, H.-J.; Lee, I.-K.; Yun, B.-S. Three undescribed dimeric ferulates from the culture broth of *Phellinus linteus* and their dengue virus type-2 inhibition activity. *Phytochem. Lett.* **2024**, *60*, 229–233. [CrossRef]
35. Li, T.J.; Lin, T.W.; Lu, T.Y.; Tseng, C.-K.; Lin, C.-K.; Chu, H.-T.; Li, I.-C.; Chen, C.-C. *Phellinus linteus* mycelia extract in COVID-19 prevention and identification of its key metabolic compounds profiling using UPLC-QTOF-MS/MS spectrometry. *Fitoterapia* **2023**, *171*, 105695. [CrossRef]
36. Feng, H.; Zhang, S.; Wan, J.M.F.; Gui, L.; Ruan, M.; Li, N.; Zhang, H.; Liu, Z.; Wang, H. Polysaccharides extracted from *Phellinus linteus* ameliorate high-fat high-fructose diet induced insulin resistance in mice. *Carbohydr. Polym.* **2018**, *200*, 144–153. [CrossRef]
37. Hwang, H.J.; Kim, S.W.; Lim, J.M.; Joo, J.-H.; Kim, H.-O.; Kim, H.-M.; Yun, J.-W. Hypoglycemic effect of crude exopolysaccharides produced by a medicinal mushroom *Phellinus baumii* in streptozotocin-induced diabetic rats. *Life Sci.* **2005**, *76*, 3069–3080. [CrossRef] [PubMed]
38. Zheng, S.; Deng, S.; Huang, Y.; Huang, M.; Zhao, P.; Ma, X.; Wen, Y.; Wang, Q.; Yang, X. Anti-diabetic activity of a polyphenol-rich extract from: *Phellinus igniarius* in KK-Ay mice with spontaneous type 2 diabetes mellitus. *Food Funct.* **2018**, *9*, 614–623. [CrossRef] [PubMed]
39. Zheng, M.; Wang, L.; Sun, Y.; Pi, X.; Zhang, W.; Gao, P.; Lu, S.; Liu, W. Hypoglycemic effect of the *Phellinus baumii* extract with α-glucosidase-inhibited activity and its modulation to gut microbiota in diabetic patients. *Biomed. Pharmacother.* **2023**, *158*, 114130. [CrossRef] [PubMed]
40. Liu, J.; Wang, Y.; Wu, J.; Georgiev, M.I.; Xu, B.; Wong, K.-H.; Bai, W.; Tian, L. Isolation, Structural Properties, and Bioactivities of Polysaccharides from Mushrooms *Termitomyces*: A Review. *J. Agric. Food Chem.* **2022**, *70*, 21–33. [CrossRef]
41. Wang, Q.; Wang, F.; Xu, Z.; Ding, Z. Bioactive mushroom polysaccharides: A review on monosaccharide composition, biosynthesis and regulation. *Molecules* **2017**, *22*, 955. [CrossRef] [PubMed]
42. Kalinowska, M.; Bielawska, A.; Lewandowska-Siwkiewicz, H.; Priebe, W.; Lewandowski, W. Apples: Content of phenolic compounds vs. variety, part of apple and cultivation model, extraction of phenolic compounds, biological properties. *Plant Physiol. Biochem.* **2014**, *84*, 169–188. [CrossRef] [PubMed]
43. Wang, T.; Hicks, K.B.; Moreau, R. Antioxidant activity of phytosterols, oryzanol, and other phytosterol conjugates. *J. Am. Oil Chem. Soc.* **2002**, *79*, 1201–1206. [CrossRef]
44. Yoshida, Y.; Niki, E. Antioxidant effects of phytosterol and its components. *J. Nutr. Sci. Vitaminol.* **2003**, *49*, 277–280. [CrossRef]
45. Zapała, L.; Ciszkowicz, E.; Kosińska-Pezda, M.; Maciołek, U.; Kozioł, A.E.; Miłoś, A.; Woźnicka, E.; Bocian, A.; Zapała, W.; Rydel-Ciszek, K.; et al. Novel silver(I) complexes with fenamates: Insights into synthesis, spectral characterization, and bioactivity. *J. Inorg. Biochem.* **2025**, *266*, 112846. [CrossRef] [PubMed]
46. Thieme, L.; Hartung, A.; Tramm, K.; Klinger-Strobel, M.; Jandt, K.D.; Makarewicz, O.; Pletz, M.W. MBEC Versus MBIC: The Lack of Differentiation between Biofilm Reducing and Inhibitory Effects as a Current Problem in Biofilm Methodology. *Biol. Proced. Online* **2019**, *21*, 18. [CrossRef]
47. Bocian, A.; Ciszkowicz, E.; Hus, K.K.; Buczkowicz, J.; Lecka-Szlachta, K.; Pietrowska, M.; Petrilla, V.; Petrillova, M.; Legáth, Ľ.; Legáth, J. Antimicrobial activity of protein fraction from Naja ashei venom against *Staphylococcus epidermidis*. *Molecules* **2020**, *25*, 293. [CrossRef]
48. Repetto, G.; del Peso, A.; Zurita, J.L. Neutral red uptake assay for the estimation of cell viability/cytotoxicity. *Nat. Protoc.* **2008**, *3*, 1125–1131. [CrossRef]
49. Geddo, F.; Antoniotti, S.; Gallo, M.P.; Querio, G. Indole-3-Propionic Acid, a Gut Microbiota-Derived Tryptophan Metabolite, Promotes Endothelial Dysfunction Impairing Purinergic-Induced Nitric Oxide Release in Endothelial Cells. *Int. J. Mol. Sci.* **2024**, *25*, 3389. [CrossRef]

50. Domański, S.; Orłaś, H.; Skirgiełło, A. *Fungi. Polyporaceae II (pileatae) Mucronoporaceae II (pileatae) Ganodermataceae, Bon darzewiaceae, Boletopsidaceae, Fistulinaceae (Grzyby)*, revised ed.; Translated from Polish; Foreign Scientific Publications Department of the National Center for Scientific, Technical, and Economic Information: Warsaw, Poland, 1973.
51. Domański, S.; Orłoś, H.; Skirgiełło, A. *Podstawczaki (Basidiomycetes), bezblaszkowe (Aphyllophorales), żagwiowate II (Polyporaceae pileatae), szczecinkowate II (Mucroporonaceae pileatae), lakownicowate (Ganodermataceae), bondarcewowate (Bondarzewiaceae), boletkowate (Boletopsidaceae), ozorkowate (Fistulinaceae)*; Flora Polska. Rośliny zarodnikowe Polski i ziem ościennych. Grzyby (Mycota); Polska Akademia Nauk, Instytut Botaniki, Państwowe Wydawnictwo Naukowe: Warsaw, Poland, 1973; Volume 3.
52. Ryvarden, L.; Gilbertson, R.L. European Polypores 1. *Synop. Fungorum* **1993**, *6*, 1–387.
53. Ryvarden, L.; Gilbertson, R.L. European Polypores 2. *Synop. Fungorum* **1994**, *7*, 394–743.
54. Larsen, M.J.; Cobb Poulle, L.A. *Phellinus* (*Hymenochaetaceae*).: A survey of the world taxa. *Synop. Fungorum* **1990**, *3*, 1.
55. Sell, I. Taxonomy of the species in the *Phellinus igniarius* group. *Mycotaxon* **2008**, *104*, 337–347.
56. Kirk, P.M.; Cannon, P.F.; Minter, D.W.; Stalpers, J.A. *Ainsworth and Bisby's Dictionary of the Fungi*; CAB International: Wallingford, UK, 2008.
57. Balouiri, M.; Sadiki, M.; Ibnsouda, S.K. Methods for in vitro evaluating antimicrobial activity: A review. *J. Pharm. Anal.* **2016**, *6*, 71–79. [CrossRef]
58. Rašeta, M.; Mišković, J.; Čapelja, E.; Zapora, E.; Fabijan, A.P.; Knežević, P.; Karaman, M. Do Ganoderma Species Represent Novel Sources of Phenolic Based Antimicrobial Agents? *Molecules* **2023**, *28*, 3264. [CrossRef]
59. Isidorov, V.A. *GC-MS of Biologically and Environmentally Significant Organic Compounds*; TMS Derivatives: Hoboken, NJ, USA, 2020.
60. Stephen, E.S. *NIST/EPA/NIH Mass Spectral Library*; National Institute of Standards and Technology: Gaithersburg, MD, USA, 2020.
61. Stenhagen, E.A.; McLafferty, F.; Abrahamsson, S. *Wiley Registry of Mass Spectral Data*, 12th ed.; John Wiley & Sons: Hoboken, NJ, USA, 2020.
62. Kedare, S.B.; Singh, R.P. Genesis and development of DPPH method of antioxidant assay. *J. Food Sci. Technol.* **2011**, *48*, 412–422. [CrossRef]
63. Re, R.; Pellegrini, N.; Proteggente, A.; Pannala, A.; Yang, M.; Rice-Evans, C. Antioxidant activity applying an improved ABTS radical cation decolorization assay. *Free. Radic. Biol. Med.* **1999**, *26*, 1231–1237. [CrossRef] [PubMed]
64. Amel, B.; Seddik, K.; Shtaywy, A.; Saliha, D.; Mussa, A.Z.; Assia, B.; Saliha, D.; Abderahmane, B.; Smain, A. Phytochemical analysis, antioxidant activity and hypotensive effect of algerian azarole (*Crataegus azarolus* L.) leaves extracts. *Res. J. Pharm. Biol. Chem. Sci.* **2014**, *5*, 286–305.
65. Karcz, D.; Starzak, K.; Ciszkowicz, E.; Lecka-Szlachta, K.; Kamiński, D.; Creaven, B.; Miłoś, A.; Jenkins, H.; Ślusarczyk, L.; Matwijczuk, A. Design, Spectroscopy, and Assessment of Cholinesterase Inhibition and Antimicrobial Activities of Novel Coumarin–Thiadiazole Hybrids. *Int. J. Mol. Sci.* **2022**, *23*, 6314. [CrossRef] [PubMed]
66. Rasband, W.S. *ImageJ*; National Institutes of Health: Bethesda, MD, USA, 2015. Available online: https://imagej.net/ij/ (accessed on 18 March 2025).

International Journal of
Molecular Sciences

MDPI

Article

Metabolomic Characterization and Bioinformatic Studies of Bioactive Compounds in Two Varieties of *Psidium guajava* L. Leaf by GC–MS Analysis

Ana Victoria Valdivia-Padilla [1], Ashutosh Sharma [2], Jorge A. Zegbe [3] and José Francisco Morales-Domínguez [1,*]

1 Centro de Ciencias Básicas, Departamento de Química, Universidad Autónoma de Aguascalientes, Aguascalientes 20131, Mexico; al97640@edu.uaa.mx

2 Centre of Bioengineering, NatProLab, School of Engineering and Sciences, Tecnologico de Monterrey, Querétaro 76130, Mexico; asharma@tec.mx

3 Campo Experimental Pabellón, Instituto Nacional de Investigaciones Forestales, Agrícolas y Pecuarias, Km 32.5 Carretera Aguascalientes-Zacatecas, Pabellón de Arteaga, Aguascalientes 20668, Mexico; zegbe.jorge@inifap.gob.mx

* Correspondence: francisco.morales@edu.uaa.mx

Abstract: The guava tree (*Psidium guajava* L.) is a tropical plant from the Myrtaceae family. Leaf extracts from this plant have been used in traditional medicine to treat gastrointestinal disorders and exhibit several functional activities that benefit human health. Different varieties of guava trees produce fruits in colors ranging from white to red and present a characteristic metabolic profile in both their leaves and fruits. This study presents a metabolomic characterization of the leaves from two guava varieties: the Caxcana cultivar with yellow fruits and the S-56 accession with pink fruits. Metabolite profiling was conducted using Gas Chromatography–Mass Spectrometry (GC–MS) on methanol extracts, followed by multivariate statistical analysis, including Principal Component Analysis (PCA), and a heat map visualization of compound concentrations in the two varieties. The results identified β-caryophyllene as the major secondary metabolite present in both varieties, with a relative abundance of 16.46% in the Caxcana variety and 23.06% in the S-56 cultivar. Furthermore, in silico analyses, such as network pharmacology and molecular docking, revealed key interactions with proteins such as CB2, PPARα, BAX, BCL2, and AKT1, suggesting potential therapeutic relevance. These findings highlight the pharmacological potential of guava leaf metabolites in natural product chemistry and drug discovery.

Keywords: medicinal plants; *Psidium guajava* L.; bioactive compounds; guava leaves; molecular docking

1. Introduction

Plants have been part of traditional medicine across various cultures over the years due to their rich content of bioactive compounds, commonly referred to as phytochemicals, primarily derived from secondary metabolism [1]. These metabolites are crucial in developing new drugs and therapeutic agents, offering individual, additive, or synergistic activities that enhance their medicinal potential [2].

The guava tree (*Psidium guajava* L.), native to South America and a member of the Myrtaceae family [3], has been particularly valued in traditional medicine. Its leaves have been extensively used in regions such as Mexico, Africa, Asia, and Central America, often prepared as tea, to address a wide range of ailments, including diarrhea, gastrointestinal disorders, dysentery, hypertension, diabetes, and dental caries, among others [4]. Rich in both macro- and micronutrients, guava leaves contain carbohydrates, carotenoids, fatty acids, polyphenols, vitamins, and minerals [5]. Furthermore, they possess a diverse array of volatile compounds [4], many of which exhibit promising therapeutic activities.

Mexico is one of the leading guava producers worldwide. This crop is characterized by significant genetic variability and diverse morphological, physical, chemical, and biological characteristics of its fruits [6]. In the "Calvillo-Cañones" region, one of the primary guava-producing regions in the country, more than 100 varieties are cultivated [7]. Among the most economically important guava varieties is the "Caxcana" cultivar, characterized by its white pulp, semi-round shape, and fruit weight ranging from 75 to 95 g [8]. This variety is distinguished by its high fruit quality, flavor, and uniformity, with over 70% classified as first-grade produce. Additionally, "Caxcana" exhibits a higher yield than other nationally cultivated guava varieties, making it a valuable option for commercial production [9].

The "S-56" accession, known for its pink pulp, produces high-quality fruits with an above-range weight of 80 to 120 g and a sweet flavor [8]. In the region, pink-fleshed guavas represent a valuable alternative for diversifying the fresh guava market and agro-industrial use [10].

Different crops and varieties of guava trees present unique metabolomic profiles in their leaves and fruits [11]. A metabolic profile provides a snapshot of the biochemical state of an individual biological sample, reflecting the genetics of each organism and the changes occurring under various physiological, environmental, and developmental conditions [12]. The metabolomic characterization of different varieties of guava tree leaves aims to identify the largest number of individual metabolites, facilitating comparison and differentiation between the varieties [13]. This process allows a detailed characterization of their phytochemical composition and the assessment of their potential activity for human health [1,2].

On the other hand, chromatographic analysis is widely recognized as one of the most important analytical chemistry tools for examining plant extracts, thanks to its simplicity, sensitivity, and effectiveness in separating components [14]. Among the various techniques available, gas chromatography coupled with a mass spectrometer (GC–MS) is particularly noteworthy for its application in studying secondary metabolites in plants [13]. This method offers excellent separation capacity, selectivity, sensitivity, and reproducibility, while also enabling the differentiation of individual plants within the same species [15].

In this study, we conducted a phytochemical characterization of two guava leaf varieties using GC–MS to identify the main components present in methanolic extracts. One of the most abundant compounds identified in both varieties was β-caryophyllene (BC). Bioinformatic analyses of pharmacological networks and molecular docking were performed with this compound to predict its potential interactions with proteins or therapeutic targets. Key metabolic pathways involving the identified bioactive compounds in guava leaves were also analyzed using databases integrating high-throughput omics data.

2. Results and Discussion

2.1. *Phytochemical Screening in Guava Leaves by GC–MS Analysis*

Twenty-eight chemical compounds were identified through GC–MS analysis of methanolic extracts from guava leaves (Figure 1). The identified compounds are listed in Table 1, with their respective retention time (RT), relative abundances expressed as percentage area, for both varieties, Caxcana and S-56 (%), molecular ion (M^+), and the mass-to-charge ratios of the most abundant ions (m/z).

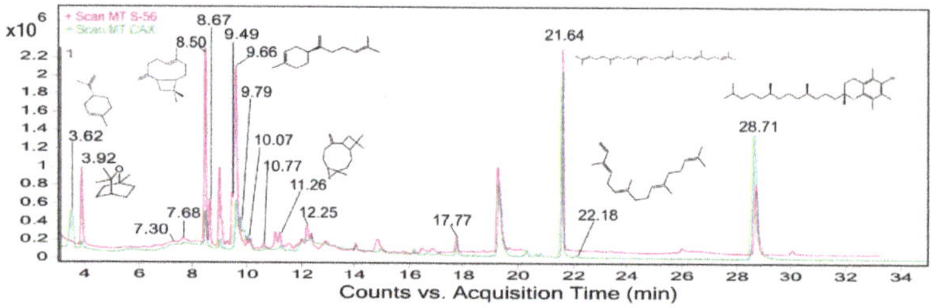

Figure 1. Total ion chromatogram obtained by analysis of methanolic extracts from guava leaves. The pink chromatogram corresponds to the S-56 variety, while the green represents the Caxcana cultivar.

Table 1. Characterization of phytochemical compounds in methanolic extracts of guava leaves varieties Caxcana and S-56.

Code	RT	CAS	Compounds	% Area Caxcana	% Area S-56	M^+	m/z (Abundance)
Primary Metabolites							
1P	14.07	112-39-0	Methyl palmitate	0.80	0.54	270.3	74.0
2P	16.21	301-00-8	Methyl linolenate	1.32	0.36	292.2	79.1
3P	21.64	7683-64-9	Squalene	10.83	13.27	410.4	69.1
Secondary Metabolites							
1S	3.62	5989-27-5	D-Limonene	9.76	0.96	136.1	68.1
2S	3.92	470-82-6	Eucalyptol	1.28	4.65	154.1	43.0
3S	7.30	3242-08-8	**γ-Elemene**	np	0.56	204.2	121.1
4S	7.68	17699-14-8	α-Cubebene	0.92	1.07	204.2	119.1
5S	8.16	1461-03-6	**β-Himachalene**	1.4	np	204.2	119.1
6S	8.31	13877-93-5	**Isocaryophyllene**	np	0.24	93.1	93.1
7S	8.50	87-44-5	β-Caryophyllene	16.46	23.06	204.2	93.1
8S	8.67	72747-25-2	**(+)-Aromadendrene**	np	6.66	204.2	161.1
9S	9.02	6753-98-6	α-Humulene	10.28	2.67	204.2	93.1
10S	9.09	6831-16-9	**(-)-Aristolene**	np	1.17	204.2	91.1
11S	9.33	39029-41-9	**γ-Cadinene**	np	0.89	204.2	161.1
12S	9.49	17066-67-0	β-Selinene	5.20	6.85	204.2	105.1
13S	9.59	22567-17-5	**γ-Gurjunene**	10.45	np	204.2	189.2
14S	9.66	495-61-4	**β-Bisabolene**	np	9.26	204.2	93.1
15S	9.79	495-60-3	**α-Zingiberene**	3.84	np	204.2	119.1
16S	10.07	483-76-1	δ-Cadinene	1.99	1.58	204.2	161.1
17S	10.77	142-50-7	Nerolidol	1.33	0.58	222.2	69.1
18S	11.10	489-41-8	(-)-Globulol	2.09	2.21	222.2	43.1
19S	11.26	1139-30-6	Caryophyllene oxide	1.66	2.22	220.2	79.1
20S	12.25	473-04-1	**Juniper camphor**	np	4.69	222.2	43.0
21S	12.42	150-86-7	Phytol	7.50	2.59	296.3	68.1
22S	17.77	1120-07-6	Nonanamide	0.94	0.62	157.1	59.0
23S	20.70	88-24-4	Antioxidant 425	0.35	1.83	368.3	191.1
24S	22.18	77898-97-6	α-Springene	0.37	2.59	418.6	149.1
25S	28.71	59-02-9	Vitamin E	11.24	8.88	430.4	165.1

RT = retention time; M^+ = molecular ion; np = not present; Compounds exclusive to the Caxcana variety are indicated in bold type. Compounds exclusive to the S-56 variety are noted in red.

Three compounds were identified as primary metabolites, of which squalene was the most abundant, with a relative abundance of 10.83% in Caxcana and 13.27% in the S-56 variety. Additionally, twenty-five compounds classified as secondary metabolites were identified. Among these, β-caryophyllene was the most abundant in both varieties, with relative abundances of 16.46% and 23.06% for Caxcana and S-56, respectively. The phytochemicals exclusive to each variety are detailed in Table 1.

The relative abundance of β-caryophyllene in the guava leaf varieties analyzed in this study was higher than previously reported in GC–MS studies on guava leaves from Pakistan (15.65%) [16], Nepal (15.80%) [17], and India (15.7%) [18], but lower than those recorded in China varieties (36.9%) [19].

The significant presence of terpenes in guava leaves can be attributed to various genetic, physiological, and anatomical factors [20]. As a result, these compounds may serve as chemophenetic markers for the Caxcana and S-56 varieties [21]. In the Myrtaceae family, terpenes primarily serve as defensive agents against herbivores and pathogens [22,23]. The evolutionary diversification of these defense mechanisms drives both qualitative and quantitative variations in the phytochemical profiles of terpenes across species, varieties, populations, and even individual plants [24,25]. It has been documented that the number of terpenes present in leaves increases in response to biotic and abiotic stress factors [22,26]. For instance, compounds such as β-Himachalene and γ-Elemene, identified in extracts from various plants, have demonstrated insecticidal properties against *Spodoptera litura* larvae (Lepidoptera: Noctuidae) and the diamondback moth *Plutella xylostella* L. (Lepidoptera: Yponomeutidae) [27,28]. The high concentration of terpene compounds in the two guava leaf varieties highlights their potential as a valuable source of therapeutic agents.

2.2. Statistical Multivariate Analysis

Principal Component Analysis (PCA) was performed to identify key factors and visualize the relationships between chemical compounds in each biological sample [29]. PCA, a widely used tool in metabolic studies, provides a two-dimensional graphical representation (Figure 2) that reveals patterns of similarity or differentiation among the identified compounds. The analysis accounted for 96.4% of the total variance of the dataset, with the first principal component (PC1) explaining 64% of the variance and the second principal component (PC2) accounting for 32.4%. The dispersion along these axes highlights the distribution of compounds with the highest concentrations in each leaf variety. In the S-56 guava variety, β-Bisabolene (14S) had the highest concentration, with an area percentage of 9.26%, followed by (+)-Aromadendrene (8S) at 6.66%. In contrast, the Caxcana variety exhibited γ-Gurjunene (13S) as the dominant compound at 10.45% and D-Limonene (1S) at 9.76%. Both guava varieties shared β-Caryophyllene (7S) and Squalene (3P) as their most representative compounds. These findings provide valuable insights into the chemical composition and variability of the guava varieties analyzed.

The representative compounds for the S-56 variety included β-Bisabolene, which has demonstrated antimicrobial activity against *Staphylococcus aureus* [30] and cytotoxic anti-proliferative effects on breast cancer cells [31]. Additionally, (+)-Aromadendrene was identified, a compound known for its various therapeutic properties, including antimicrobial activity against *Streptococcus pyogenes* [32] and inhibition effects on the growth of skin cancer cells and HaCaT precancerous cells [33].

The representative compounds for the Caxcana variety were γ-Gurjunene and D-Limonene, a monoterpene with multiple therapeutic activities, predominantly found in plants of the *Citrus* spp. Genus. D-Limonene exhibits notable antioxidant activity, demonstrated by its protective effects against oxidative stress in lymphocyte cells induced by the exogenous addition of H_2O_2 [34]. This antioxidant property underscores the molecule's potential role in several chronic degenerative disease treatments, including diabetes, cancer, chronic inflammation, and cardiovascular and gastrointestinal disorders, among others [35].

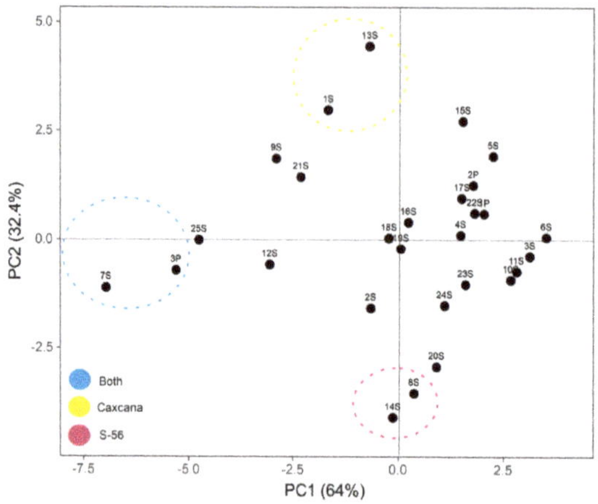

Figure 2. Principal Component Analysis (PCA) biplot, based on the variability of phytochemical profiles in the leaves of Caxcana and S-56 guava varieties. Created using ClustVis. (https://biit.cs.ut.ee/clustvis/, accessed on 10 February 2025).

A heatmap graph was also generated to evaluate the metabolite concentrations in each sample. Figure 3 displays the combined hierarchical clustering, highlighting the relative abundance of chemical compounds across the analyzed varieties. The color scale ranges from 14, represented in intense red, to 0, depicted in dark blue, providing a clear visual representation of the variability in compound concentrations.

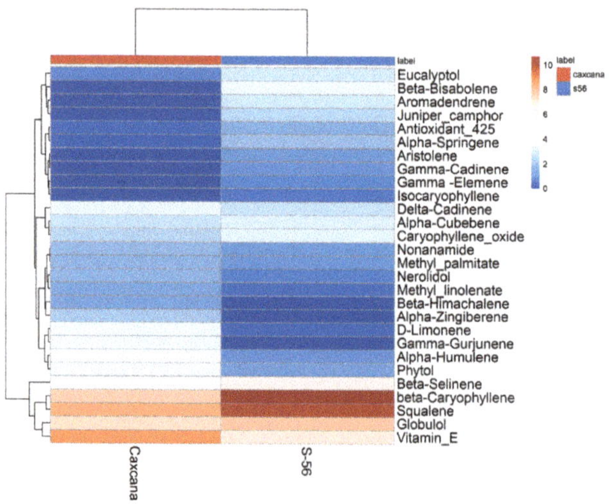

Figure 3. Heatmap with a color gradient from red to blue illustrates the relative abundance of chemical compounds from high to low in Caxcana and S-56 varieties of *P. guajava* L. leaves. Created using ClustVis.

As highlighted in the heatmap, β-caryophyllene (BC) stands out for its high concentration in the two analyzed guava leaf varieties. This compound is also present in extracts from other plants, including *Cannabis sativa*, *Syzygium aromaticum*, *Piper nigrum*, *Rosmarinus officinalis*, *Origanum vulgare* L. [36], and species within the genus *Veronica* L. [21]. BC is a bicyclic sesquiterpene belonging to the phytocannabinoids family, recognized for its extensive range of biological activities, including anti-inflammatory, anticancer, and antioxidant properties. Furthermore, BC is used in the treatment of nervous system disorders, and atherosclerosis [37], as well as being a potent analgesic [38]. The notable presence of this compound in guava leaves further validates the traditional use of this plant for the treatment of stomach pain, underscoring its medicinal relevance.

2.3. Network Pharmacology and Target Protein Prediction

For this analysis, BC was selected due to its high abundance in the two guava leaf varieties. BC was used as a ligand in various databases to elucidate its pharmacological action by evaluating its binding affinity to potential human target proteins. The analysis identified several target proteins likely to interact with BC: Cannabinoid Receptor 2 (CB2, ID: 1269), Peroxisome Proliferator-Activated Receptor Alpha (PPARα, ID: 5465), Bcl2-Associated Protein X (BAX, ID: 581), B-cell Lymphoma 2 (BCL2, ID: 596), and Protein Kinase B (AKT1, ID: 207). These findings provide insights into potential mechanisms of action of BC at a molecular level.

2.4. Molecular Docking Analysis

The pharmacological network analysis was complemented by molecular docking studies involving BC and its target proteins. The interactions between BC and the identified proteins—CB2, PPARα, BAX, BCL2, and AKT1—are summarized in Table 2. This table provides the Vina scores, which quantify the binding affinity of BC to each target protein, alongside detailed information on the ligand-protein binding sites.

Table 2. Results of molecular docking studies: binding interactions of β-Caryophyllene with key ligands of target proteins CB2, PPARα, BAX, BCL2, and AKT1.

Compound (Ligand)	PubChem ID	Origin	Binding Affinity (kcal/mol)	Cavity Size	Aminoacidic Interaction	Reference
CB2 Protein						
β-caryophyllene	5281515	Natural	−8.6	1654	87–288	
Quercetin	5280343	Natural	−8.8	1654	24–288	[39]
Codeine	5284371	Natural	−9.1	1654	110–288	[40]
Cannabidiol	644019	Natural	−9.3	1654	87–288	[41]
HU-308	5311172	Synthetic	−9.5	1654	87–288	[41]
JWH-015	4273754	Synthetic	−10.6	1654	87–285	[41]
Levonantradol	5361881	Synthetic	−11.0	1654	87–288	[40]
PPARα protein						
β-caryophyllene	5281515	Natural	−7.2	3640	272–464	
Arachidonic Acid	444899	Natural	−7.0	3640	220–464	[42]
Resveratrol	445154	Natural	−8.4	3640	269–464	[43]
Fenofibric acid	64929	Synthetic	−7.8	3640	269–464	[44]
Bezafibrate	39042	Synthetic	−8.8	3640	272–464	[44]
Pemafibrate	11526038	Synthetic	−9.2	3640	218–355	[44]

Table 2. *Cont.*

CB2 Protein						
Compound (Ligand)	**PubChem ID**	**Origin**	**Binding Affinity (kcal/mol)**	**Cavity Size**	**Aminoacidic Interaction**	**Reference**
			BAX protein			
β-caryophyllene	5281515	Natural	−5.5	183	13–159	
Thymoquinone	10281	Natural	−5.3	148	13–159	[45]
Kaempferol	5280863	Natural	−6.3	326	13–159	[46]
Quercetin	5280343	Natural	−6.4	180	13–159	[45]
BAM7	3101542	Synthetic	−7.2	148	28–61	[47]
SMBA1	6070109	Synthetic	−7.4	148	29–61	[47]
BTSA1	3857348	Synthetic	−7.8	183	29–61	[47]
			BCL2 protein			
β-caryophyllene	5281515	Natural	−7.1	124	60–167	
Maytansine	5281828	Natural	−7.2	272	65–114	[48]
Annocatalin	44566987	Natural	−7.3	272	69–114	[48]
Ginsenoside Rg1	441923	Natural	−8.4	272	60–167	[49]
Obatoclax	11404337	Synthetic	−7.5	94	65–114	[50]
Navitoclax	24978538	Synthetic	−9.2	272	65–114	[50]
Sonrotoclax	149553242	Synthetic	−9.4	91	65–114	[50]
			AKT1 protein			
β-caryophyllene	5281515	Natural	−7.1	1673	164–291	
Tehranolide	6711941	Natural	−7.3	1673	156–438	[51]
Artemisinin	68827	Natural	−7.8	1673	156–292	[51]
Shogaol	5281794	Natural	−7.5	1673	156–438	[51]
XM1	46870040	Synthetic	−8.7	1668	156–442	[52]
Ipatasertib	24788740	Synthetic	−8.7	1668	156–442	[52]
Uprosertib	51042438	Synthetic	−9.4	1668	156–438	[52]

The binding prediction of BC to the CB2 receptor yielded an energy value of −8.6 kcal/mol. This interaction is characterized by polar bonds with the amino acids Thr114, Ser285, and Cys288 (Figure 4a). According to the PDBe database, 15 specific ligands have been reported for the CB2 receptor, 11 synthetic, and 4 of natural origin. The interactions observed with reported ligands, such as cannabidiol and other synthetic compounds, align with the active side interaction between Phe87 and Cys288. The activation of the CB2 receptor in humans is associated with a range of therapeutic benefits, including pain relief and the treatment of diseases linked to chronic inflammation [53]. This suggests that BC could be a promising natural remedy with both anticancer and analgesic properties [54]. Additionally, BC may offer a potential treatment for neuroinflammation related to psychiatric conditions such as depression and anxiety [53,55]. A notable example of its therapeutic potential applications is the development of BC-based ophthalmic nanoemulsions, designed to treat Acanthamoeba keratitis and relieve retinal pain in pigs [56].

The interaction between the PPARα protein and BC obtained a binding energetic value of −7.2 kcal/mol. This affinity is higher than that of other natural ligands, such as palmitic acid and arachidonic acid, but lower than resveratrol (Table 2). In the PDBe database, 43 ligands have been reported for PPARα, including 31 synthetic compounds and 12 of natural origin. Polar interactions between BC and PPARα occur at the amino acids Cys276, Gln277, Ser280, Tyr314, His440, and Tyr264 (Figure 4b), with the binding site located between residues Phe218 and Tyr 464. These results are consistent with those reported by Kamata et al., 2020 [57] who analyzed both synthetic and natural PPARα ligands. PPARα is a member of the nuclear hormone receptor superfamily and plays a crucial role in regulating oxidative stress, energy homeostasis, fatty acid metabolism, and inflammatory processes [58,59]. The anti-inflammatory effects of PPARα ligands, such as fatty acids and fibrates, have been demonstrated in various mouse models. These effects are primarily attributed to the inhibition of pro-inflammatory cytokines like IL-7 and IFNγ [60,61], as well as the reduction in leukocyte recruitment to sites of inflammation [62]. Consequently, PPARα is considered a potential therapeutic target for the treatment of neurodegenerative diseases [59], diabetes, cardiovascular diseases [63], obesity (as a hypolipidemic agent) [64], and chronic lymphocytic leukemia [65], among other conditions.

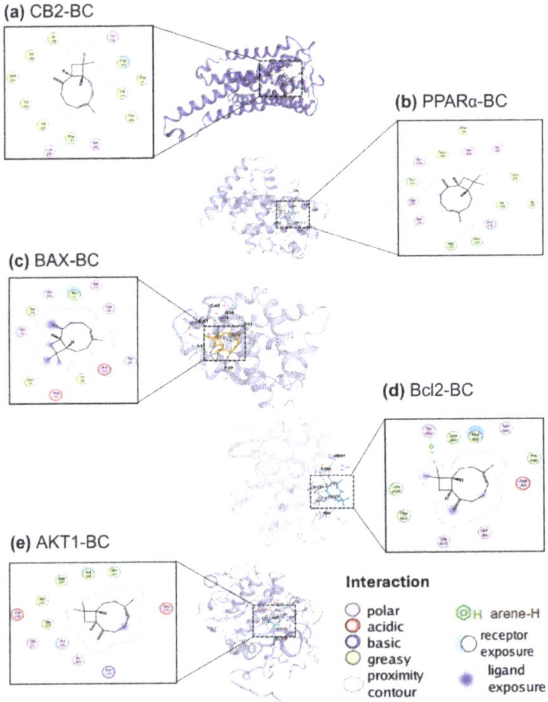

Figure 4. Two-dimensional and 3D models of protein–ligand active site interactions: molecular docking of BC with the following proteins: (**a**) CB2, (**b**) PPARα, (**c**) BAX, (**d**) BCL2, (**e**) AKT1.

Another protein identified in this analysis as a predicted interaction partner of BC was the BAX protein. The binding energy of this interaction was calculated to be −5.5 kcal/mol, comparable to the synthetic ligand K6G and higher than those naturally occurring ligands such as dodecane and thymoquinone. According to the PDBe database, 18 ligands are associated with BAX protein including 8 synthetic ligands and 10 of natural origin. The BC binding site on BAX spans residues Pro13 to Asp159, featuring polar interactions with the amino acids Gln18, Thr22, Ser55, Gly156, and Gly157, in addition to an acidic interaction with Asp53 and Asp159 (Figure 4c). BAX, a member of the BCL2 protein family, plays a proapoptotic role and is crucial in regulating cell death [66,67]. This protein is widely expressed across various cell types, including cancer cells. In most cancer cells, such as those from lungs, colon, and breast cancer, among others; BAX is predominantly found in an inactive state [68]. Consequently, the direct activation of BAX to induce apoptosis in tumor cells presents a promising target for pharmacological cancer therapies [69].

On the other hand, the interaction between BC and the BCL2 protein displayed a binding energy of −7.1 kcal/mol, comparable to that of natural ligands but lower than synthetic ones. This suggests that BC could act as a potential antagonist of BCL2. In the PDBe database, 30 ligands are reported for BCL2, including 3 of natural origin and 27 synthetic ligands. The BC–BCL2 interaction involves the active site residues Gln60 and Met167, which aligns with the interactions observed in other reported ligands. Specifically, BC forms polar interactions with Gln60, Gly135, Tyr163, and Ser166, an acidic interaction with Asp64, and a hydrophobic interaction with Tyr163 (Figure 4d). Both BCL2 and BAX are key regulators of apoptotic pathways and belong to the family of anti-apoptotic proteins [69]. The BCL2 gene is classified as an oncogene due to its overexpression in various cancers and its association with chemotherapy resistance in tumors [70]. Inhibitory molecules targeting BCL2 have been shown to induce apoptosis in chronic lymphocytic leukemia [71] and lung cancer cell lines [72]. Pharmaceutical therapies targeting BCL2 hold promise as potential cancer treatments, either as standalone agents or in combination with chemotherapeutics to enhance their efficacy [73].

Finally, for the AKT1 protein, the binding energy interaction with BC was calculated to be −7.1 kcal/mol, a value comparable to those observed for natural ligands. In contrast, synthetic ligands exhibited binding energies ranging from −8.7 to 9.4 kcal/mol. As reported in the PDBe database, AKT1 has been associated with 40 ligands, including 9 natural and 31 synthetic ligands. Notably, these share a common binding site on AKT1, from residues Val164 to Thr291, as is the case for BC artemisinin. Detailed analysis of the BC–AKT1 interaction revealed specific polar contacts with Thr211, Tyr229, and Thr291, an acidic interaction with Glu228, and a basic interaction with Lys179 (Figure 4e). These interactions collectively stabilize the binding of BC within the AKT1 binding site. AKT1 plays a critical role in regulating cell proliferation and survival [74], serving as a key component of the PI3K/AKT signaling pathway. Dysregulation of AKT1, including abnormal activation and overexpression, has been implicated in the progression of various cancers [75], such as breast, ovarian, lung, and pancreatic cancer [76,77]. Inhibitory molecules targeting AKT1 have demonstrated potential therapeutic benefits, including the induction of apoptosis in prostate cancer cells and the inhibition of breast cancer cell growth [78,79]. These findings underscore the potential of AKT1 as a therapeutic target, particularly for interventions aimed at suppressing tumorigenesis during the early stages of cancer development [80].

2.5. Interaction of Bioactive Compounds

In this analysis, we identified key proteins and genes involved in the synthesis of phytochemicals with therapeutic activity, providing valuable insights into the genetic improvement of guava tree varieties (Figure 5). Among the identified proteins, the SQS1 protein, encoded by the Squalene Synthase (SQS) gene, exhibited the highest number of interactions with bioactive compounds. This enzyme plays a vital role in the squalene and triterpenoid biosynthesis pathway [81,82]. The SQS gene is present in all plant organisms, though its expression varies significantly among species [83]. Molecular characterization of the SQS gene has been conducted in various medicinal plants to facilitate phylogenetic studies, metabolic engineering, and the production of squalene and triterpenoids in bacterial systems. Overexpression of the SQS gene has been achieved in plants such as Withania somnifera to enhance withanolide synthesis [84], in Eleutherococcus senticosus to increase phytosterol and triterpenic saponin production [85], and in hairy roots cultures of Centella asiatica to boost centelloside triterpenoid synthesis [86].

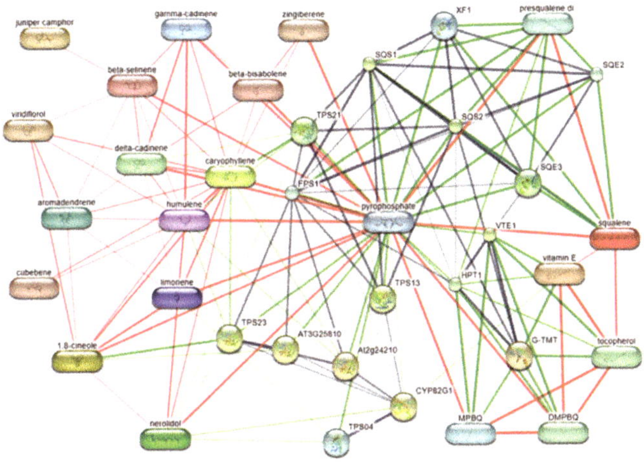

Figure 5. Interaction network of bioactive metabolites in *Psidium guajava*. Stronger associations are represented by thicker lines: protein-protein interactions are shown in gray, compound–protein interactions in green, and compound–compound interactions in red.

Other proteins with the highest number of interactions identified in this study were TPS23 and TPS21, members of the Terpene Synthases (TPS) gene family [87]. These proteins are precursors in the biosynthesis of thousands of terpenes found in plants, contributing to the distinctive terpene compositions observed across different taxa [88,89]. Among plant families, the Myrtaceae family is noted for having the highest concentrations and diversity of foliar terpenes. Recent studies have shown that species such as *Eucalyptus grandis*, *E. globulus*, and *Corymbia citriodora* possess the largest number of TPS genes [90]. In Brazilian guava cultivars, variations in TPS genes have been analyzed, enabling the mapping of functional genes involved in terpene biosynthesis and their use as molecular markers for genetic improvement [91]. The diverse range of terpene compounds identified in this study, along with the variability among guava varieties highlight the potential for future research on TPS genes in *Psidium guajava* L. Such investigations could deepen our understanding of metabolic profiles, elucidate their roles in adaptation mechanisms, and identify molecular targets for enhancing the production of therapeutic compounds through genetic improvement.

3. Materials and Methods

3.1. Plant Material

Young, green, firm, and spot-free leaves were collected from guava trees maintained in the germplasm bank of the National Institute of Forestry, Agricultural, and Livestock Research (INIFAP), located at the geographical coordinates 22°03′21.2″ N, 102°52′53.6″ W. Two varieties were selected for analysis: the yellow-fruited Caxcana and the pink-fruited

S-56 (Figure 6). Collected leaves were washed with distilled water to remove contaminants and subsequently freeze-dried using a Labconco Freeze Dry System/Freezone 4.5.

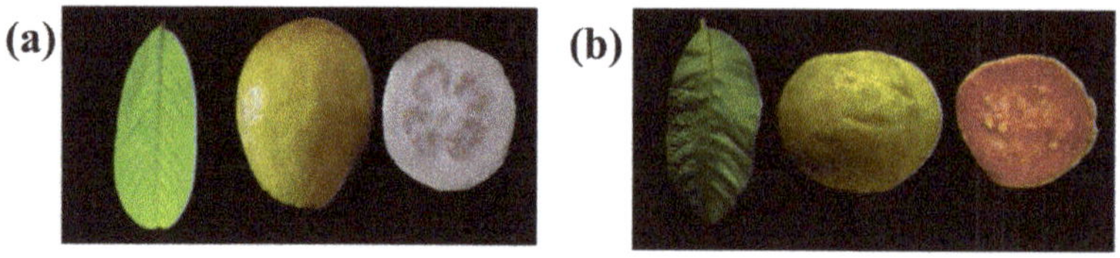

Figure 6. Guava fruit and leaf varieties from the "Calvillo-Cañones" region. (**a**) Caxcana cultivar. (**b**) S-56 accession.

3.2. Preparation of Methanol Extracts

A total of 500 mg of freeze-dried leaf material was weighed and transferred into a 50 mL Falcon tube. To each sample, 5 mL of HPLC-grade methanol (Karal, Leon, Guanajuato, Mexico) was added. The samples were sonicated at 40 °C for 20 min using a Bransonic 1800 ultrasonic cleaner (Emerson, Allen Park, MI, USA), followed by vortex mixing for 30 s. Subsequently, the samples were centrifuged at 6000 rpm for 10 min, and the supernatant was carefully collected for analysis by gas chromatography-mass spectrometry (GC–MS) using an Agilent 7890A GC system. Each extraction process was performed in triplicate for every leaf sample, and three technical replicates were analyzed for each biological replicate.

3.3. Analysis and Compound Identification by GC–MS Chromatography

Gas chromatography (GC) analyses were performed using an Agilent 7890A GC system coupled with a 5975C single quadrupole mass spectrometer (MS). The system was equipped with an Agilent DB-17HT capillary column (30 m × 0.25 mm × 0.15 μL), and helium served as the carrier gas at a constant flow rate of 1.29 mL/min. Samples (1 μL) were injected in splitless mode, with the injector temperature set at 250 °C. The oven temperature program was initiated at 50 °C, followed by a 10 °C /min ramp to reach 250 °C, and held for 15 min. Data acquisition and processing were conducted using Agilent MassHunter Quantitative Analysis software B.07.00, which allowed mass spectral deconvolution and integration. Compound identification was achieved by matching the acquired spectra against the NIST2011 library, considering compounds present in at least 70% of the total reads analyzed, and exhibiting a minimum similarity score of 80% for reliable identification

3.4. GC–MS Statistical Multivariate Analysis

Principal Component Analysis (PCA) was conducted on a data matrix comprising the percentage area of each chemical component identified in the guava leaf varieties. Data normalization and scaling were carried out using the Pareto scaling method [92]. Additionally, hierarchical clustering analysis (HCA) was performed, accompanied by a heat map to visualize the metabolite concentrations across the guava varieties. Euclidean distance was used as the metric for clustering. Data matrices for PCA and HCA were exported to the ClustVis (http://biit.cs.ut.ee/clustvis/ (accessed on 10 February 2025)), for graph generation and visualization.

3.5. Molecular Docking Analysis

The 3D structures of the target proteins CB2 (ID:6PT0), PPARα (ID:6KXY), BAX (ID:6EB6), BCL2 (ID:5FCG), and AKT1 (ID:4EKL) were retrieved from Protein Data Bank (www.pdb.org). Protein preparation for molecular docking analysis was carried out using MOE 2019 software. The chemical structure of BC (ID:6PT0) was obtained from PubChem (https://pubchem.ncbi.nlm.nih.gov/ (accessed on 18 January 2025)). Docking studies were conducted using the cb-Dock2 platform (https://cadd.labshare.cn/cb-dock2/index.php/ (accessed on 18 January 2025)) following standard procedures. Interaction diagrams and 2D images depicting the ligand-target protein interactions were generated using MOE 2019 based on the docking simulation results.

3.6. Interaction of Bioactive Compounds

The identified phytochemicals were analyzed in silico using the STITCH 4 database (http://stitch.embl.de/ (accessed on 18 January 2025)), which integrates protein–chemical interactions with user-provided data.

4. Conclusions

In this study, we identified several metabolites with therapeutic potential, mainly from the terpene family, with β-caryophyllene (BC) being the main component in the leaves of the two varieties analyzed. This study provides valuable insights into the metabolomic composition of guava leaves from the "Calvillo-Cañones" region. PCA, pharmacological network bioinformatics, and molecular docking approaches uncovered specific metabolites with potential therapeutic relevance. Notably, the predicted interactions of BC with key target proteins (CB2, PPARα, BAX, BCL2, and AKT1) suggest its involvement in crucial biochemical pathways related to inflammation, apoptosis, and cellular survival. These findings bridge an important gap in understanding guava leaf metabolites and their pharmacological potential. The identification of cultivar-specific compounds further refines our knowledge of the phytochemical diversity within *Psidium guajava* L. Additionally, our results provide a foundation for future research on the biomedical applications of guava-derived terpenes, particularly in drug discovery and therapeutic development. Further experimental validation of these interactions will be essential to fully elucidate their pharmacological significance and clinical applicability.

Author Contributions: A.V.V.-P.: methodology, formal analysis, investigation, writing—original draft preparation; A.S.: methodology, formal analysis, software, resources, supervision, review and editing; J.A.Z.: resources, review and editing; J.F.M.-D.: conceptualization, validation, visualization, supervision, funding acquisition, project administration, writing—review and editing. All authors have read and agreed to the published version of the manuscript.

Funding: This research was founded by the Autonomous University of Aguascalientes under project grant number PIBT23-1.

Institutional Review Board Statement: Not applicable.

Informed Consent Statement: Not applicable.

Data Availability Statement: Data are contained in the article.

Acknowledgments: The authors are grateful for the grant awarded by the SECIHTI with the number 860444. Special thanks are also extended to Jorge Artemio Zegbe Dominguez and the entire team at the Cañones Experimental Site for providing the plant material used in this study.

Conflicts of Interest: The authors declare no conflicts of interest.

Abbreviations

The following abbreviations are used in this manuscript:

GC–MS	Gas Chromatography coupled with Mass Spectrometry
PCA	Principal Component Analysis

BC	β-caryophyllene
CB2	Cannabinoid Receptor 2
PPARα	Peroxisome Proliferator-Activated Receptor Alpha
BAX	Bcl2-Associated Protein X
BCL2	B-cell Lymphoma 2
AKT1	Protein Kinase B

References

1. Marrelli, M. Medicinal Plants. *Plants* **2021**, *10*, 1355. [CrossRef] [PubMed]
2. Abdelhafez, O.H.; Othman, E.M.; Fahim, J.R.; Desoukey, S.Y.; Pimentel-Elardo, S.M.; Nodwell, J.R.; Schirmeister, T.; Tawfike, A.; Abdelmohsen, U.R. Metabolomics Analysis and Biological Investigation of Three Malvaceae Plants. *Phytochem. Anal.* **2020**, *31*, 204–214. [CrossRef]
3. Angulo-López, J.E.; Flores-Gallegos, A.C.; Torres-León, C.; Ramírez-Guzmán, K.N.; Martínez, G.A.; Aguilar, C.N. Guava (*Psidium Guajava* L.) Fruit and Valorization of Industrialization By-Products. *Processes* **2021**, *9*, 1075. [CrossRef]
4. Naseer, S.; Hussain, S.; Naeem, N.; Pervaiz, M.; Rahman, M. The Phytochemistry and Medicinal Value of *Psidium Guajava* (Guava). *Clin. Phytoscience* **2018**, *4*, 32. [CrossRef]
5. Kumar, M.; Tomar, M.; Amarowicz, R.; Saurabh, V.; Nair, M.S.; Maheshwari, C.; Sasi, M.; Prajapati, U.; Hasan, M.; Singh, S.; et al. Guava (*Psidium Guajava* L.) Leaves: Nutritional Composition, Phytochemical Profile, and Health-Promoting Bioactivities. *Foods* **2021**, *10*, 752. [CrossRef] [PubMed]
6. Suárez-Toledo, J.R.; Hernández-Aguilar, C.; Domínguez-Pacheco, F.A.; Aceves-Hernandez, F.J. Caracterización de La Guayaba Cultivada En México. *Rev. Mex. Cienc. Agric.* **2022**, *13*, 1233–1245. [CrossRef]
7. Padilla-Ramírez, J.S. Caracterización Morfológica de Fruto de la Colección Ex Situ de *Psidium Guajava* L. *Agro Product.* **2016**, *9*, 4.
8. Pérez-Barraza, M.H.; Osuna-García, J.A.; Padilla-Ramírez, J.S.; Sánchez-Lucio, R.; Nolasco-González, Y.; González-Gaona, E. Fenología, productividad y calidad de fruto de guayaba pulpa crema y rosa en clima tropical en México. *Interciencia* **2015**, *40*, 198–203.
9. Padilla Ramírez, J.S.; González Gaona, E.; Perales de la Cruz, M.Á. Nuevas variedades de guayaba (*Psidium guajava* L.). In *Folleto Técnico 42*; Instituto Nacional de Investigaciones Forestales, Agrícolas y Pecuarias: Coyoacán, Mexican, 2010; ISBN 978-607-37-0634-6.
10. Padilla-Ramírez, J.S.; González-Gaona, E.; Esquivel-Villagrana, F.; Mercado-Silva, E.; Hernández-Delgado, S.; Mayek-Pérez, N. Caracterización de Germoplasma Sobresaliente de Guayabo de la Región Calvillo-Cañones, México. *Rev. Fitotec. Mex.* **2002**, *25*, 393–399. [CrossRef]
11. Wang, S.; Liu, G.; Xie, C.; Zhou, Y.; Yang, R.; Wu, J.; Xu, J.; Tu, K. Metabolomics Analysis of Different Quinoa Cultivars Based on UPLC-ZenoTOF-MS/MS and Investigation into Their Antioxidant Characteristics. *Plants* **2024**, *13*, 240. [CrossRef]
12. Hill, C.B.; Roessner, U. Metabolic Profiling of Plants by GC–MS. In *The Handbook of Plant Metabolomics*; Wiley Online Books: Wiley, NJ, USA, 2013; pp. 1–23.
13. Abadie, C.; Lalande, J.; Tcherkez, G. Exact Mass GC-MS Analysis: Protocol, Database, Advantages, and Application to Plant Metabolic Profiling. *Plant Cell Environ.* **2022**, *45*, 3171–3183. [CrossRef] [PubMed]
14. Al-Rubaye, A.F.; Hameed, I.H.; Kadhim, M.J. A Review: Uses of Gas Chromatography-Mass Spectrometry (GC-MS) Technique for Analysis of Bioactive Natural Compounds of Some Plants. *Int. J. Toxicol. Pharmacol. Res.* **2017**, *9*, 81–85. [CrossRef]
15. Sobrinho, A.C.G.; Corpes, R.S.; dos Santos, K.I.P.; Barra, I.M.M.; Miyagawa, H.K.; Santos, A.S.S. Untargeted GC-MS Metabolomics Applied to Wild Leaves and Callus Produced by Plant Tissue Culture of *Hibiscus Sabdariffa* L. *Arab. J. Chem.* **2022**, *15*, 104103. [CrossRef]
16. Afzal, M.; Iqbal, R.; Mahmood, Z.; Zeshan, B.; Wattoo, J.I. Study of GC-MS and HPLC Characterized Metabolic Compounds In Guava (*Psidium guajava* L.) Leaves. *Pak. J. Agric. Sci.* **2019**, *56*, 3.
17. Satyal, P.; Paudel, P.; Lamichhane, B.; Setzer, W.N. Leaf essential oil composition and bioactivity of *Psidium guajava* from Kathmandu, Nepal. *Am. J. Essent. Oils Nat. Prod.* **2015**, *3*, 11–14.
18. Chaturvedi, T.; Singh, S.; Nishad, I.; Kumar, A.; Tiwari, N.; Tandon, S.; Saikia, D.; Verma, R.S. Chemical composition and antimicrobial activity of the essential oil of senescent leaves of guava (*Psidium guajava* L.). *Nat. Prod. Res.* **2021**, *35*, 1393–1397. [CrossRef]
19. Xu, C.; Liang, Z.; Tang, D.; Xiao, T.; Tsunoda, M.; Zhang, Y.; Zhao, L.; Deng, S.; Song, Y. Gas Chromatography-Mass Spectrometry (GC-MS) Analysis of Volatile Components from Guava Leaves. *J. Essent. Oil Bear. Plants* **2017**, *20*, 1536–1546. [CrossRef]
20. Brosset, A.; Blande, J.D. Volatile-Mediated Plant–Plant Interactions: Volatile Organic Compounds as Modulators of Receiver Plant Defence, Growth, and Reproduction. *J. Exp. Bot.* **2022**, *73*, 511–528. [CrossRef]
21. Dunkić, V.; Nazlić, M.; Ruščić, M.; Vuko, E.; Akrap, K.; Topić, S.; Milović, M.; Vuletić, N.; Puizina, J.; Grubešić, R.J.; et al. Hydrodistillation and Microwave Extraction of Volatile Compounds: Comparing Data for Twenty-One *Veronica* Species from Different Habitats. *Plants* **2022**, *11*, 902. [CrossRef]

22. Montejano-Ramírez, V.; Ávila-Oviedo, J.L.; Campos-Mendoza, F.J.; Valencia-Cantero, E. Microbial Volatile Organic Compounds: Insights into Plant Defense. *Plants* **2024**, *13*, 2013. [CrossRef]
23. Rosenkranz, M.; Chen, Y.; Zhu, P.; Vlot, A.C. Volatile Terpenes–Mediators of Plant-to-plant Communication. *Plant J.* **2021**, *108*, 617–631. [CrossRef] [PubMed]
24. Padovan, A.; Keszei, A.; Külheim, C.; Foley, W.J. The Evolution of Foliar Terpene Diversity in Myrtaceae. *Phytochem. Rev.* **2014**, *13*, 695–716. [CrossRef]
25. Padovan, A.; Keszei, A.; Wallis, I.R.; Foley, W.J. Mosaic Eucalypt Trees Suggest Genetic Control at a Point That Influences Several Metabolic Pathways. *J. Chem. Ecol.* **2012**, *38*, 914–923. [CrossRef] [PubMed]
26. Portillo-Estrada, M. Limitations of Plant Stress Tolerance upon Heat and CO_2 Exposure in Black Poplar: Assessment of Photosynthetic Traits and Stress Volatile Emissions. *Plants* **2024**, *13*, 1165. [CrossRef]
27. Reddy, S.G.; Dolma, S.K.; Koundal, R.; Singh, B. Chemical Composition and Insecticidal Activities of Essential Oils against Diamondback Moth, *Plutella Xylostella* (L.) (Lepidoptera: Yponomeutidae). *Nat. Prod. Res.* **2016**, *30*, 1834–1838. [CrossRef]
28. Benelli, G.; Govindarajan, M.; AlSalhi, M.S.; Devanesan, S.; Maggi, F. High Toxicity of Camphene and γ-Elemene from *Wedelia Prostrata* Essential Oil against Larvae of *Spodoptera litura* (Lepidoptera: Noctuidae). *Environ. Sci. Pollut. Res.* **2018**, *25*, 10383–10391. [CrossRef]
29. Tanabe, K.; Hayashi, C.; Katahira, T.; Sasaki, K.; Igami, K. Multiblock Metabolomics: An Approach to Elucidate Whole-Body Metabolism with Multiblock Principal Component Analysis. *Comput. Struct. Biotechnol. J.* **2021**, *19*, 1956–1965. [CrossRef]
30. Li, D.S.; Shi, L.L.; Guo, K.; Luo, S.H.; Liu, Y.C.; Chen, Y.G.; Liu, Y.; Li, S.H. A New Sesquiterpene Synthase Catalyzing the Formation of (R)-β-Bisabolene from Medicinal Plant Colquhounia Coccinea Var. Mollis and Its Anti-Adipogenic and Antibacterial Activities. *Phytochemistry* **2023**, *211*, 113681. [CrossRef]
31. Yeo, S.K.; Ali, A.Y.; Hayward, O.A.; Turnham, D.; Jackson, T.; Bowen, I.D.; Clarkson, R. β-Bisabolene, a Sesquiterpene from the Essential Oil Extract of Opoponax (*Commiphora Guidottii*), Exhibits Cytotoxicity in Breast Cancer Cell Lines. *Phytother. Res.* **2016**, *30*, 418–425. [CrossRef]
32. Mulyaningsih, S.; Sporer, F.; Zimmermann, S.; Reichling, J.; Wink, M. Synergistic Properties of the Terpenoids Aromadendrene and 1,8-Cineole from the Essential Oil of *Eucalyptus Globulus* against Antibiotic-Susceptible and Antibiotic-Resistant Pathogens. *Phytomedicine* **2010**, *17*, 1061–1066. [CrossRef]
33. Pavithra, P.S.; Mehta, A.; Verma, R.S. Aromadendrene Oxide 2, Induces Apoptosis in Skin Epidermoid Cancer Cells through ROS Mediated Mitochondrial Pathway. *Life Sci.* **2018**, *197*, 19–29. [CrossRef]
34. Roberto, D.; Micucci, P.; Sebastian, T.; Graciela, F.; Anesini, C. Antioxidant Activity of Limonene on Normal Murine Lymphocytes: Relation to H2O2 Modulation and Cell Proliferation. *Basic Clin. Pharmacol. Toxicol.* **2010**, *106*, 38–44. [CrossRef]
35. Anandakumar, P.; Kamaraj, S.; Vanitha, M.K. D-limonene: A Multifunctional Compound with Potent Therapeutic Effects. *J. Food Biochem.* **2021**, *45*, e13566. [CrossRef] [PubMed]
36. Mishra, G.; Singh, P.; Molla, M.; Yimer, Y.; Ewunetie, A.; Tadesse, T.Y.; Ayele, T.M.; Kefale, B. Nutraceuticals: A Source of Benefaction for Neuropathic Pain and Fibromyalgia. *J. Funct. Foods* **2022**, *97*, 105260. [CrossRef]
37. Francomano, F.; Caruso, A.; Barbarossa, A.; Fazio, A.; La Torre, C.; Ceramella, J.; Mallamaci, R.; Saturnino, C.; Iacopetta, D.; Sinicropi, M.S. β-Caryophyllene: A Sesquiterpene with Countless Biological Properties. *Appl. Sci.* **2019**, *9*, 5420. [CrossRef]
38. Paula-Freire, L.I.G.; Andersen, M.L.; Gama, V.S.; Molska, G.R.; Carlini, E.L.A. The Oral Administration of Trans-Caryophyllene Attenuates Acute and Chronic Pain in Mice. *Phytomedicine* **2014**, *21*, 356–362. [CrossRef] [PubMed]
39. Ettitaou, A.; Kabdy, H.; Oubella, K.; Raoui, K.; Oubahmane, M.; Aboufatima, R.; Elyazouli, L.; Garzoli, S.; Chait, A. Molecular docking of quercetin: A promising approach for the development of new anti-inflammatory and analgesic drugs. *Nat. Prod. Res.* **2024**, *38*, 1–10. [CrossRef]
40. Pertwee, R.G.; Thomas, A. Therapeutic applications for agents that act at CB1 and CB2 receptors. In *The Cannabinoid Receptors*; Reggio, P.H., Ed.; The Receptors; Humana Press: Totowa, NJ, USA, 2009. [CrossRef]
41. Turcotte, C.; Blanchet, M.R.; Laviolette, M.; Flamand, N. The CB_2 receptor and its role as a regulator of inflammation. *Cell. Mol. Life Sci.* **2016**, *73*, 4449–4470. [CrossRef]
42. Rigano, D.; Sirignano, C.; Taglialatela-Scafati, O. The potential of natural products for targeting PPARα. *Acta Pharm. Sin. B* **2017**, *7*, 427–438. [CrossRef]
43. Wang, L.; Waltenberger, B.; Pferschy-Wenzig, E.M.; Blunder, M.; Liu, X.; Malainer, C.; Blazevic, T.; Schwaiger, S.; Rollinger, J.M.; Heiss, E.H.; et al. Natural product agonists of peroxisome proliferator-activated receptor gamma (PPARγ): A review. *Biochem. Pharmacol.* **2014**, *92*, 73–89. [CrossRef]
44. Honda, A.; Kamata, S.; Akahane, M.; Machida, Y.; Uchii, K.; Shiiyama, Y.; Habu, Y.; Miyawaki, S.; Kaneko, C.; Oyama, T.; et al. Functional and Structural Insights into Human PPARα/δ/γ Subtype Selectivity of Bezafibrate, Fenofibric Acid, and Pemafibrate. *Int. J. Mol. Sci.* **2022**, *23*, 4726. [CrossRef] [PubMed]

45. Alam, S.; Mohammad, T.; Padder, R.A.; Hassan, M.I.; Husain, M. Thymoquinone and quercetin induce enhanced apoptosis in non-small cell lung cancer in combination through the Bax/Bcl2 cascade. *J. Cell. Biochem.* **2022**, *123*, 259–274. [CrossRef]
46. Zhang, Q.; Chen, L.; Gao, M.; Wang, S.; Meng, L.; Guo, L. Molecular docking and in vitro experiments verified that *kaempferol* induced apoptosis and inhibited human HepG2 cell proliferation by targeting BAX, CDK1, and JUN. *Mol. Cell. Biochem.* **2023**, *478*, 767–780. [CrossRef]
47. Feng, G.; Zhang, X.; Li, Y.; Wang, R. Analysis of the Binding Sites on BAX and the Mechanism of BAX Activators through Extensive Molecular Dynamics Simulations. *J. Chem. Inf. Model.* **2022**, *62*, 5208–5222. [CrossRef] [PubMed]
48. Rosdi, M.N.M.; Arif, S.M.; Abu Bakar, M.H.; Razali, S.A.; Zulkifli, R.M.; Ya'akob, H. Molecular docking studies of bioactive compounds from *Annona muricata* Linn as potential inhibitors for Bcl-2, Bcl-w and Mcl-1 antiapoptotic proteins. *Apoptosis* **2018**, *23*, 27–40. [CrossRef]
49. Sathishkumar, N.; Sathiyamoorthy, S.; Ramya, M.; Yang, D.U.; Lee, H.N.; Yang, D.C. Molecular docking studies of anti-apoptotic BCL-2, BCL-XL, and MCL-1 proteins with ginsenosides from *Panax ginseng*. *J. Enzym. Inhib. Med. Chem.* **2012**, *27*, 685–692. [CrossRef]
50. Liu, W.; Bulgaru, A.; Haigentz, M.; Stein, C.A.; Perez-Soler, R.; Mani, S. The BCL2-family of protein ligands as cancer drugs: The next generation of therapeutics. *Curr. Med. Chem.-Anti-Cancer Agents* **2003**, *3*, 217–223. [CrossRef] [PubMed]
51. Mirza, Z.; Karim, S. Structure-Based Profiling of Potential Phytomolecules with AKT1 a Key Cancer Drug Target. *Molecules* **2023**, *28*, 2597. [CrossRef]
52. Mahajan, P.; Wadhwa, B.; Barik, M.R.; Malik, F.; Nargotra, A. Combining ligand- and structure-based in silico methods for the identification of natural product-based inhibitors of Akt1. *Mol. Divers.* **2020**, *24*, 45–60. [CrossRef]
53. Whiting, Z.M.; Yin, J.; de la Harpe, S.M.; Vernall, A.J.; Grimsey, N.L. Developing the Cannabinoid Receptor 2 (CB2) Pharmacopoeia: Past, Present, and Future. *Trends Pharmacol. Sci.* **2022**, *43*, 754–771. [CrossRef]
54. Scandiffio, R.; Geddo, F.; Cottone, E.; Querio, G.; Antoniotti, S.; Gallo, M.P.; Maffei, M.E.; Bovolin, P. Protective Effects of (E)-β-Caryophyllene (BCP) in Chronic Inflammation. *Nutrients* **2020**, *12*, 3273. [CrossRef]
55. Porto, D.S.; Port, B.D.C.B.; Conte, J.; Argenta, D.F.; Balleste, M.P.; Micke, G.A.; Campos, Â.M.M.; Caumo, K.S.; Caon, T. Development of Ophthalmic Nanoemulsions of β-Caryophyllene for the Treatment of *Acanthamoeba Keratitis*. *Int. J. Pharm.* **2024**, *659*, 124252. [CrossRef] [PubMed]
56. Ricardi, C.; Barachini, S.; Consoli, G.; Marazziti, D.; Polini, B.; Chiellini, G. Beta-Caryophyllene, a Cannabinoid Receptor Type 2 Selective Agonist, in Emotional and Cognitive Disorders. *Int. J. Mol. Sci.* **2024**, *25*, 3203. [CrossRef] [PubMed]
57. Kamata, S.; Oyama, T.; Saito, K.; Honda, A.; Yamamoto, Y.; Suda, K.; Ishikawa, R.; Itoh, T.; Watanabe, Y.; Shibata, T.; et al. PPARα Ligand-Binding Domain Structures with Endogenous Fatty Acids and Fibrates. *iScience* **2020**, *23*, 101727. [CrossRef]
58. Bougarne, N.; Weyers, B.; Desmet, S.J.; Deckers, J.; Ray, D.W.; Staels, B.; De Bosscher, K. Molecular Actions of PPARα in Lipid Metabolism and Inflammation. *Endocr. Rev.* **2018**, *39*, 760–802. [CrossRef] [PubMed]
59. Wójtowicz, S.; Strosznajder, A.K.; Jeżyna, M.; Strosznajder, J.B. The Novel Role of PPAR Alpha in the Brain: Promising Target in Therapy of Alzheimer's Disease and Other Neurodegenerative Disorders. *Neurochem. Res.* **2020**, *45*, 972–988. [CrossRef]
60. Xu, J.; Chavis, J.A.; Racke, M.K.; Drew, P.D. Peroxisome Proliferator-Activated Receptor-α and Retinoid X Receptor Agonists Inhibit Inflammatory Responses of Astrocytes. *J. Neuroimmunol.* **2006**, *176*, 95–105. [CrossRef]
61. Straus, D.S.; Glass, C.K. Anti-Inflammatory Actions of PPAR Ligands: New Insights on Cellular and Molecular Mechanisms. *Trends Immunol.* **2007**, *28*, 551–558. [CrossRef]
62. Han, L.; Shen, W.-J.; Bittner, S.; Kraemer, F.B.; Azhar, S. PPARs: Regulators of Metabolism and As Therapeutic Targets in Cardiovascular Disease. Part I: PPAR-α. *Futur. Cardiol.* **2017**, *13*, 259–278. [CrossRef]
63. Cheng, A.Y.Y.; Leiter, L.A. PPAR-alpha: Therapeutic Role in Diabetes-related Cardiovascular Disease. *Diabetes Obes. Metab.* **2008**, *10*, 691–698. [CrossRef]
64. Al-Harbi, L.N. Morin Prevents Non-Alcoholic Hepatic Steatosis in Obese Rats by Targeting the Peroxisome Proliferator-Activated Receptor Alpha (PPARα). *Life* **2024**, *14*, 945. [CrossRef]
65. Spaner, D.E.; Lee, E.; Shi, Y.; Wen, F.; Li, Y.; Tung, S.; McCaw, L.; Wong, K.; Gary-Gouy, H.; Dalloul, A.; et al. PPAR-Alpha Is a Therapeutic Target for Chronic Lymphocytic Leukemia. *Leukemia* **2013**, *27*, 1090–1099. [CrossRef] [PubMed]
66. Spitz, A.Z.; Gavathiotis, E. Physiological and Pharmacological Modulation of BAX. *Trends Pharmacol. Sci.* **2022**, *43*, 206–220. [CrossRef]
67. Liu, Z.; Ding, Y.; Ye, N.; Wild, C.; Chen, H.; Zhou, J. Direct Activation of Bax Protein for Cancer Therapy. *Med. Res. Rev.* **2016**, *36*, 313–341. [CrossRef] [PubMed]
68. Zhang, Z.; Hou, L.; Liu, D.; Luan, S.; Huang, M.; Zhao, L. Directly Targeting BAX for Drug Discovery: Therapeutic Opportunities and Challenges. *Acta Pharm. Sin. B* **2024**, *14*, 2378–2401. [CrossRef]
69. Reed, J.C. Proapoptotic Multidomain Bcl-2/Bax-Family Proteins: Mechanisms, Physiological Roles, and Therapeutic Opportunities. *Cell Death Differ.* **2006**, *13*, 1378–1386. [CrossRef] [PubMed]

70. Vogler, M.; Dinsdale, D.; Dyer, M.J.S.; Cohen, G.M. Bcl-2 Inhibitors: Small Molecules with a Big Impact on Cancer Therapy. *Cell Death Differ.* **2009**, *16*, 360–367. [CrossRef]
71. Campàs, C.; Cosialls, A.M.; Barragán, M.; Iglesias-Serret, D.; Santidrián, A.F.; Coll-Mulet, L.; de Frias, M.; Domingo, A.; Pons, G.; Gil, J. Bcl-2 Inhibitors Induce Apoptosis in Chronic Lymphocytic Leukemia Cells. *Exp. Hematol.* **2006**, *34*, 1663–1669. [CrossRef]
72. Tahir, S.K.; Yang, X.; Anderson, M.G.; Morgan-Lappe, S.E.; Sarthy, A.V.; Chen, J.; Warner, R.B.; Ng, S.-C.; Fesik, S.W.; Elmore, S.W.; et al. Influence of Bcl-2 Family Members on the Cellular Response of Small-Cell Lung Cancer Cell Lines to ABT-737. *Cancer Res.* **2007**, *67*, 1176–1183. [CrossRef]
73. D'Aguanno, S.; Del Bufalo, D. Inhibition of Anti-Apoptotic Bcl-2 Proteins in Preclinical and Clinical Studies: Current Overview in Cancer. *Cells* **2020**, *9*, 1287. [CrossRef]
74. Hill, M.M.; Hemmings, B.A. Inhibition of Protein Kinase B/Akt. Implications for cancer therapy. *Pharmacol. Ther.* **2002**, *93*, 243–251. [CrossRef] [PubMed]
75. Barnett, S.F.; Bilodeau, M.T.; Lindsley, C.W. The Akt/PKB Family of Protein Kinases: A Review of Small Molecule Inhibitors and Progress Towards Target Validation. *Curr. Top. Med. Chem.* **2010**, *10*, 458–477. [CrossRef]
76. Hinz, N.; Jücker, M. Distinct Functions of AKT Isoforms in Breast Cancer: A Comprehensive Review. *Cell Commun. Signal.* **2019**, *17*, 154. [CrossRef] [PubMed]
77. Song, M.; Bode, A.M.; Dong, Z.; Lee, M.-H. AKT as a Therapeutic Target for Cancer. *Cancer Res.* **2019**, *79*, 1019–1031. [CrossRef]
78. Kaur, P.; Shukla, S.; Gupta, S. Plant Flavonoid Apigenin Inactivates Akt to Trigger Apoptosis in Human Prostate Cancer: An in Vitro and in Vivo Study. *Carcinogenesis* **2008**, *29*, 2210–2217. [CrossRef]
79. Lakshmi, T.P.; Kumar, A.; Vijaykumar, V.; Natarajan, S.; Krishna, R. Identification of Natural Allosteric Inhibitor for Akt1 Protein through Computational Approaches and in Vitro Evaluation. *Int. J. Biol. Macromol.* **2017**, *96*, 200–213. [CrossRef] [PubMed]
80. Alwhaibi, A.; Verma, A.; Adil, M.S.; Somanath, P.R. The Unconventional Role of Akt1 in the Advanced Cancers and in Diabetes-Promoted Carcinogenesis. *Pharmacol. Res.* **2019**, *145*, 104270. [CrossRef]
81. Lee, M.H.; Jeong, J.H.; Seo, J.W.; Shin, C.G.; Kim, Y.S.; In, J.G.; Yang, D.C.; Yi, J.S.; Choi, Y.E. Enhanced Triterpene and Phytosterol Biosynthesis in *Panax Ginseng* Overexpressing Squalene Synthase Gene. *Plant Cell Physiol.* **2004**, *45*, 976–984. [CrossRef]
82. Aminfar, Z.; Tohidfar, M. In Silico Analysis of Squalene Synthase in Fabaceae Family Using Bioinformatics Tools. *J. Genet. Eng. Biotechnol.* **2018**, *16*, 739–747. [CrossRef]
83. Hazra, A.; Dutta, M.; Dutta, R.; Bhattacharya, E.; Bose, R.; Biswas, S.M. Squalene Synthase in Plants–Functional Intricacy and Evolutionary Divergence While Retaining a Core Catalytic Structure. *Plant Gene* **2023**, *33*, 100403. [CrossRef]
84. Patel, N.; Patel, P.; Kendurkar, S.V.; Thulasiram, H.V.; Khan, B.M. Overexpression of Squalene Synthase in *Withania somnifera* Leads to Enhanced Withanolide Biosynthesis. *Plant Cell Tissue Organ Cult. (PCTOC)* **2015**, *122*, 409–420. [CrossRef]
85. Seo, J.W.; Jeong, J.H.; Shin, C.G.; Lo, S.C.; Han, S.S.; Yu, K.W.; Harada, E.; Han, J.Y.; Choi, Y.E. Overexpression of Squalene Synthase in *Eleutherococcus senticosus* Increases Phytosterol and Triterpene Accumulation. *Phytochemistry* **2005**, *66*, 869–877. [CrossRef]
86. Alcalde, M.A.; Palazon, J.; Bonfill, M.; Hidalgo-Martinez, D. Enhancing Centelloside Production in *Centella asiatica* Hairy Root Lines through Metabolic Engineering of Triterpene Biosynthetic Pathway Early Genes. *Plants* **2023**, *12*, 3363. [CrossRef] [PubMed]
87. Nagegowda, D.A.; Gupta, P. Advances in Biosynthesis, Regulation, and Metabolic Engineering of Plant Specialized Terpenoids. *Plant Sci.* **2020**, *294*, 110457. [CrossRef]
88. Zhang, A.; Xiong, Y.; Fang, J.; Jiang, X.; Wang, T.; Liu, K.; Peng, H.; Zhang, X. Diversity and Functional Evolution of Terpene Synthases in Rosaceae. *Plants* **2022**, *11*, 736. [CrossRef]
89. Ashour, M.; Wink, M.; Gershenzon, J. Biochemistry of Terpenoids: Monoterpenes, Sesquiterpenes and Diterpenes. In *Annual Plant Reviews Volume 40: Biochemistry of Plant Secondary Metabolism*; Wink, M., Ed.; Wiley-Blackwell: Oxford, UK, 2010; pp. 258–303.
90. Canal, D.; Escudero, F.L.G.; Mendes, L.A.; da Silva Ferreira, M.F.; Turchetto-Zolet, A.C. Genome-Wide Identification, Expression Profile and Evolutionary Relationships of TPS Genes in the Neotropical Fruit Tree Species *Psidium cattleyanum*. *Sci. Rep.* **2023**, *13*, 3930. [CrossRef] [PubMed]
91. Canal, D.; dos Santos, P.H.D.; de Avelar Carpinetti, P.; Silva, M.A.; Fernandes, M.; Brustolini, O.J.B.; Ferreira, A.; da Silva Ferreira, M.F. Exploring the Versatility of Sesquiterpene Biosynthesis in Guava Plants: A Comparative Genome-Wide Analysis of Two Cultivars. *Sci. Rep.* **2024**, *14*, 574. [CrossRef] [PubMed]
92. Okada, T.; Afendi, F.M.; Altaf-Ul-Amin, M.; Takahashi, H.; Nakamura, K.; Kanaya, S. Metabolomics of medicinal plants: The importance of multivariate analysis of analytical chemistry data. *Curr. Comput.-Aided Drug Des.* **2010**, *6*, 179–196. [CrossRef]

International Journal of
Molecular Sciences

MDPI

Article

Expanding Knowledge about the Influence of Citral on Cognitive Functions—In Vitro, In Vivo and Ex Vivo Studies

Karolina Wojtunik-Kulesza [1,*], Monika Rudkowska [2], Katarzyna Klimek [3], Monika Agacka-Mołdoch [4], Jarosław Mołdoch [5] and Agnieszka Michalak [2]

1 Department of Inorganic Chemistry, Medical University of Lublin, Chodźki 4a, 20-093 Lublin, Poland
2 Independent Laboratory of Behavioral Studies, Medical University of Lublin, 4A Chodźki, 20-093 Lublin, Poland; monika.rudkowska@umlub.pl (M.R.); agnieszka.michalak@umlub.pl (A.M.)
3 Department of Biochemistry and Biotechnology, Medical University of Lublin, 1 Chodźki Street, 20-093 Lublin, Poland; katarzyna.klimek@umlub.pl
4 Department of Plant Breeding and Biotechnology, Institute of Soil Science and Plant Cultivation, State Research Institute, Czartoryskich 8, 24-100 Puławy, Poland; magacka@iung.pulawy.pl
5 Department of Biochemistry and Crop Quality, Institute of Soil Science and Plant Cultivation, State Research Institute, Czartoryskich 8, 24-100 Puławy, Poland; jmoldoch@iung.pulawy.pl
* Correspondence: karolina.wojtunik@umlub.pl

Abstract: Citral, a common monoterpene found in numerous plants, is an interesting compound that has been shown to have various biological activities. Although it is widely distributed in nature and there are many studies presenting its biological activities, its anti-neurodegenerative activity, especially under in vivo conditions, is very poorly understood. Thus, this paper aimed to deepen knowledge about citral activity towards factors and symptoms of neurodegeneration. To accomplish this, several comprehensive tests were conducted, including the estimation of butyrylcholinesterase inhibition, the evaluation of hepatotoxicity and the detection of oxidative stress and lipid peroxidation in vitro, as well as an in vivo behavioral assessment using mice models. Additionally, ex vivo determination of level of the compound in the brain and blood of a tested animal was undertaken. The results obtained revealed that citral is able to inhibit butyrylcholinesterase activity and protect hepatic cells against oxidative stress and lipid peroxidation in vitro. Moreover, behavioral tests in vivo indicated that citral (50 mg/kg) improves memory processes associated with acquisition (passive avoidance test), both in acute and subchronic administration. Additionally, we found that the administration of citral at 25 mg/kg and 50 mg/kg did not significantly affect the locomotor activity. Beyond the aforementioned, gas chromatography-mass spectrometry analysis revealed the presence of the compound in the blood and brain after subchronic administration of citral. Taken together, the results obtained in vitro, in vivo and ex vivo clearly indicate that citral is a promising monoterpene that can potentially be used towards cognition improvement.

Keywords: citral; memory; mice; brain; neurodegeneration; butyrylcholinesteraze; hepatotoxicity; passive avoidance test; locomotor activity; lipid peroxidation

1. Introduction

Citral (3,7-dimethyl-2-6-octadienal) (Figure 1) is a volatile unsaturated aldehyde belonging to the largest group of secondary plant metabolites, namely terpenes. This monoterpene has been subjected to detailed study and is commonly used as a fragrance and flavoring, due to its characteristic odor. It can be found in numerous plants, e.g, A. citriodora, P. citriodora and Cymbopogon varieties [1,2]. The open-chain monoterpenoid, both natural and synthetic, is a mixture of isomers: cis-neral and trans-geranial. It is known that geranial is structurally similar to all-trans retinaldehyde. There are study results which indicate that citral is an antagonist of vitamin A, acting by inhibiting the oxidation of retinal to retinoic acid [3].

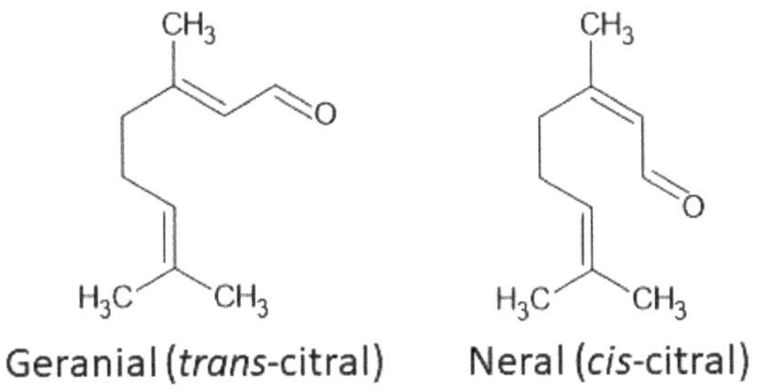

Figure 1. Structure of citral isomers: geranial (*trans*-citral) and neral (*cis*-citral).

This compound has been revealed to be involved in numerous pro-health activities (antioxidant, antimicrobial, acetylcholinesterase (AChE) inhibitory, anti-inflammatory, chemo-preventive, learning improvement properties, etc.) [3–5]. Despite the significant pro-health side effects of the monoterpenoid, there are also results presenting pathological processes which are associated with citral activity, such as cutaneous anaphylaxis and prostatic hyperplasia [6]. Nevertheless, it is worth mentioning that citral is Generally Regarded as Safe (GRAS) by the US Food and Drug Administration, hence, this compound is considered safe for human and animal consumption [7]. Detailed in vivo studies based on rat embryos revealed that citral is embryotoxic only after topical, intraperitoneal or oral treatment at dose levels that were also toxic for mothers [3].

The presented paper is focused on the determination of citral's influence on neurodegeneration symptoms (which, in most cases, are associated with behavioral disturbances). It is known that dementia, as well as disorders strictly associated with it (i.e., Alzheimer's disease), may manifest themselves as minor disturbances related to memory and orientation in space, which deepen along with disease progress. Most of the neurological changes are multifactorial and therefore it is extremely important to find a substance that will be active in several possible directions [8]. In the case of Alzheimer's disease (AD), the most important factors are oxidative stress, low level of neurotransmitters (i.e., acetylcholine), neurofibrillary tangles and amyloid β accumulation, disturbances to the levels of ion metals including Fe(II), Fe(III), Cu(II), Ca(II) and neuroinflammation [9].

Increasing attention is being paid to disease prevention based on following a healthy lifestyle and practicing proper nutrition by consuming a diet rich in plant products and natural compounds. To date, several drugs have been approved for the treatment of neurodegenerative diseases. Among them are drugs accepted for AD treatment, namely galantamine, donepezil, rivastigmine, memantine and aducanumab [10]. Nevertheless, their use is also associated with serious side effects, including hepatotoxicity [11]. Thus, alternatives are desired.

An analysis of available study results revealed that citral can be a good multi-directional agent against neurodegeneration and/or prevent the changes. Citral has been studied both in vitro and in vivo towards various neurodegeneration factors, but in vivo study results are limited. Among the in vitro assays, there are results pointing towards free radical scavenging activity, Fe(III) and Cu(II) reduction and Fe(II) chelation. Studies have revealed satisfactory activity towards all of the aforementioned factors [12–14]. Additionally, in vitro and in silico analyses of the monoterpene towards AChE have indicated the ability of citral to counteract low levels of neurotransmiter acetylcholine [15]. Despite research into the interesting and significant activity of citral in in vitro conditions, studies based on animal models are limited [16–18].

The presented paper is intended to enrich our current knowledge of the anti-neurodegenerative activity of citral as evidenced through mice models, as well as its butyrylcholinesterase (BChE) inhibitory activity, hepatotoxicity, detection of oxidative stress and lipid peroxidation using hepatic cells.

2. Results

2.1. BChE Inhibitory Activity

Thin-layer chromatography was employed to evaluate citral's BChE inhibitory activity. The use of a slightly modified Marston method [19] allowed us to determine the activity of various concentrations of citral and galantamine-a known AChE and BChE inhibitor. Figure 2 presents the obtained results.

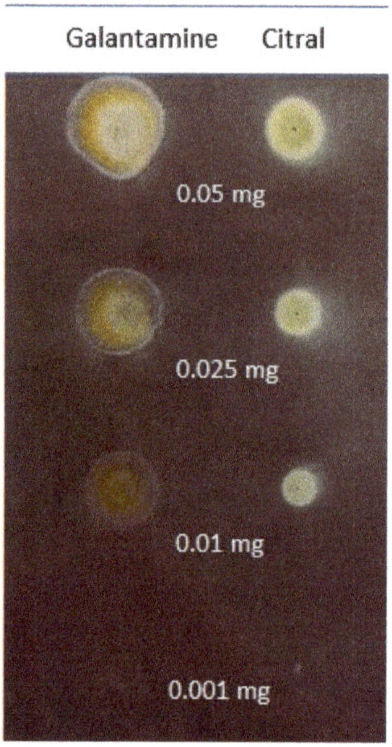

Figure 2. Inhibitory activity of citral and standard (galantamine) at various amounts: 0.05 mg, 0.025 mg, 0.01 mg and 0.001 mg.

The results were analyzed using the Sorbfil TLC Videodensitometer program which allows the activity of compounds to be assessed based on changes in the color of the spots with the applied compound compared to the background of the plate that does not contain active compounds. Additionally, comparative analysis allows the obtained results to be compared to the activity of the reference substance, which, in this case, is galantamine. Table 1 shows the BChE inhibitory activity (%) of citral in comparison to galantamine (activity assumed as 100%) for selected amounts of terpene.

Table 1. BChE inhibitory activity [%] of citral and galantamine.

Compound	Inhibitory Activity [%] for Selected Amount of Compounds			
	0.05 mg	**0.025 mg**	**0.01 mg**	**0.001 mg**
Galantamine *	100	100	100	100
citral	44	63	131	35

* galantamine activity was assumed to be 100%.

In accordance with Table 1, citral was revealed to show satisfactory activity towards BChE inhibition. Interestingly, its activity did not change linearly with changing concentration, although it was expected that the activity would decrease as the amount of citral applied was reduced. Instead, the performed assay revealed a non-linear concentration–

activity relationship for which the highest activity turned out to be for 0.01 mg of applied compound, whereas the highest and the lowest amounts of citral revealed the weakest activity.

2.2. In Vitro Hepatotoxicity of Citral

In this study, the HepG2 cell line was used, as it is considered to be a model cell line for hepatotoxicity evaluation in searching for new potential drugs for treating neurodegeneration [20,21]. After 48 h of incubation, the 3-[4,5-dimethylthiazol-2-yl]-2,5 diphenyl tetrazolium bromide (MTT) assay [22] showed that citral possessed hepatotoxic activity at some tested concentrations (Figure 3A). Thus, the CC_{50} value for citral was determined at approx. 0.57 mM. This result was confirmed by CLSM observations (Figure 3B). In this case, cells treated only with selected concentrations were observed and photographed. Nevertheless, it is clearly evident that the viability of hepatocytes incubated with citral at concentration of 1.25 mM and 5 mM is significantly decreased. In turn, galantamine (reference compound, positive control) exhibited a lower hepatotoxocity compared to citral, with a CC_{50} value close to 5.0 mM (Figure 3A,B).

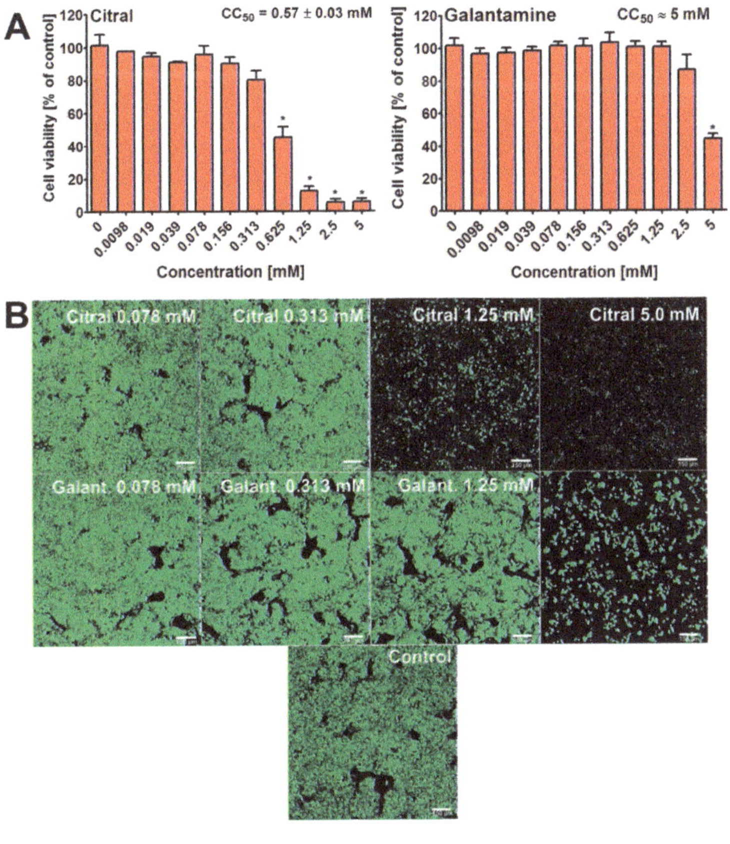

Figure 3. Viability of human hepatocellular carcinoma cells (HepG2 cell line, HB-8065) after 48 h incubation with citral and galantamine. MTT assay was used for quantitative evaluation of cytotoxicity (**A**); * statistically significant differences between cell viability after treatment with citral/galantamine and cell viability after incubation without tested substance (control, 0 mM), Student *t*-test, $p < 0.05$. Moreover, Live/Dead Double staining was used for visualization of live (green fluorescence) and dead cells (red fluorescence) (**B**); magnification 100×, scale bar = 150 μm.

2.3. *In Vitro Hepatoprotective Activity of Citral*

The hepatoprotective activity of citral against H_2O_2-induced oxidative stress and lipid peroxidation was assessed via CLSM observations (Figures 4 and 5). Based on data obtained from the MTT assay (previous experiment—in vitro hepatotoxicity of citral), a citral at concentration of 0.078 mM was used. Firstly, CLSM observations showed that citral significantly decreased the generation of ROS by H_2O_2 in HepG2 cells (Figure 4). As can be seen, H_2O_2 possessed a strong oxidative potential in HepG2 cells pre-incubated only with culture medium (positive control). These observations were also confirmed during staining (Figure 5), where ROS generated by H_2O_2 induced strong lipid peroxidation in HepG2 cells pre-incubated only with culture medium (positive control). In turn, citral protected HepG2 cells against lipid peroxidation. Hence, these observations indicated that this monoterpene has hepatoprotective potential in vitro.

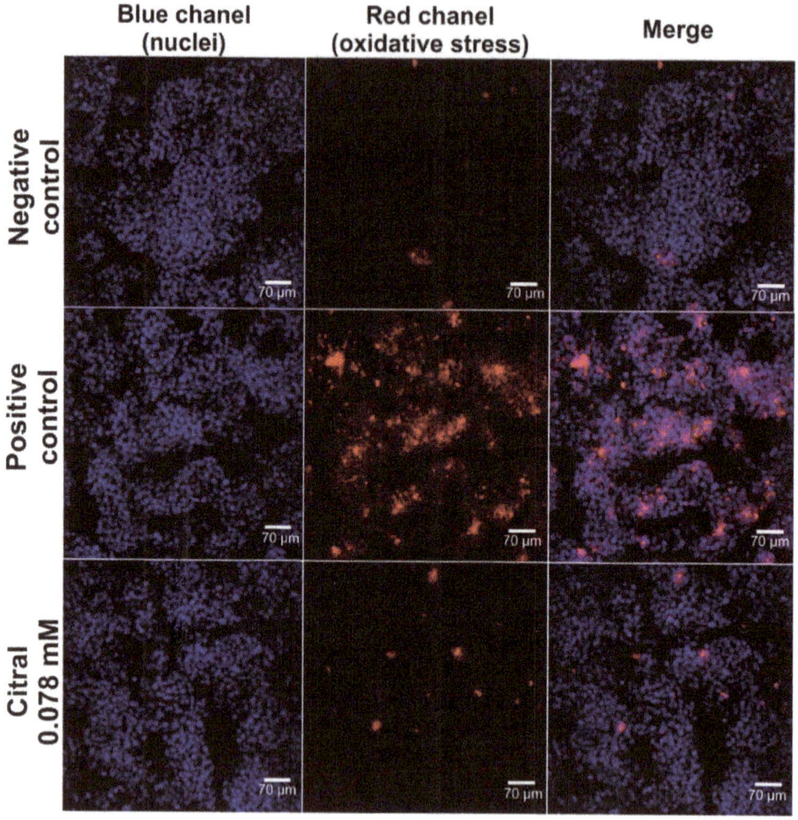

Figure 4. The CLSM images presenting hepatoprotective activity of citiral against H_2O_2-induced oxidative stress. Human hepatocellular carcinoma cells (HepG2 cell line, HB-8065) were incubated with citral for 48 h, followed by 1-h incubation with 500 µM H_2O_2. In turn, cells incubated only with culture medium are served as negative control, while cells incubated with culture medium, followed by incubation with 500 µM H_2O_2 are served as positive control. Cells were stained with Hoechst 33342 for nuclei visualization (blue fluorescence) and with CellROX^{TM} Deep Red Reagent for visualization of oxidative stress (red fluorescence); magnification 200×, scale bar = 70 µm.

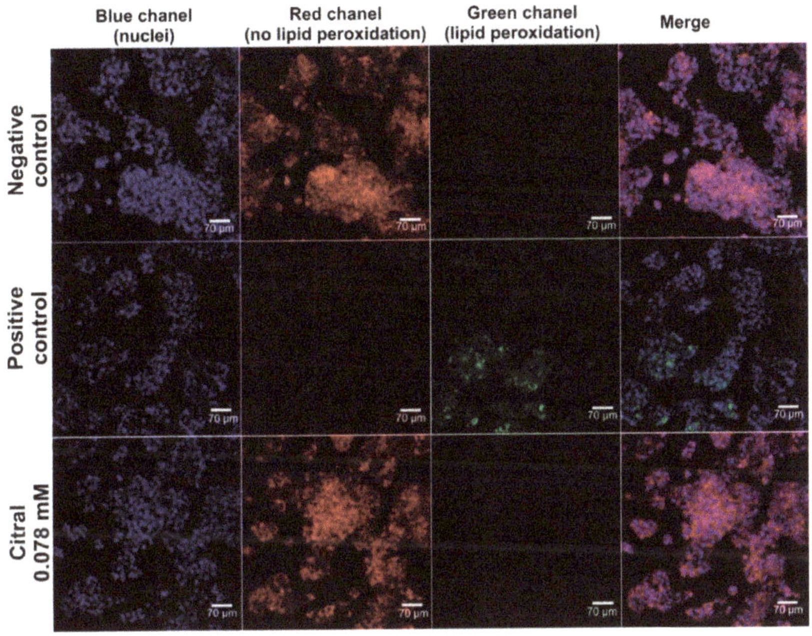

Figure 5. The CLSM images presenting hepatoprotective activity of citiral against H2O2-induced lipid peroxidation. Human hepatocellular carcinoma cells (HepG2 cell line, HB-8065) were incubated with citral for 48 h, followed by 1-h incubation with 500 µM H_2O_2. In turn, cells incubated only with culture medium served as negative control, while cells incubated with culture medium, followed by incubation with 500 µM H_2O_2 served as positive control. Cells were stained with Hoechst 33342 for nuclei visualization (blue fluorescence) and with Image-IT^{TM} Lipid Peroxidation for visualization of lipid peroxidation (green fluorescence) or lack of lipid peroxidation (red fluorescence); magnification 200×, scale bar = 70 µm.

2.4. The Impact of Acute and Subchronic Administration of Citral on Memory Acquisition Processes in the Passive Avoidance (PA) Test

One-way ANOVA revealed that acute administration of citral influenced the IL in the context of memory acquisition processes in the PA test [F (2, 24) = 5.085; p = 0.0144]. The Dunnett's post hoc test showed a significant increase in IL after the injection of citral at a dose of 50 mg/kg compared to the control group receiving saline solution, indicating an improvement in memory processes associated with acquisition [$p < 0.05$]. Citral at a dose of 25 mg/kg did not change the IL compared to the saline-treated control group (Figure 6).

The nonparametric Kruskal–Wallis ANOVA test revealed that subchronic administration of citral influenced the IL in the context of memory acquisition processes in the PA [F (2, 26) = 2.387; p = 5.040]. The Dunn's post hoc test showed a significantly increased IL after the repeated injection of citral at a dose of 50 mg/kg, as compared to the control group receiving saline solution, indicating an improvement in memory processes associated with acquisition [$p < 0.05$]. Citral at a dose of 25 mg/kg increased the IL value, but it did not reach the level of statistical significance (Figure 7).

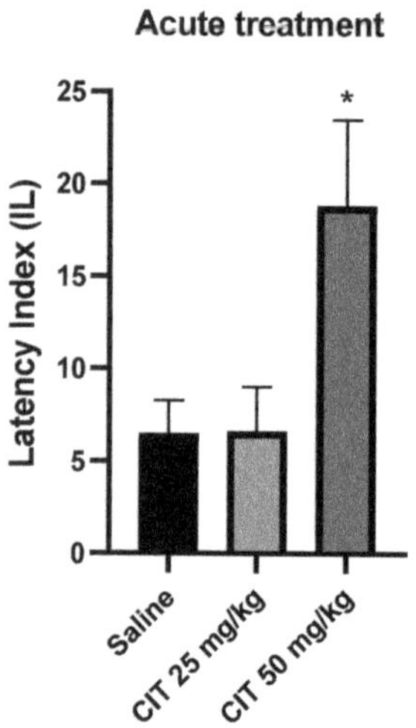

Figure 6. The effect of acute administration of citral (CIT) on the memory acquisition trials using the PA test in mice. CIT (25 or 50 mg/kg i.p.) or saline solution (i.p.) were administered 30 min before training. Subsequently, 24 h after training, the animals were re-tested. Results are expressed as latency index (IL) and presented as mean ± SEM, n = 8–10. Statistical analysis of data was performed using one-way analysis of variance (ANOVA), followed by the post hoc Dunnett's test. * $p < 0.05$ vs. saline-treated control group.

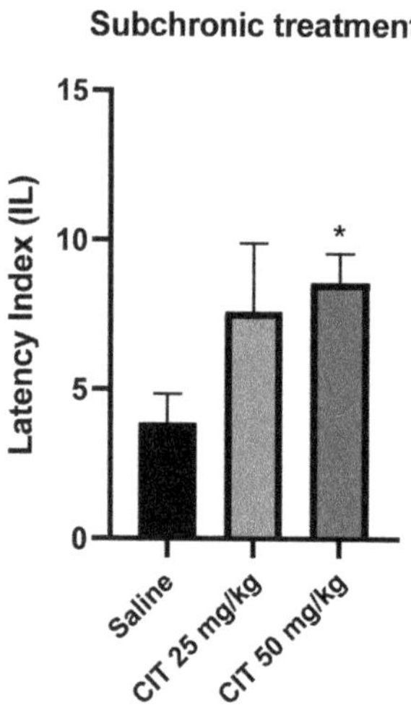

Figure 7. The effect of subchronic administration of citral (CIT) on the memory acquisition trials using the PA test in mice. CIT (25 or 50 mg/kg, i.p.) or saline solution (i.p.) were administered for 6 days, twice daily. On the seventh day, the appropriate group of mice received CIT (25 or 50 mg/kg, i.p.) or saline solution (i.p.) 30 min before training. Twenty-four hours after training, the animals were tested again. Results are expressed as latency index (IL) and presented as mean ± SEM, n = 9–10. Statistical analysis of data was performed via nonparametric Kruskal–Wallis ANOVA test, followed by the post hoc Dunn's test. * $p < 0.05$ vs. saline-treated control group.

2.5. The Impact of Acute and Subchronic Administration of Citral on Memory Acquisition Processes Impairment by an Acute Injection of Scopolamine in the Passive Avoidance (PA) Test

Two-way analysis of variance (ANOVA) showed statistically significant changes after scopolamine pretreatment [F (1, 46) = 50.87; $p < 0.0001$] and citral treatment [F (2, 46) = 3.253; $p = 0.0477$], while interactions between scopolamine and citral were observed [F (2, 46) = 4.850; $p = 0.0123$] (Figure 8). The post hoc Tukey's test confirmed that scopolamine (1 mg/kg) treatment impaired the acquisition of memory [$p < 0.01$]. An improvement in memory acquisition induced by citral at a dose of 50 mg/kg was observed [$p < 0.05$]. There was a statistically significant difference in the IL values between the group treated with citral 50 mg/kg and the group treated with citral 50 mg/kg with scopolamine [$p < 0.0001$]. However, no significant statistical differences in the IL values were observed between the groups receiving citral (25 or 50 mg/kg) with scopolamine compared to the group receiving scopolamine alone, indicating that citral at doses of 25 and 50 mg/kg did not prevent the amnestic action of scopolamine.

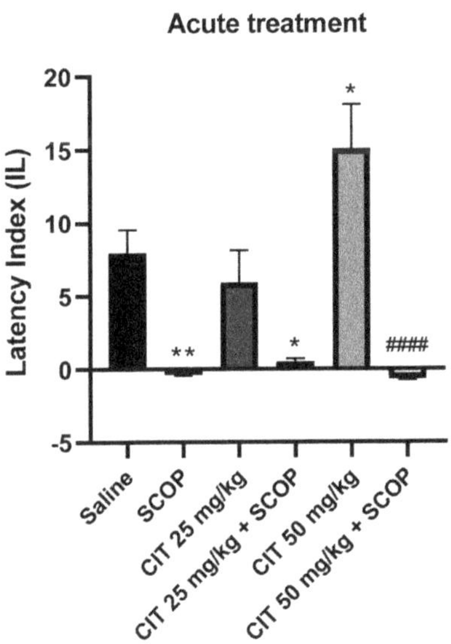

Figure 8. The effect of acute administration of citral (CIT) in scopolamine–induced impairment of memory acquisition trial using the PA test in mice. CIT (25 or 50 mg/kg, i.p.) or saline solution (i.p.) or scopolamine (1 mg/kg, s.c.) were administered 30 min before training. Subsequently, 24 h after training, the animals were tested again. Results are expressed as latency index (IL) and presented as mean ± SEM, $n = 7–10$. Statistical analysis of data was performed by employing two–way analysis of variance (ANOVA), followed by the post hoc Bonferroni's test. * $p < 0.05$; ** $p < 0.01$ vs. saline-treated control group; #### $p < 0.0001$ vs. CIT 50.

Two-way ANOVA showed statistically significant changes after scopolamine pretreatment [F (1, 50) = 45.59; $p < 0.0001$] and citral treatment [F (2, 50) = 3.226; $p = 0.0481$], while no interactions were observed between scopolamine and citral [F (2, 50) = 1.416; $p = 0.2523$] (Figure 9). The post hoc Bonferroni's test confirmed that scopolamine (1 mg/kg) treatment impaired the acquisition of memory [$p < 0.05$]. An improvement in memory acquisition induced by citral at a dose of 50 mg/kg was observed [$p < 0.05$]. Significantly different results were obtained for the groups receiving citral (25 or 50 mg/kg) compared to animals receiving citral (25 or 50 mg/kg) with scopolamine [$p < 0.01$; $p < 0.001$, respectively]. However, no statistically significant differences were found between the groups receiving citral (25 or 50 mg/kg) with scopolamine and the groups receiving only scopolamine, indicating that citral at these doses does not reverse the adverse effects of scopolamine on the acquisition of memory.

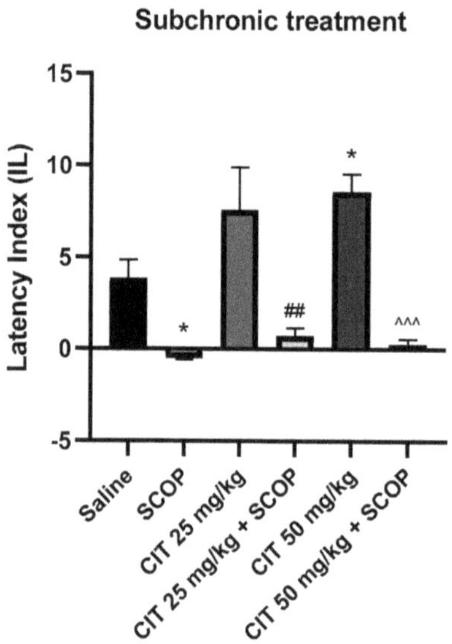

Figure 9. The effect of subchronic administration of citral (CIT) in scopolamine−induced impairment of memory acquisition trial using the PA test in mice. CIT (25 or 50 mg/kg, i.p.) or saline solution (i.p.) were administered for 6 days, twice daily. On the seventh day, the appropriate group of mice received CIT (25 or 50 mg/kg, i.p.) or saline solution (i.p.) or scopolamine (1 mg/kg, s.c.) 30 min before training. The animals were retested 24 h after training. Results are expressed as latency index (IL) and presented as mean ± SEM, n = 9–10. Statistical analysis of data was performed using two−way analysis of variance (ANOVA), followed by the post hoc Bonferroni's test. * $p < 0.05$ vs. saline−treated control group; ## $p < 0.01$ vs. CIT 25; ^^^ $p < 0.001$ vs. CIT 50.

2.6. The Effect of Citral on Locomotor Activity in Mice

One-way ANOVA analysis showed that neither single [F (2, 21) = 0.9401; $p = 0.4064$] nor repeated [F (2, 21) = 0.6018; $p = 0.5570$] administration of citral at doses of 25 and 50 mg/kg affected the locomotor activity of mice measured for 30 min after injections. Similarly, during 60 min, acute [F (2, 20) = 0.8448; $p = 0.4444$] and repeated [F (2, 21) = 1.543; $p = 0.2371$] administration of citral in these doses did not significantly affect the tested parameter (Table 2).

Table 2. The effect of citral on locomotor activity in mice. In acute treatment, CIT (25 or 50 mg/kg, i.p.) or saline solution (i.p.) were administered once immediately before the spontaneous locomotor activity test. In repeated treatment, the same compounds were administered for 5 days, twice daily. On the sixth day, the appropriate group of mice received CIT (25 or 50 mg/kg, i.p.) or saline solution (i.p.) and were then immediately placed in actimeters. Locomotor activity (number of interruptions of light beams) was recorded for 30 and 60 min. Data are presented as means ± SEM, n = 7–8.

	Acute Treatment			Repeated Treatment		
	Control	**CIT 25 mg/kg**	**CIT 50 mg/kg**	**Control**	**CIT 25 mg/kg**	**CIT 50 mg/kg**
Photocell beam breaks ± SEM (30 min)	4950 ± 235.8	4260 ± 447.7	4865 ± 442.4	4686 ± 443.5	3784 ± 367.0	3799 ± 401.1
Photocell beam breaks ± SEM (60 min)	7938 ± 527.9	6860 ± 494.9	7946 ± 860.8	7721 ± 693.1	6143 ± 717.5	6312 ± 682.0

2.7. GC-MS Analysis of Citral in Biological Material–Plasma, Hippocampus and the Rest of the Brain

GC-MS analysis showed the presence of cis-citral in all analyzed samples of biological material collected from mice after subchronic administration of the monoterpene at a dose of 50 mg/kg, which turned out to be most efficient in behavioral studies (Table 3). It should be noted that trans-citral was not detected in the analyzed tissues.

Table 3. Quantitative GC-MS analysis of cis-citral in tissues collected from mice after subchronic administration of the monoterpene at a dose of 50 mg/kg. Data are presented as the mean concentration ± SD [μg] in 1 mL of plasma or 1 mg of tissue from which citral was extracted. RSD%—relative standard deviation (n = 3). Statistical analysis of data was performed using t-test. *** $p < 0.001$ vs. remnant of the brain.

	Tissues		
	Plazma [μg/mL]	Hippocampus [μg/mg of Tissue]	Remnant of the Brain [μg/mg of Tissue]
Average content	13.105	0.117 ***	0.02
SD	0.12	0.01	0.0003
RSD%	0.89	7.42	1.37

The obtained results confirm that citral is able to enter the brain and simultaneously cross the blood–brain barrier. Citral was detected in all analyzed tissues after subchronic administration. It should be emphasized that the amount of the monoterpene in hippocampus was much greater than in the rest of the brain. Analysis with use of t-test showed that the difference is statistically significant ($p < 0.0001$). Unfortunately, the monoterpene was not identified in biological samples after acute administration. However, if citral was detected after semi-chronic administration, it is highly probable that chronic administration will increase the amount of citral in the examined tissues.

3. Discussion

The neurological activity of citral was evaluated using various methods, both in vitro, in silico, in vivo and ex vivo. Due to the compound showing a wide range of biological activity, it is highly probable that it can act as an anti-neurodegenerative agent [23]. Despite a wide range of activity, in current literature, detailed in vivo analysis is limited, thus any approach to research with an animal model is important to expand our knowledge about this interesting natural compound.

Considering the results obtained for assays associated with neurodegeneration, citral can be considered as a potential candidate for a more detailed analysis of its effect upon dementia and neurodegeneration. In our work, a detailed in vitro Marston assay [19] revealed the strong ability of the monoterpene to inhibit BChE—one of the enzymes strictly associated with neurodegeneration. Interestingly, a non-linear activity–concentration relationship was observed. The highest activity, 131% in comparison to reference galantamine (100% activity), was observed for 0.01 mg of compound being applied onto the plate, whereas 0.05 mg induced a 44% inhibition of the enzyme.

Potential drugs should primarily be effective in non-cytotoxic doses. In our previous study, we preliminary evaluated the significance of citral as a potential drug for the treatment of Alzheimer's disease [15]. Among many analyses, we also assessed cytotoxicity towards two model normal cell lines. Our research indicated that CC_{50} values for citral were 0.076 ± 0.012 mM (normal skin fibroblasts) and 0.136 ± 0.012 mM (normal kidney cells) [15]. Nevertheless, it is also important to evaluate the influence of drugs on hepatocytes, as they play crucial role in drug metabolism. It is known that drugs may become less toxic or even more toxic for cells after metabolic transformation [20]. The research conducted by other researchers indicated that even low concentrations of citral can induce cytotoxicity and genotoxicity in hepatocytes [24]. Our experiment, however, showed that this monoterpene did not exhibit hepatotoxicity (cell viability $\geq$ 95%) at low concentrations (0.0098–0.078) mM. It is also worth noting that although citral showed higher in vitro hepatotoxicity than galantamine, this monoterpene, unlike galantamine, has a broader spectrum of biological activities [12–14].

Effective drugs, in many cases, may cause hepatotoxicity. However, the mechanisms of drug-induced hepatotoxicity are poorly understood. Published research demonstrates that drug metabolism may lead to the creation of reactive metabolites that may cause oxidative damage to cells induced by generated ROS [25]. According to available literature data, citral not only does not cause oxidative damages in cells, but it also has protective properties against the effects of ROS [26,27]. This phenomenon is crucial and allows citral to be considered as a promising agent for counteracting and treating neurodegeneration [28].

In order to evaluate the influence of citral on locomotor activity and memory, in vivo tests based on a mice model were conducted. It should be emphasized that some studies were conducted to assess the effect of citral on cholinergic transmission using scopolamine. Scopolamine is a known blocker of muscarinic cholinergic receptors (mAChRs) (subtypes; M1 and M2) which manifests its effects in short-term and long-term memory disturbances [29]. In accordance with Giridharm et al. [30], scopolamine administration leads to significant reduction of acetylcholine level in mice. It should be recalled that scopolamine-induced amnesia is linked with oxidative stress and disturbance in the metabolism of low molecular weight antioxidants including glutathione. Such changes can have a significant impact on the high levels of lipid peroxidation that occur within the brain [31]. The mice model based on scopolamine-induced amnesia is preferred in studies of memory dysfunction, especially in testing new substances for their anti-AD properties.

Detailed analysis of passive avoidance test results revealed that citral is able to improve memory both in acute and subchronic administration. In both cases, the most effective dose was found to be 50 mg/kg—which improved memory with statistical significance. The lower dose (25 mg/kg) was also revealed to have a positive impact on IL value in subchronic administration; nevertheless, it did not reach the level of statistical significance. It should be emphasized that citral did not change the effects of scopolamine, and scopolamine eliminated the beneficial effect of citral on memory processes. In the context of the study and knowledge of the scopolamine mechanism, the active substance in this test should have an impact by increasing the ACh levels in the brain, along with strong antioxidant activity, which together would reverse the effect of scopolamine.

It is difficult to clearly answer the question of whether citral can act in the mechanism of AChE and BuChE inhibition. It should be taken into account that the scopolamine used in the studies is one of the most potent muscarinic receptor antagonists. The affinity of scopolamine for muscarinic receptors is <1 nM and is similar to ACh. Our study demonstrated the activity of citral in inhibiting AChE [15] and BuChE in vitro. We then found a procognitive effect of the tested compound, which was abolished by the scopolamine injection. These data suggest that the effect of citral on memory may be related to cholinergic transmission, but its effects at the doses used are not strong enough to reverse the adverse effects of scopolamine on memory acquisition. The results obtained in the passive avoidance test for citral and for citral in combination with scopolamine may also suggest a mechanism of action in the field of memory other than cholinergic.

A significant aspect of the activity in in vivo conditions is the influence of the examined compound on locomotor activity. Our work demonstrated that the administration of citral in 25 mg/kg and 50 mg/kg doses did not significantly affect the tested parameter.

GC-MS analysis of citral (sum of isomers) allowed us to detect only cis-citral. Citral was detected in all tissues collected from mice after subchronic administration of the monoterpene at dose of 50 mg/kg. The results indicate that citral crosses the blood–brain barrier and accumulates in the brain, especially in the hippocampus, which is involved in memory, learning and emotions [32]. Taking into account that the 'trans' isomer was not detected in the GC-MS analysis, it is important to assume that the 'cis' isomer is responsible for the positive effect on memory observed in the passive avoidance test.

In vitro activity towards various enzymes linked with dementia was evaluated. Among these, there is acetylcholinesterase (AChE), for which citral revealed 26.02% of inhibition at concentration 3.5 mM (Ellman assay), and satisfactory activity in the Marston assay (activity obtained for 0.01 mg of compound applied onto the plate) [15]. Additionally, citral activity was evaluated using molecular docking simulation (AutoDock and Molegro Virtual Docker), which revealed the compound's interactions (H-bonds) with amino acid residues in close contact (up to 4 Å) with the AChE active site [15]. What is more, citral and essential oils rich in this monoterpene were analyzed in detail for their role in AChE inhibition. An example of which is the extract from leaves of C. limon that revealed in vitro and in vivo (Swiss mice) activity towards enzyme inhibition (approx. 30%), in comparison to neostigmine [28].

Besides AChE inhibitory ability, citral also reveals various other activities against neurodegeneration, i.e., antioxidant, Fe(III) reduction ability, impacts on the concentration of brain-derived neurotrophic factor (BDNF) and malondialdehyde (MDA), etc. [33]. Previous research points to the various possible mechanisms of action of citral, which are based on the in vivo results obtained. A representative case is that presented by Beniwal et al. [33], who presented an improvement in behavioral responses and a decrease in anxiety and depression in rats as a result of citral administration. In this work, the activity of the monoterpene was found to result from an increase in BDNF and an improvement in the antioxidant–oxidant status. The inhibitory activity of citral in retinoic acid synthesis is equally interesting [3]. It is known that retinoic acid is responsible for a strong enhancement effect on acetylcholine (ACh) transmitter functions in brain cholinergic neurons [34]. It has been shown that low concentrations of citral almost completely inhibits retinoic acid formation. The sensitivity is demonstrated by all three isozymes of human aldehyde dehydrogenase [35].

In analyzing the activity of citral, various concentrations were studied. Of significance is that the biphasic effect of the monoterpene was found to depend on its concentration when observed in animal models. The effect was first observed by Yang et al., who analyzed the impact of citral on spatial learning and memory in rats [6]. In our work, the results revealed opposing activity of the monoterpene depending on dosage. In the case of 0.1 mg/kg, citral had an explicit influence on the spatial learning capabilities and enhanced the spatial reference memory of rats. The opposite effect was observed for a dose of 1 mg/kg. The effect has been associated with the retinoic acid concentration in the hippocampus, namely, a low dose of citral brought about an increase in the retinoic acid concentration, whereas a high dose decreased it.

One of the common symptoms of cognitive impairment is anxiety, which can be observed in most patients with AD and 10% to 45% of all patients with mild cognitive impairment (MCI) [36]. The activity of citral towards this symptom was also analyzed in our study. In the paper presented by Moghaddam et al. [7], intraperitoneal administration of citral resulted in a positive effect on anxiety (elevated plus maze test, open field test), whereas the highest activity was observed for a dose of 20 mg/kg, but, simultaneously, lower dosages of 5 and 10 mg/kg also caused significant increase in anxiolytic effect in comparison to control. It is probable that citral, besides having impact on retinoic acid, can act via GABAA and 5-HT1A receptor modulation [7].

It is worth mentioning that citral can be quickly eliminated and does not bioaccumulate in the body; hence, a direct evaluation of its impact on spatial memory by determining the level of the monoterpene in organs is very difficult to undertake [6]. Citral is an unstable and hydrophobic compound under normal storage conditions, so it can easily be degraded and lose its pharmaceutical properties [37]. This known effect can explain the trace amounts of citral in the tested ex vivo samples.

In analyzing the in vivo activity of citral, it is important to underline that an extremely small number of studies have been aimed at determining the concentration of citral or its metabolites in various organs in animal models. In most cases, the activity of the monoterpene in animal models is estimated based on behavioral tests, enzyme activity or level of selected parameters, and only a few papers have reported on the evaluation of level of citral (neral/geranial) or its potential metabolites in serum or selected organs. One was presented by Phillips et al. in 1976, who considered the issue of absorption, distribution and excretion of citral in rat and mouse models [18]. Herein, [14C] citral was administered to mice in the form of a compound dissolved in corn oil via oral intubation, and we provided a dose equal to 100 mg/kg. The results showed that the greatest proportion of radioactivity was absorbed after 12 h, but activity was not detected in the brain, thymus, eye lens and skeleton at this time, whereas activity in these parts was observed at 72 h. Moreover, nearly 90% of all radioactivity was eliminated within 72 h, and complete excretion was observed after 120 h.

4. Materials and Methods

4.1. In Vitro Assays

4.1.1. Materials

The following reagents: citral (≥95%), butyrylcholinesterase, type V-S from Electrophorus electricus, albumin from bovine serum, 1-naphthyl acetate, Trizma® (2-amino-2-(hydroxymethyl)-1,3-propanediol) hydrochloride solution (1 M, pH 7.8), acetylthiocholine iodide (≥99%), DTNB (5,5′-dithiobis(2-nitrobenzoic acid)), galantamine hydrobromide from Lycoris sp. (>94%) and Fast Blue B salt 95%, penicillin, streptomycin, Live/Dead Cell Double Staining Kit, Hoechst 33342 dye, layer chromatography plates (HPTLC silica gel 60), scopolamine hydrobromide, citral and Tween 20 were purchased from Sigma Aldrich (now, Merck) (St. Louis, MO, USA). HepG2 cells (HB-8065, human hepatocellular carcinoma cell line) and Eagle's Minimum Essential Medium (EMEM) were supplied by ATCC (London, UK), while fetal bovine serum (FBS) was obtained from Pan-Biotech (Aidenbach, Germany). Additionally, CellROXTM Deep Red Reagent and Image-ITTM Lipid Peroxidation Kit were purchased from Invitrogen, ThermoFisher Scientific (Waltham, MA, USA). Solvents (analytical purity grade) were obtained from Polish Reagents (Gliwice, Poland).

4.1.2. Methods

Butyrylcholinesterase (BChE) Inhibitory Assay on TLC Plates According to the Marston MethodThe assay was performed based on a slightly modified method first presented by Marston et al. [19]. The experiment was performed using chromatographic HPTLC silica gel 60 (F254) plates. The plates were first prepared by elution with ethanol and activation for 40 min at 105 °C. The next step was terpene application onto the plate by means of an automatic applicator, Desaga AS-30 (Biostep GmbH, Burkhardtsdorf, Germany). Due to the high volatility of the monoterpene, before application, citral was diluted 10 times in ethanol. The following amounts of citral were used [mg]: 0.05; 0.025; 0.01 and 0.001. The amount of the compound employed was decided on the basis of previously conducted tests for acetylcholinesterase [15]. Galantamine, a known AChE inhibitor, was chosen as a reference standard. Solution of the standard (1 mg/mL in ethanol) was applied onto the plates in the same amount as the investigated terpene. The next step of assay was performed in accordance with [15]. The activity was determined based on the discoloration of the dark background where the compounds were spotted. Obtained results were analyzed via the Sorbfil TLC Videodensitometer program (ver. 2.0).

Evaluation of Hepatotoxicity In Vitro

The experiments were performed using human hepatocellular carcinoma cells (HepG2 cell line). The cells were maintained according to manufacturer recommendations. Briefly, the cells were cultured in Eagle's Minimum Essential Medium (EMEM) supplemented with 100 U/mL penicillin, 100 µg/mL streptomycin and 10% fetal bovine serum (FBS). The appropriate conditions for cell culture were 37 °C, 5% CO_2 and 95% humidity. Before the experiment, the HepG2 cells were seeded onto 96-well plates at concentration of

3×10^4 cells/well and incubated for 24 h at 37 °C (5% CO_2, 95% humidity). Subsequently, serial dilutions (0.0098–5 mM) of citral or galantamine (reference compound, positive control) were prepared and added to the cells. After 48 h of incubation, the cell viability was evaluated using MTT assay, according to procedure described previously in detail [38]. For this reason, four independent experiments were conducted. The results were expressed as percentage of control (the OD for control cells incubated without tested compounds was considered as 100%). The results were presented as mean value ± standard deviation (SD). Student *t*-test was used to determine statistical differences between samples. Moreover, 4-parametric logarithmic curves were created in order to assess concentration that caused 50% decrease of cell viability (CC_{50} value) compared to control cells. In turn, to confirm quantitative results, the cells were stained with Live/Dead Cell Double Staining Kit, observed under confocal laser scanning microscope (CLSM), and photographed.

Detection of Oxidative Stress and Lipid Peroxidation

The experiment was carried out according to the same procedure as described in Section Evaluation of hepatotoxicity in vitro, but using citral at concentration of 0.078 mM. After 48 h of incubation, the solution was removed and oxidative stress was induced by 1 h treatment with 500 μM H_2O_2 prepared in culture medium. Cells incubated with a culture medium for 48 h, followed by 1 h without H_2O_2 served as negative control, while cells incubated with culture medium by 48 h followed by 1 h treatment with H_2O_2 served as a positive control. Subsequently, the cells were stained using CellROX^{TM} Deep Red Reagent or Image-IT^{TM} Lipid Peroxidation Kit, while cell nuclei were stained with Hoechst 33342 dye. CellROXTM Deep Red Reagent is a non-fluorescent dye in a reduced state, which after oxidation by reactive oxygen species (ROS), exhibits strong red fluorescence. In turn, Image-ITTM Lipid Peroxidation is a fluorescent dye enabling the highlighting of lipid peroxidation cells. After oxidation by ROS, this dye exhibits green fluorescence indicating lipid peroxidation, and gives red fluorescence when lipid peroxidation did not occur. The cells were observed under CLSM and photographed.

4.2. *In Vivo Assays*

4.2.1. Preparation and Administration of Drugs

Scopolamine was dissolved in saline solution (0.9% NaCl) and administered subcutaneously (s.c.), while citral was suspended in a 1% solution of Tween 20 and administered intraperitoneally (i.p.). All substances were administered at a volume of 10 mL/kg, 30 min prior to memory tests or directly before locomotor activity assessment. Drug solutions were prepared daily immediately before use. The control group received injections of saline solution in the same volume and at the same time as the corresponding drugs. The doses of citral and scopolamine were selected based on published literature [16,20].

4.2.2. Animals

The experiments were conducted on naive adult male Swiss mice, maintained in the Centre of Experimental Medicine of the Medical University of Lublin. The mice weighed between 20 and 25 g at the beginning of the study. Experiments were carried out between 8.30 a.m. and 4.00 p.m. The animals were housed in colony cages under standardized laboratory conditions: a natural 12/12 h light–dark cycle, temperature maintained at 21 ± 1 °C, air humidity between 50 ± 5%, with air exchange occurring at a rate of 15/h, and with free access to tap water and food (Altromin, Animalab, Lage, Germany). After a 7-day acclimatization period, the mice were randomly assigned to experimental groups consisting of 8–10 animals.

4.2.3. Methods

Passive Avoidance Task

The passive avoidance (PA) task is based on the natural aversion of rodents to bright places and is considered a measure of long-term memory [39]. Depending on the testing protocol used, the PA test offers the opportunity to explore different memory stages based on the timing of drug administration. In our research, compounds were given prior to the initial trial (before the pretest), impeding the process of information acquisition.

The apparatus consisted of an acrylic box divided by sliding doors into two compartments: illuminated and darkened. The floor of the apparatus is made of stainless-steel bars serving as conductors for electric current.

On the first day of the experiment, the tested mice, after drug administration, were placed in the illuminated part of the apparatus and allowed to freely explore the illuminated chamber for 30 s with the doors closed. After this time, the sliding doors were raised, and they had access to the darkened chamber. Upon entry into the dark chamber with all four paws, the sliding doors were closed, and a weak electrical current (2 s exposure to shock, 0.2 mA) was delivered through the floor (Figure 10). The latency time for entry into the dark chamber was recorded as TL1. The observation period lasted for a maximum of 300 s. If the mouse did not enter the dark box during this time, it was placed in the dark box, the doors were closed, and a weak electrical current was delivered to the animal's feet. In such cases, the TL1 value was recorded as 300 s.

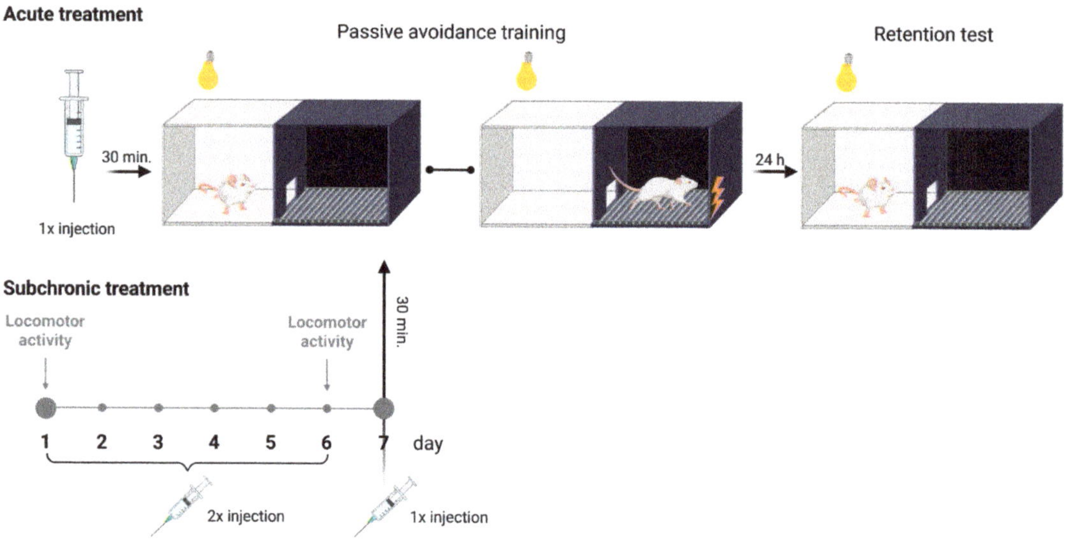

Figure 10. Graphical diagram of the experimental protocol used in the undertaken behavioral research created with BioRender.com.

On the following day of the test (24 h later), the same mice were again placed in the illuminated compartment of the apparatus with the sliding doors closed. After a 30 s adaptation period, the doors between the compartments were raised, and the time at which the animal entered the dark chamber was recorded as TL2. No electric shock stimulus was delivered in this trial (Figure 10). If the animal did not enter the dark chamber within 300 s, the test was terminated, and TL2 was recorded as 300 s [40].

Locomotor Activity

The locomotor activity scores were assessed using an Opto-Varimex-4 Auto-Track actimeter (Columbus Instruments, Columbus, OH, USA). The apparatus consists of a square plexiglass cage measuring (43 × 43 × 32 cm). Photoresistors and sensors are mounted on the wall at 2 heights (2.5 and 1.2 cm). When a moving animal crosses the beam, the change is recorded by the respective counter as one movement. The difference in counter readings determines the spontaneous mobility of the animals. Immediately after compound administration, the mice were individually placed in the actimeter for 60 min. The number of beam breaks by mice was recorded as locomotor activity at 30 and 60 min.

Treatment of Behavioral Research

The behavioral component of the experiment aimed to assess the influence of acute and subchronic administration of citral on memory acquisition in mice using the PA test. In the subsequent stage, an attempt was made to evaluate the mechanism of action of citral by examining its impact on memory acquisition processes following impairment by scopolamine [41].

To assess the impact of a single dose of citral on memory processes, animals were randomly assigned to 3 groups and administered the following substances: (1) saline solution (0.9% NaCl) i.p., (2) citral 25 mg/kg i.p. and (3) citral 50 mg/kg i.p. To evaluate the potential mechanism of action of citral, animals were randomly allocated to 6 groups and given the following compounds: (1) saline solution (0.9% NaCl) i.p., (2) scopolamine 1 mg/kg s.c., (3) citral 25 mg/kg i.p., (4) citral 25 mg/kg i.p. and scopolamine 1 mg/kg s.c., (5) citral 50 mg/kg i.p., (6) citral 50 mg/kg i.p. and scopolamine 1 mg/kg s.c. On the first day of the experiment, after administration of the test compounds, mice were trained in the PA test. All compounds were introduced 30 min before the pretest (assessment of memory acquisition processes). 24 h later, mice were retested in the PA test.

To assess the impact of subchronic administration of citral on memory processes, animals were randomly allocated to 3 groups: (1) control, (2) citral (25 mg/kg), (3) citral (50 mg/kg). For 6 days, twice daily (at 8:30 a.m. and 6:00 p.m.), and on the seventh day in the morning, animals received (1) saline solution (0.9% NaCl) i.p., (2) citral 25 mg/kg i.p. and (3) citral 50 mg/kg i.p., and 30 min after injection, animals were trained in the PA test. On the eighth day, mice were retested in the PA test, without injections.

To evaluate the potential mechanism of action of citral following subchronic administration, subsequent animals were randomly allocated into 6 groups: (1) control, (2) scopolamine (1 mg/kg), (3) citral (25 mg/kg), (4) citral (25 mg/kg) with scopolamine, (5) citral (50 mg/kg), (6) citral (50 mg/kg) with scopolamine. For the first 6 days of the experiment, animals from each group received the following compounds twice daily (at 8:30 a.m. and 6:00 p.m.): (1) and (2) saline solution (0.9% NaCl) i.p., (3) and (4) citral 25 mg/kg i.p., (5) and (6) citral 50 mg/kg i.p. On the seventh day in the morning, animals received the following injections: (1) saline solution (0.9% NaCl) i.p., (2) scopolamine 1 mg/kg s.c., (3) citral 25 mg/kg i.p., (4) citral 25 mg/kg i.p. and scopolamine 1 mg/kg s.c., (5) citral 50 mg/kg i.p., (6) citral 50 mg/kg i.p. and scopolamine 1 mg/kg s.c. (Table 4). On the seventh day of the experiment, 30 min after injection, animals were trained in the PA test. On the eighth day, the PA test was reapplied. Additionally, locomotor activity was measured in animals assigned to groups 1, 3 and 5. Animals were tested on the first and sixth days of the experiment following the morning injection.

Table 4. Substance administration scheme to assess the potential mechanism of action of citral after subchronic administration. SCOP—scopolamine; CIT—citral.

Groups	Substances Administered from Day 1 to 6	Substances Administered on Day 7
(1) Control	saline solution (0.9% NaCl)	saline solution (0.9% NaCl)
(2) SCOP	saline solution (0.9% NaCl)	scopolamine 1 mg/kg s.c.
(3) CIT 25 mg/kg	Citral 25 mg/kg i.p.	Citral 25 mg/kg i.p.
(4) CIT 25 mg/kg + SCOP	Citral 25 mg/kg i.p.	Citral 25 mg/kg i.p. + scopolamine 1 mg/kg s.c.
(5) CIT 50 mg/kg	Citral 50 mg/kg i.p.	Citral 50 mg/kg i.p.
(6) CIT 50 mg/kg + SCOP	Citral 50 mg/kg i.p.	Citral 50 mg/kg i.p. + scopolamine 1 mg/kg s.c.

Tissue Collection and Preparation

Immediately after the passive avoidance test (24 h after the last injection), the animals were decapitated. Blood and brains were collected. Blood was centrifuged at 13.000 rpm,

10 min., at +4 °C and the supernatant was collected. Brains were rinsed with an ice-cold saline solution, then hippocampi were isolated from the brains. The obtained biological material was stored at −80°C for further analysis. Tissues were homogenized in ultrapure water (1:2 *w/v* for remnants of the brains; 1:4 *w/v* for hippocampi) using a bead mill homogenizer (BeadBug 6, Benchmark Scientific, Sayreville, NJ USA) at 4000 rpm for 15 s. The homogenate was centrifuged at 10.000 rpm for 15 min at +4°C, and the supernatant was collected for further analysis.

4.2.4. Ethical Declaration

All experiments were conducted following the ARRIVE guidelines and the European Community Council Directive for the Care and Use of Laboratory Animals of 22 September 2010 (2010/63/EU) and approved by the Local Ethics Committee in Lublin, Poland (Permission No: 42/2022).

4.3. Ex Vivo Assay

GC-MS Analysis of Citral in Biological Material–Plasma, Hippocampus and the Rest of the Brain

The GC–MS system consisted of a model 7890 A series gas chromatograph coupled with Detector Agilent Technologies 5975 C inert XL MSD with Triple–Axis Detector (Agilent, Wood Dale, IL, USA). Separation was performed on a HP–5MS column (30 m × 0.25 mm, 0.25 µm film thickness) supplied by Agilent Technologies (Wood Dale, IL, USA). The carrier gas was 99.99% high purity helium with a flow rate of 1.2 mL min-1. The sample volume was 1 µL. The injection port and the detector temperatures were 270 °C and 280 °C, respectively. Oven temperature program was initially set at 50 °C for 1 min, then ramped at 3 °C min-1 to 110 °C and held for 1.0 min, then ramped at 20 °C min^{-1} to 200 °C and held for 5 min. The total run time was 31.5 min with a solvent delay of 10 min. Ionization was performed in electron impact ionization (EI) mode at 70 eV. Selective ion monitoring (SIM) was set for quantification with dwell time 100 ms ion^{-1}. Data were collected using Agilent MassHunter Quantitative Analysis Software–Perlan (ver. 05.02). Retention times of cis-citral, trans-citral and thymol (IS—internal standard) were 22.13 min, 23.10 min and 21.08 min, respectively. The characteristic ions (m/z) cis–citral and trans–citral were 152, 69, 41 and for thymol (m/z) z 150, 135, 91. The final concentration of the IS was 30 µg mL^{-1}. Cis- and trans-citral contents were calculated from a commercial standard (Sigma-Aldrich, C83007-5ML) based on the area of the peaks obtained during chromatographic separation.

In this part of the experiment, 200 µL of hexane–ethyl acetate mixture (1:1) was added to the supernatant (from the collected biological material-plasma, hippocampus and remnant of the brain after separation of the hippocampus), with an internal standard so as to produce a final concentration of 30 ug/mL). The sample was shaken for 10 min to extract the active compounds by utilizing a vortex mixer. The sample was then centrifuged in a high-speed cryogenic centrifuge at 12,000 rpm for 10 min and the supernatant was transferred to a 1.5 mL clean autosample tube. The, 1 µL of the supernatant was injected into GC–MS system for analysis. Four samples of biological material were used for each analysis and tested three times. Concentrations of cis- and trans-citral in mice were calculated on the basis of calibration curves (Table 5).

Table 5. Parameters of the calibration curve for cis- and trans-citral.

No.	Compound	Calibration Curve	R^2
1	*cis*-citral	y = −0.0254426 x2 + 1.46612 x + 0.0137605	0.997
2	*trans*-citral	y = −0.0116753 x2 + 1.43904 x + 0.164956	0.997

4.4. Statistical Analysis

The data were presented as the mean ± SEM. Outliers were identified and excluded from the analysis by employing the ROUT method. The following tests were applied for statistical analysis of the obtained results: one-way analysis of variance ANOVA, followed by the post hoc Dunnett's test; two-way analysis of variance ANOVA, followed by the post hoc Bonferroni's test; nonparametric Kruskal–Wallis ANOVA test, followed by the post hoc Dunn's test. The significance level was set at $p < 0.05$. The statistical analysis was conducted based on the nature of the data, with respect to best statistic practice in biomedical research [42].

To assess memory processes, a latency index (IL) was utilized, which was calculated for each animal. IL serves for the quantitative assessment of PA test performance as the disparity between retention and training latencies, and was computed as follows:

$$IL = (TL2 - TL1)/TL1$$

where: TL1 signifies the duration taken to enter the dark compartment during the training phase TL2 indicates the period taken to re-enter the dark compartment during retention [20,23].

In order to determine statistical significance in the difference in citral content in the hippocampus and the rest of the brain, *t*-test was used.

5. Conclusions

Scientific studies have explicitly indicated the various biological activities of citral. Among these, some activities can be associated with neurodegeneration, and the positive effects of citral administration on this condition suggests that citral can be considered for use in preventing and treating neurodegeneration and dementia. The in vitro studies presented in this paper explicitly demonstrated the activity of citral towards BuChE inhibition, along with hepatoprotective properties. Moreover, citral did not reveal hepatotoxicity at low dose and simultaneously protected HepG2 cells against oxidative stress and lipid peroxidation. Hence, these observations indicate that this monoterpene has hepatoprotective potential in vitro. It is also worth mentioning that citral positively influenced memory after its acute and subchronic administration in vivo. The most effective rate of intake turned out to be a dose equal to 50 mg/kg, which simultaneously did not impact locomotor activity.

It is difficult to answer the question of whether citral can act in the mechanism of AChE and BuChE inhibition. It is known that the pro-cognitive activity of the monoterpene, with a high degree of probability, results from its cholinergic activity. However, based on the obtained results, this activity was not high enough to reverse the effect of the muscarinic receptor antagonist scopolamine. A mechanism of action of citral other than cholinergic is also possible. Additionally, the presence of cis-citral in the brain of the tested animals after subchronic administration of the monoterpene (mix of isomers) has confirmed the ability of citral to cross the blood-brain barrier. It should be emphasized that the presented analysis of citral in tissues collected from tested animals is one of the few available studies that showed the presence of this compound using direct tests.

Author Contributions: Conceptualization, K.W.-K.; methodology, K.W.-K., M.R., K.K., J.M., M.A.-M.; software, A.M. and K.W.-K.; validation K.W.-K., M.R., K.K., J.M., M.A.-M. and A.M.; formal analysis, K.K. and A.M.; investigation, K.W.-K. and K.K.; resources, J.M., K.W.-K. and M.R.; data curation, M.A.-M.; writing—original draft preparation, K.W.-K., M.R., K.K. and J.M.; writing—review and editing, K.W.-K., M.R., K.K. and J.M.; visualization, K.W.-K.; supervision, A.M. and M.R.; project administration, K.W.-K.; funding acquisition, K.W.-K. All authors have read and agreed to the published version of the manuscript.

Funding: This research was conducted as part of project: Internal Grants for Young Scientists (ID:PBmb65) funded by Medical University of Lublin, Poland.

Institutional Review Board Statement: All experiments were (a) performed in compliance with ARRIVE guidelines, (b) carried out in accordance with the National Institute of Health Guidelines for the Care and Use of Laboratory Animals and with the European Community Council Directive for the Care and Use of Laboratory Animals of 22 September 2010 (2010/63/EU) and (c) approved by the Local Ethics Committee in Lublin, Poland (Permission No: 42/2022).

Informed Consent Statement: Not applicable.

Data Availability Statement: The research data are available from the authors.

Conflicts of Interest: The authors declare no conflicts of interest.

References

1. Hirai, M.; Ota, Y.; Ito, M. Diversity in Principal Constituents of Plants with a Lemony Scent and the Predominance of Citral. *J Nat. Med.* **2022**, *76*, 254–258. [CrossRef] [PubMed]
2. Rana, V.S.; Das, M.; Blazqeuz, M.A. Essential Oil Yield, Chemical Composition, and Total Citral Content of Nine Cultivars of Cymbopogon Species from Western India. *J. Herbs Spices Med. Plants* **2016**, *22*, 289–299. [CrossRef]
3. Di Renzo, F.; Broccia, M.L.; Giavini, E.; Menegola, E. Citral, an Inhibitor of Retinoic Acid Synthesis, Attenuates the Frequency and Severity of Branchial Arch Abnormalities Induced by Triazole-Derivative Fluconazole in Rat Embryos Cultured in Vitro. *Reprod. Toxicol.* **2007**, *24*, 326–332. [CrossRef] [PubMed]
4. Ganjewala, D.; Gupta, A.K.; Muhury, R. An Update on Bioactive Potential of a Monoterpene Aldehyde Citral. *J. Biol. Act. Prod. Nat.* **2012**, *2*, 186–199. [CrossRef]
5. Shi, C.; Song, K.; Zhang, X.; Sun, Y.; Sui, Y.; Chen, Y.; Jia, Z.; Sun, H.; Sun, Z.; Xia, X. Antimicrobial Activity and Possible Mechanism of Action of Citral against *Cronobacter sakazakii*. *PLoS ONE* **2016**, *11*, e0159006. [CrossRef] [PubMed]
6. Yang, Z.; Xi, J.; Li, J.; Qu, W. Biphasic Effect of Citral, a Flavoring and Scenting Agent, on Spatial Learning and Memory in Rats. *Pharmacol. Biochem. Behav.* **2009**, *93*, 391–396. [CrossRef]
7. Hajizadeh Moghaddam, A.; Mashayekhpour, M.A.; Tabari, M.A. Anxiolytic-like Effects of Citral in the Mouse Elevated plus Maze: Involvement of GABAergic and Serotonergic Transmissions. *Naunyn Schmiedebergs Arch. Pharmacol.* **2023**, *396*, 301–309. [CrossRef]
8. Gupta, G.L.; Samant, N.P. Current Druggable Targets for Therapeutic Control of Alzheimer's Disease. *Contemp. Clin. Trials* **2021**, *109*, 106549. [CrossRef]
9. Wojtunik-Kulesza, K.; Oniszczuk, A.; Waksmundzka-Hajnos, M. An Attempt to Elucidate the Role of Iron and Zinc Ions in Development of Alzheimer's and Parkinson's Diseases. *Biomed. Pharmacother.* **2019**, *111*, 1277–1289. [CrossRef]
10. Bokhoven, P.; de Wilde, A.; Vermunt, L.; Leferink, P.; Heetveld, S.; Cummings, J.; Scheltens, P.; Vijverberg, E. The Alzheimer's Disease Drug Development Landscape. *Alzheimer's Res. Ther.* **2021**, *13*, 186. [CrossRef]
11. Sedky, K.; Nazir, R.; Joshi, A.; Kaur, G.; Lippmann, S. Which Psychotropic Medications Induce Hepatotoxicity? *Gen. Hosp. Psychiatry* **2012**, *34*, 53–61. [CrossRef] [PubMed]
12. Wojtunik, K.A.; Ciesla, L.M.; Waksmundzka-Hajnos, M. Model Studies on the Antioxidant Activity of Common Terpenoid Constituents of Essential Oils by Means of the 2,2-Diphenyl-1-Picrylhydrazyl Method. *J. Agric. Food Chem.* **2014**, *62*, 9088–9094. [CrossRef] [PubMed]
13. Wojtunik-Kulesza, K.A.; Oniszczuk, A. Ability of Selected Monoterpenes to Reduce Fe(III) Ions Being Pro-Neurodegenerative Factors: Tests Based on a FRAP Reaction Extended to 48 Hours. *Int. J. Mol. Sci.* **2024**, *25*, 2191. [CrossRef] [PubMed]
14. Wojtunik-Kulesza, K.A.; Wiśniewska, R. Interactions of Selected Monoterpenes with Iron and Copper Ions Based on Ferrozine and CUPRAC Methods—The Preliminary Studies. *Chem. Biodivers.* **2022**, *19*, e202200461. [CrossRef] [PubMed]
15. Wojtunik-Kulesza, K.A.; Targowska-Duda, K.; Klimek, K.; Ginalska, G.; Jóźwiak, K.; Waksmundzka-Hajnos, M.; Cieśla, Ł. Volatile Terpenoids as Potential Drug Leads in Alzheimer's Disease. *Open Chem.* **2017**, *15*, 332–343. [CrossRef]
16. do Vale, T.G.; Furtado, E.C.; Santos, J.G.; Viana, G.S.B. Central Effects of Citral, Myrcene and Limonene, Constituents of Essential Oil Chemotypes from *Lippia Alba* (Mill.) n.e. Brown. *Phytomedicine* **2002**, *9*, 709–714. [CrossRef] [PubMed]
17. Charret, T.S.; Pereira, M.T.M.; Pascoal, V.D.B.; Lopes-Cendes, I.; Cristina Rheder Fagundes Pascoal, A. Citral Effects on the Expression Profile of Brain-Derived Neurotrophic Factor and Inflammatory Cytokines in Status Epilepticus-Induced Rats Using the Lithium–Pilocarpine Model. *J. Med. Food* **2021**, *24*, 916–924. [CrossRef] [PubMed]
18. Phillips, J.C.; Kingsnorth, J.; Gangolli, S.D.; Gaunt, I.F. Studies on the Absorption, Distribution and Excretion of Citral in the Rat and Mouse. *Food Cosmet. Toxicol.* **1976**, *14*, 537–540. [CrossRef] [PubMed]
19. Marston, A.; Kissling, J.; Hostettmann, K. A Rapid TLC Bioautographic Method for the Detection of Acetylcholinesterase and Butyrylcholinesterase Inhibitors in Plants. *Phytochem. Anal.* **2002**, *13*, 51–54. [CrossRef]
20. Cetin, S.; Knez, D.; Gobec, S.; Kos, J.; Pišlar, A. Cell Models for Alzheimer's and Parkinson's Disease: At the Interface of Biology and Drug Discovery. *Biomed. Pharmacother.* **2022**, *149*, 112924. [CrossRef]
21. Ruoß, M.; Vosough, M.; Königsrainer, A.; Nadalin, S.; Wagner, S.; Sajadian, S.; Huber, D.; Heydari, Z.; Ehnert, S.; Hengstler, J.G.; et al. Towards Improved Hepatocyte Cultures: Progress and Limitations. *Food Chem. Toxicol.* **2020**, *138*, 111188. [CrossRef] [PubMed]
22. van Meerloo, J.; Kaspers, G.J.L.; Cloos, J. Cell Sensitivity Assays: The MTT Assay. *Methods Mol. Biol.* **2011**, *731*, 237–245. [CrossRef] [PubMed]
23. Bouzenna, H.; Hfaiedh, N.; Giroux-Metges, M.-A.; Elfeki, A.; Talarmin, H. Biological Properties of Citral and Its Potential Protective Effects against Cytotoxicity Caused by Aspirin in the IEC-6 Cells. *Biomed. Pharmacother.* **2017**, *87*, 653–660. [CrossRef] [PubMed]

24. Souza, A.C.S.; Silva, L.K.; Queiroz, T.B.; Marques, E.S.; Hiruma-Lima, C.A.; Gaivão, I.O.M.; Maistro, E.L. Citral Presents Cytotoxic and Genotoxic Effects in Human Cultured Cells. *Drug Chem. Toxicol.* **2020**, *43*, 435–440. [CrossRef]
25. Donato, M.T.; Tolosa, L. High-Content Screening for the Detection of Drug-Induced Oxidative Stress in Liver Cells. *Antioxidants* **2021**, *10*, 106. [CrossRef]
26. Uchida, N.S.; Silva-Filho, S.E.; Cardia, G.F.E.; Cremer, E.; de Souza Silva-Comar, F.M.; Silva, E.L.; Bersani-Amado, C.A.; Cuman, R.K.N. Hepatoprotective Effect of Citral on Acetaminophen-Induced Liver Toxicity in Mice. *Evid. -Based Complement. Altern. Med.* **2017**, *2017*, e1796209. [CrossRef]
27. Asle-Rousta, M.; Amini, R.; Aghazadeh, S. Carvone Suppresses Oxidative Stress and Inflammation in the Liver of Immobilised Rats. *Arch. Physiol. Biochem.* **2023**, *129*, 597–602. [CrossRef]
28. Feitosa, C.M.; de Freitas, R.M.; Silva, V.L.; da Silva Araújo, L.; de Melo, C.H.S.; Santos, F.P.D.S. Citrus: A Perspective for Developing Phytomedicines for Neurodegenerative Diseases. In *Citrus Pathology*; IntechOpen: London, UK, 2017; ISBN 978-953-51-3072-7.
29. Noroozi, N.; Shayan, M.; Maleki, A.; Eslami, F.; Rahimi, N.; Zakeri, R.; Abdolmaleki, Z.; Dehpour, A.R. Protective Effects of Dapsone on Scopolamine-Induced Memory Impairment in Mice: Involvement of Nitric Oxide Pathway. *Dement. Geriatr. Cogn. Disord. Extra* **2022**, *12*, 43–50. [CrossRef]
30. Giridharan, V.V.; Thandavarayan, R.A.; Sato, S.; Ko, K.M.; Konishi, T. Prevention of Scopolamine-Induced Memory Deficits by Schisandrin B, an Antioxidant Lignan from Schisandra Chinensis in Mice. *Free Radic. Res.* **2011**, *45*, 950–958. [CrossRef]
31. Budzynska, B.; Boguszewska-Czubara, A.; Kruk-Slomka, M.; Skalicka-Wozniak, K.; Michalak, A.; Musik, I.; Biala, G. Effects of Imperatorin on Scopolamine-Induced Cognitive Impairment and Oxidative Stress in Mice. *Psychopharmacology* **2015**, *232*, 931–942. [CrossRef]
32. Opitz, B. Memory Function and the Hippocampus. *Front. Neurol. Neurosci.* **2014**, *34*, 51–59. [CrossRef] [PubMed]
33. Beniwal, S.; Chatterjee, I.; Gohil, N.; Ghalami, F.; Lorenzo-Villegas, D.; Moradikor, N. Evaluation of the Neuroprotective Activity of Citral Nanoemulsion on Alzheimer's Disease-Type Dementia in a Preclinical Model: The Assessment of Cognitive and Neurobiochemical Responses. *Life Neurosci.* **2023**, *1*, 9–17. [CrossRef]
34. Szutowicz, A.; Bielarczyk, H.; Jankowska-Kulawy, A.; Ronowska, A.; Pawełczyk, T. Retinoic Acid as a Therapeutic Option in Alzheimer's Disease: A Focus on Cholinergic Restoration. *Expert Rev. Neurother.* **2015**, *15*, 239–249. [CrossRef] [PubMed]
35. Kikonyogo, A.; Abriola, D.P.; Dryjanski, M.; Pietruszko, R. Mechanism of Inhibition of Aldehyde Dehydrogenase by Citral, a Retinoid Antagonist. *Eur. J. Biochem.* **1999**, *262*, 704–712. [CrossRef]
36. Gomoll, B.P.; Kumar, A. Managing Anxiety Associated with Neurodegenerative Disorders. *F1000Prime Rep.* **2015**, *7*, 05. [CrossRef]
37. Fourtaka, K.; Christoforides, E.; Tzamalis, P.; Bethanis, K. Inclusion of Citral Isomers in Native and Methylated Cyclodextrins: Structural Insights by X-Ray Crystallography and Molecular Dynamics Simulation Analysis. *J. Mol. Struct.* **2021**, *1234*, 130169. [CrossRef]
38. Pitucha, M.; Woś, M.; Miazga-Karska, M.; Klimek, K.; Mirosław, B.; Pachuta-Stec, A.; Gładysz, A.; Ginalska, G. Synthesis, Antibacterial and Antiproliferative Potential of Some New 1-Pyridinecarbonyl-4-Substituted Thiosemicarbazide Derivatives. *Med. Chem. Res.* **2016**, *25*, 1666–1677. [CrossRef]
39. Venault, P.; Chapouthier, G.; de Carvalho, L.P.; Simiand, J.; Morre, M.; Dodd, R.H.; Rossier, J. Benzodiazepine Impairs and β-Carboline Enhances Performance in Learning and Memory Tasks. *Nature* **1986**, *321*, 864–866. [CrossRef] [PubMed]
40. Allami, N.; Javadi-Paydar, M.; Rayatnia, F.; Sehhat, K.; Rahimian, R.; Norouzi, A.; Dehpour, A.R. Suppression of Nitric Oxide Synthesis by L-NAME Reverses the Beneficial Effects of Pioglitazone on Scopolamine-Induced Memory Impairment in Mice. *Eur. J. Pharmacol.* **2011**, *650*, 240–248. [CrossRef]
41. Skalicka-Wozniak, K.; Budzynska, B.; Biala, G.; Boguszewska-Czubara, A. Scopolamine-Induced Memory Impairment Is Alleviated by Xanthotoxin: Role of Acetylcholinesterase and Oxidative Stress Processes. *ACS Chem. Neurosci.* **2018**, *9*, 1184–1194. [CrossRef] [PubMed]
42. Yan, F.; Robert, M.; Li, Y. Statistical Methods and Common Problems in Medical or Biomedical Science Research. *Int. J. Physiol. Pathophysiol. Pharmacol.* **2017**, *9*, 157–163. [PubMed]

International Journal of
Molecular Sciences

MDPI

Article

Preventive Effect of the Total Polyphenols from *Nymphaea candida* on Sepsis-Induced Acute Lung Injury in Mice via Gut Microbiota and NLRP3, TLR-4/NF-κB Pathway

Chenyang Li [1], Xinxin Qi [1], Lei Xu [2], Yuan Sun [3], Yan Chen [2], Yuhan Yao [2] and Jun Zhao [1,2,*]

[1] School of Public Health, Xinjiang Medical University, Urumqi 830011, China; licy0609@126.com (C.L.); xjmuqxx@163.com (X.Q.)
[2] Xinjiang Key Laboratory for Uighur Medicine, Institute of Materia Medica of Xinjiang, Urumqi 830004, China; xulei19820218@126.com (L.X.); chenmaomao158@163.com (Y.C.); 17726840515@163.com (Y.Y.)
[3] School of Pharmacy, Xinjiang Medical University, Urumqi 830011, China; sym5658@163.com
* Correspondence: zhaojun21cn@163.com

Abstract: This study aimed to investigate the preventive effects of the total polyphenols from *Nymphaea candida* (NCTP) on LPS-induced septic acute lung injury (ALI) in mice and its mechanisms. NCTP could significantly ameliorate LPS-induced lung tissue pathological injury in mice as well as lung wet/dry ratio and MPO activities ($p < 0.05$). NCTP could significantly decrease the blood leukocyte, neutrophil, monocyte, basophil, and eosinophil amounts and LPS contents in ALI mice compared with the model group ($p < 0.05$), improving lymphocyte amounts ($p < 0.05$). Moreover, compared with the model group, NCTP could decrease lung tissue TNF-α, IL-6, and IL-1β levels ($p < 0.05$) and downregulate the protein expression of TLR4, MyD88, TRAF6, IKKβ, IκB-α, p-IκB-α, NF-κB p65, p-NF-κB p65, NLRP3, ASC, and Caspase1 in lung tissues ($p < 0.05$). Furthermore, NCTP could inhibit ileum histopathological injuries, restoring the ileum tight junctions by increasing the expression of ZO-1 and occludin. Simultaneously, NCTP could reverse the gut microbiota disorder, restore the diversity of gut microbiota, increase the relative abundance of Clostridiales and Lachnospiraceae, and enhance the content of SCFAs (acetic acid, propionic acid, and butyric acid) in feces. These results suggested that NCTP has preventive effects on septic ALI, and its mechanism is related to the regulation of gut microbiota, SCFA metabolism, and the TLR-4/NF-κB and NLRP3 pathways.

Keywords: *Nymphaea candida*; polyphenols; acute lung injury; inflammatory response; NLRP3; TLR-4/NF-κB pathway; gut microbiota; short-chain fatty acids

1. Introduction

Acute lung injury (ALI) is a life-threatening disease, causing high morbidity and mortality worldwide, which is a clinical syndrome caused by various direct or indirect causative factors, including pathogenic microbial infections, and mainly manifested by acute progressive respiratory distress and intractable hypoxemia [1]. Acute inflammation of the lung interstitium in ALI can produce neutrophil infiltration and pro-inflammatory cell aggregation and, ultimately, impede the lung gas exchange. It is reported that the mortality rate of acute respiratory distress syndrome (ARDS) can be up to 40% [2,3]. For example, novel coronavirus (SARS-CoV-2) infection can trigger a cytokine storm, which, in turn, leads to severe ALI, ARDS, and even multi-organ failure until death [4,5]. Moreover, sepsis is a leading cause of death in intensive care unit (ICU) patients [6]. The progress of sepsis is rapid and causes damage to multiple organs, of which lung injury and failure are the earliest and most frequent occurrences. In recent years, Chinese medicines have played a greater role in the prevention and therapy of ALI induced by various causes, especially in

the treatment of ALI/ARDS caused by SARS-CoV-2 infection, such as Qingfei Paidu Tang, Ma Xing Shi Gan Tang, and total flavonoids from Sanyeqing and baicalein [1,7–9].

It is reported that there is a close interaction between gastrointestinal health and the severity of sepsis-induced lung disease. The lungs and intestines can influence each other through the flora and the immune system to achieve a bidirectional regulatory effect [10]. The intestinal epithelium is a natural physical barrier that prevents the transfer of harmful intestinal bacteria and their metabolic byproducts, including lipopolysaccharide (LPS), into the bloodstream [11]. LPS-mediated pathological responses have been a hot spot; exogenous LPS can enter the portal circulation, causing an imbalance in the intestinal flora in vivo. Imbalanced and disturbed flora causes the exudation of endogenous LPS, which in turn destroys the intestinal mucosa, increases intestinal permeability, and permits the bioproducts of intestinal microbiota to be released in the intestinal tract, which activates the systemic immune system, where a large amount of LPS enters the lungs along with blood and exacerbates the immune damage to the lungs [12]. Animal models of sepsis usually use infection to elicit a systemic inflammatory response, such as intraperitoneal injection of LPS, to mimic the mechanisms of sepsis onset and progression in humans [5]. Serum-increased LPS binds to the toll-like receptor 4 (TLR4) complex, which has been shown to be the main signaling receptor for LPS-mediated injury in the body. NF-κB is an important signaling molecule in the signaling pathway downstream of TLR4 [13] and is also a key hub in the regulation of inflammatory pathways, which can increase the inflammatory response and activate inflammatory cascade reactions in ALI [14].

Nymphaea candida flower is a traditional Chinese medicine, with efficacy in moisturizing the lungs and relieving coughs, clear heat, and strengthening the heart, mainly distributed in the Bosten Lake, Yili, and Altay regions in Xinjiang, China [15,16]. *N. candida* is also one of the main medicinal materials in the Uyghur medicine compound "Zukamu granules", and this compound preparation has played a significant role in the treatment of COVID-19 [17,18]. Zukamu granules could significantly reduce the apoptosis rate of alveolar macrophages, enhance phagocytosis of alveolar macrophages, and inhibit high expression of the TLR4/MyD88/NF-κB signaling pathway [19]. *N. candida* has a variety of biological activities, such as antibacterial, anti-inflammatory, hypotensive, hepatoprotective, and neuroprotective effects, and polyphenols are its main characteristic compounds [16,20,21]. Total polyphenols from *N. candida* (NCTP) have been previously prepared and showed better anti-inflammatory, analgesic, antipyretic, and cough activities [22]. However, whether it has the effect of preventing pneumonia has not been reported in the literature. Therefore, this study intends to investigate the preventive effects of NCTP on LPS-induced septic acute lung injury in mice via the gut microbiota, metabolism of SCFAs, and NLRP3 and TLR-4/NF-κB pathways. The result may provide a new way of thinking for the prevention of ALI.

2. Results

2.1. Compound Analysis of NCTP

Chemical compositions of NCTP were confirmed by UHPLC-HRMS based on their retention time (Rt), accurate molecular mass (mass error of less than 5 ppm), and major MS/MS fragment ions. As shown in Figure 1 and Table 1, a total of 14 compounds were identified and deduced, including 10 hydrolyzable tannins and 4 flavonoid glycoside compounds. Some compounds were identified compared with reference substances, such as isostrictiniin, ellagic acid, and nicotiflorin [22].

2.2. Protective Effect of NCTP on Lung Tissue in ALI Mice

In the experiment, the control group had a healthy fur status, good spirit, and normal defecation. After modeling, a large number of secretions appeared in the corners of the eyes in the model group, meaning that mice were unable to open their eyes, followed by messy fur, laziness, slowness, and poorly formed feces. The mice in the NCTP (50 mg/kg) group had a small amount of secretions in their eyes, which resulted in their eyes being slightly open, with messy fur and unshaped feces, while the mice in the NCTP (100, 200 mg/kg) group had occasional secretions in the eyes, and their fur and other appearances were close to the control group. The DEX (3 mg/kg) had occasional secretions in their eyes, messy fur, and symptoms of slow reaction. All groups showed weight loss after modeling, but there were no differences between all groups in terms of weight (Table 2). There were no deaths in any groups of mice. As a major feature of ALI, pulmonary oedema of lung tissue was reflected by the lung wet-to-dry (W/D) ratio. Compared with the control group, the lung W/D ratio in the model group was significantly increased after LPS stimulation ($p < 0.01$). In contrast, NCTP (100, 200 mg/kg) and DEX (3 mg/kg) could decrease the lung W/D ratio in LPS-induced ALI mice ($p < 0.05$ or $p < 0.01$) (Table 2).

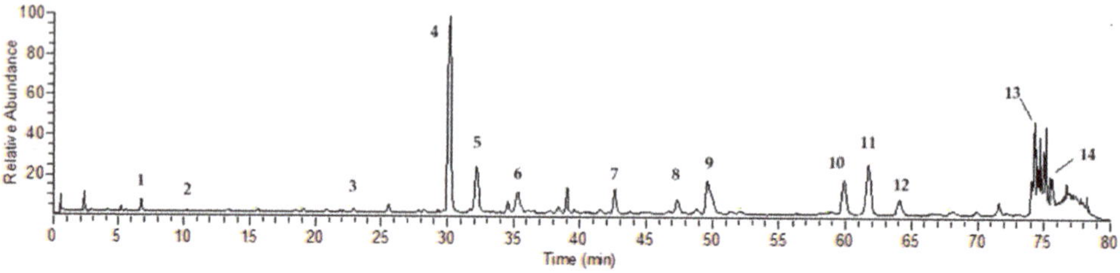

Figure 1. UHPLC-Q-Orbitrap-HRMS chromatogram of NCTP (ESI-negative mode). (The number in the figure caption means peak number).

Table 1. MS and MS/MS data of NCTP.

Peak No	Rt (min)	$[M-H]^-$	Error	MS/MS m/z (% of Base Peak)	Formula	Compound Identity
1 [a]	6.75	169.01385	4.17	125 (100)	$C_7H_6O_5$	Gallic acid
2	10.29	483.07758	1.33	331 (20), 169 (100), 125 (80)	$C_{20}H_{20}O_{14}$	Digalloyl-glucoside
3	21.87	635.08838	0.77	483 (40), 313 (20), 169 (100), 125 (80)	$C_{27}H_{24}O_{18}$	Trigalloyl-glucoside
4 [a]	30.19	633.07281	0.91	463 (20), 301 (100), 275 (20), 185 (10)	$C_{27}H_{22}O_{18}$	Isostrictiniin
5 [a]	32.15	197.04503	0.77	169 (100), 125 (50)	$C_9H_{10}O_5$	Gallic acid ethyl ester
6	35.44	785.08282	−0.48	633 (10), 301 (100), 275 (50), 257 (10), 169 (40), 125 (30)	$C_{34}H_{26}O_{22}$	Digalloyl-ellagitannoyl-glucoside
7	42.54	787.09906	0.26	635 (20), 617 (70), 465 (40), 313 (30), 169 (100), 125 (60)	$C_{34}H_{28}O_{22}$	Tetragalloyl-glucoside
8	47.28	433.04095	1.85	301 (100), 271 (5), 243 (5)	$C_{19}H_{13}O_{12}$	Quercetin-O-arabinoside
9 [a]	49.69	300.99802	0.41	283 (20), 229 (10), 185 (20)	$C_{14}H_6O_8$	Ellagic acid
10 [a]	59.92	939.10931	−0.53	787 (20), 769 (100), 465 (10), 313 (10), 169 (90), 125 (80)	$C_{41}H_{32}O_{26}$	Pentagalloyl-glucoside
11 [a]	61.66	593.15015	0.08	285 (80), 255 (60), 227 (40), 151 (10)	$C_{27}H_{30}O_{15}$	Nicotiflorin
12 [a]	64.07	447.09293	1.65	285 (60), 255 (70), 227 (50), 151 (10)	$C_{21}H_{20}O_{11}$	Kaempferol O-glucoside
13	74.66	447.09305	1.92	300 (100), 271 (70), 227 (50), 169 (100), 125 (20)	$C_{21}H_{20}O_{11}$	Ellagitannoyl-rhamnoside
14 [a]	75.47	285.03970	1.18	257 (10), 229 (10), 151 (10)	$C_{15}H_{10}O_6$	Kaempferol

Rt, retention time; [a], compared with reference substance.

Table 2. Influence of NCTP on body weight in ALI mice.

Group	Pre-Modeling Weight (g)	Post-Modeling Weight (g)	W/D Ratio
Control	24.72 ± 0.58	22.54 ± 0.51	3.96 ± 0.46
Model	23.38 ± 1.32	20.48 ± 1.29	5.62 ± 0.42 ##
NCTP, 50 mg/kg	23.08 ± 0.82	21.28 ± 0.66	4.37 ± 0.77
NCTP, 100 mg/kg	22.90 ± 0.69	20.76 ± 0.48	4.26 ± 0.66 *
NCTP, 200 mg/kg	23.01 ± 1.25	21.14 ± 1.10	4.38 ± 0.33 **
DEX, 3 mg/kg	23.04 ± 1.56	21.22 ± 1.10	4.12 ± 0.30 **

Data are presented as mean ± SD, $n = 6$. ## $p < 0.01$ versus control group, * $p < 0.05$ and ** $p < 0.01$ versus model group.

We explored lung injury in each group, as shown in Figure 2. The lung tissues in the control group represented normal structure without histopathologic changes. Compared to the control group, the lung tissues in the model group exhibited severe injury as follows: alveoli, alveolar tubes, and bronchi were distorted; the connective tissue showed hyperplasia; a large amount of inflammatory cell infiltration was observed in the pulmonary interstitium. Compared with the model group, the lung tissues in the NCTP (50 mg/kg) group showed hyperplasia in a small amount of the connective tissue, inflammatory cell

infiltration, and thickening of the alveolar wall; the lung tissues in the NCTP (100 mg/kg) group showed significant improvement, with a small amount of connective tissue hyperplasia and occasional mild inflammatory cell infiltration; the lung tissues in the NCTP (200 mg/kg) and DEX (3 mg/kg) groups did not show significant inflammatory infiltration and connective tissue hyperplasia. Non-parametric tests of pathological grading showed that the mean rank value of the model group was significantly higher compared with the control group ($p < 0.01$). Compared with the model group, the mean rank value of NCTP (100, 200 mg/kg) decreased significantly ($p < 0.05$, Table 3). The results showed that NCTP could improve pathologic changes in acute inflammation attenuating lung tissue injury.

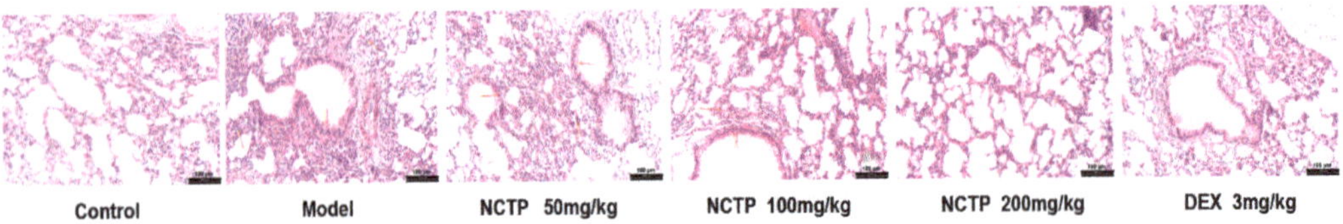

Figure 2. Representative images of H&E-stained lung sections in mice (n = 6) at 16 h after LPS administration, demonstrating alveolar structure. Scale bar, 100 μm.

Table 3. The pathological grading results of lung tissues in ALI mice.

Group	n	The Degree of Lung Injury				Rank Mean
		0	I	II	III	
Control	6	6	0	0	0	9.50
Model	6	0	0	6	0	33.50 ##
NCTP, 50 mg/kg	6	0	6	0	0	24.50
NCTP, 100 mg/kg	6	3	3	0	0	17.00 *
NCTP, 200 mg/kg	6	4	2	0	0	14.50 **
DEX, 3 mg/kg	6	5	1	0	0	12.00 **

n = 6. ## $p < 0.01$ versus control group, * $p < 0.05$ and ** $p < 0.01$ versus model group

As an indicator of inflammatory cell extravasation, the lung tissue MPO activities in the model group mice were significantly upregulated compared with the control group ($p < 0.01$). However, NCTP (100, 200 mg/kg) and DEX (3 mg/kg) could markedly reverse this change induced by LPS ($p < 0.05$ or $p < 0.01$). The results indicated that NCTP could inhibit the LPS-induced recruitment of inflammatory cells (Figure 3 and Table S1).

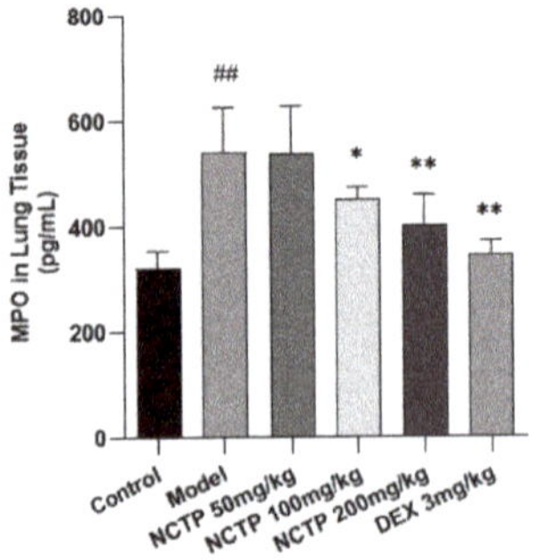

Figure 3. Effect of NCTP on lung tissues MPO levels in ALI mice. Data were presented as mean ± SD, n = 6. ## $p < 0.01$, versus control group; * $p < 0.05$, ** $p < 0.01$ versus model group.

2.3. Effect of NCTP on BALF BCA and LDH Activities in ALI Mice

As shown in Figure 4 and Table S2, the BALF BCA and LDH activities were significantly enhanced in LPS-induced septic ALI mice compared to the control group ($p < 0.05$

or $p < 0.01$). Nevertheless, treatment of NCTP (200 mg/kg) could reverse LPS-induced increases in BCA ($p < 0.05$). NCTP (50, 100 and 200 mg/kg) and DEX (3 mg/kg) could markedly decrease the LDH activity in ALI mice ($p < 0.01$).

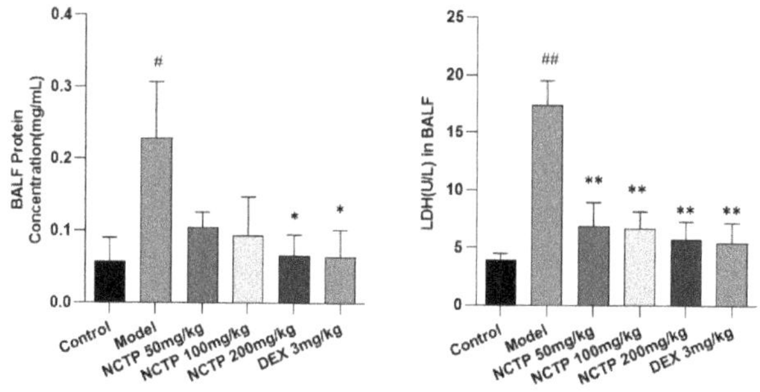

Figure 4. Effect of NCTP on BALF total protein; LDH contents in ALI mice. Data were presented as mean ± SD, $n = 6$. # $p < 0.05$, ## $p < 0.01$, versus control group; * $p < 0.05$, ** $p < 0.01$ versus model group.

2.4. *Effect of NCTP on Inflammatory Cells and Pro-Inflammatory Cytokines in ALI Mice*

The white blood cell (WBC) count is a well-known basic parameter to reflect the host immune response, which would be increased under chronic or acute inflammation conditions [23]. In this study, there was a significant increase in the WBC count in the model group when compared to the control group ($p < 0.01$). Moreover, compared with the model group, treatments with NCTP (100, 200 mg/kg) displayed a significant decrease in the WBC count ($p < 0.01$). From the results of the routine blood test, compared with the control group, the levels of neutrophil count (NEU), monocyte count (MON), basophil count (BAS), and eosinophil count (EOS) were significantly increased in the model group ($p < 0.05$ or $p < 0.01$). NCTP (100, 200 mg/kg) improved the trend of the above indicators after intervention ($p < 0.05$ or $p < 0.01$). The level of lymphocyte count (LYM) decreased in the model group ($p < 0.05$), and the NCTP (200 mg/kg) could improve the decreasing trend ($p < 0.05$) (Figure 5A–F and Table S3).

Next, we explored the plasma LPS content. The plasma LPS level in the model group was significantly increased compared to the control group ($p < 0.01$). The NCTP (50, 100 and 200 mg/kg) and DEX (3 mg/kg) groups could significantly decrease the plasma LPS level in ALI mice compared to the model group ($p < 0.05$ or $p < 0.01$) and showed a dose-dependent effect (Figure 6A and Table S4). The lung tissue inflammatory levels of cytokines TNF-α, IL-6, and IL-1β 16 h post-LPS-treatment are shown in Figure 6B–D. The results showed that LPS induced significant increases in TNF-α, IL-6, and IL-1β production ($p < 0.01$), while the NCTP (50, 100 and 200 mg/kg) and DEX (3 mg/kg) groups reduced TNF-α, IL-6, and IL-1β levels in lung tissue ($p < 0.01$).

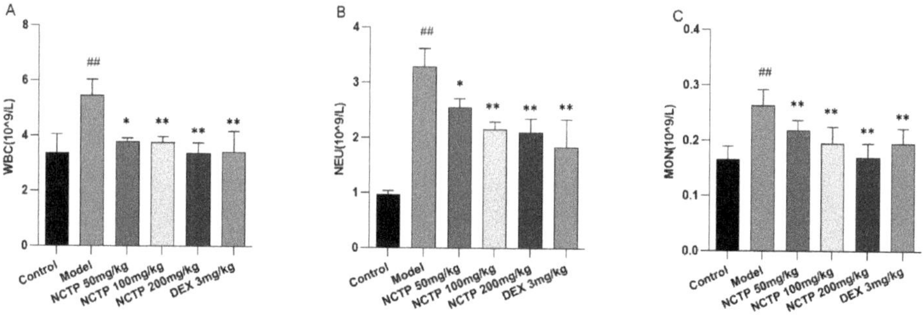

Figure 5. *Cont.*

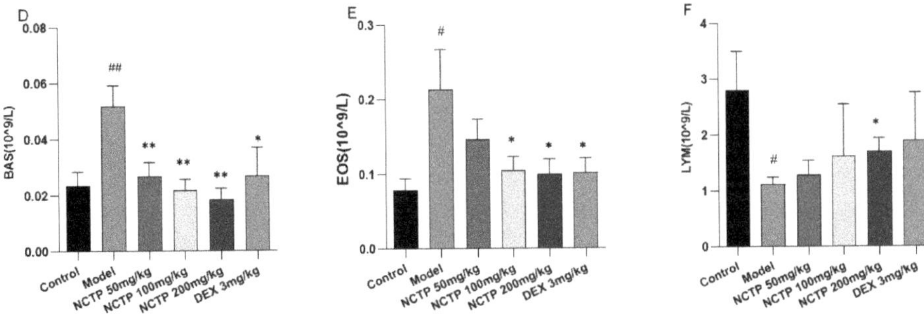

Figure 5. Effects of inflammatory factors in blood in ALI mice. (**A**) WBC count in blood in mice. (**B**) NEU count in blood in mice. (**C**) MON count in blood in mice. (**D**) BAS count in blood in mice. (**E**) EOS count in blood in mice. (**F**) LYM count in blood in mice. Data were presented as mean ± SD, $n = 6$. $^{\#}$ $p < 0.05$, $^{\#\#}$ $p < 0.01$, versus control group; * $p < 0.05$, ** $p < 0.01$ versus model group.

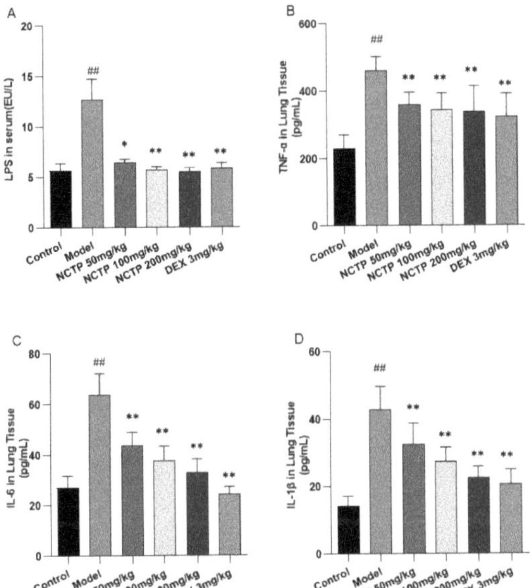

Figure 6. Effects of NCTP on inflammatory factors in ALI mice. (**A**) The levels of LPS in plasma of LPS-induced ALI mice. (**B–D**) The levels of TNF-α, IL-6, IL-1β in lung tissues of LPS-induced ALI mice. Data were presented as mean ± SD, $n = 6$. $^{\#\#}$ $p < 0.01$, versus control group; * $p < 0.05$, ** $p < 0.01$, versus model group.

2.5. Effect of NCTP on NF-κB and NLRP3 Signaling Pathway Activation

Intestinal microflora dysfunction may induce G-bacteria to produce a large amount of LPS that potentially enters the circulation through intestinal leakage, upregulating the lung immune response and activating the TLR4 signaling pathway [12]. Therefore, we next explored how the gut flora regulated immunity in animals with LPS-induced ALI by regulating the TLR4/NF-κB signaling pathway [24].

In this study, the expressions of the TLR4/NF-κB pathways in lung tissues were detected by Western blotting. As shown in Figures 7 and S1 and Table S5, the levels of TLR4, MyD88, TRAF6, and IKKβ proteins showed obvious increases in the model group ($p < 0.01$), which were attenuated after treatment of NCTP (200 mg/kg) ($p < 0.05$ or $p < 0.01$). In addition, LPS stimulation apparently induced the phosphorylation of NF-κB p65 and

IκB-α ($p < 0.01$), whereas NCTP (200 mg/kg) ($p < 0.05$ or $p < 0.01$) administration reversed this result.

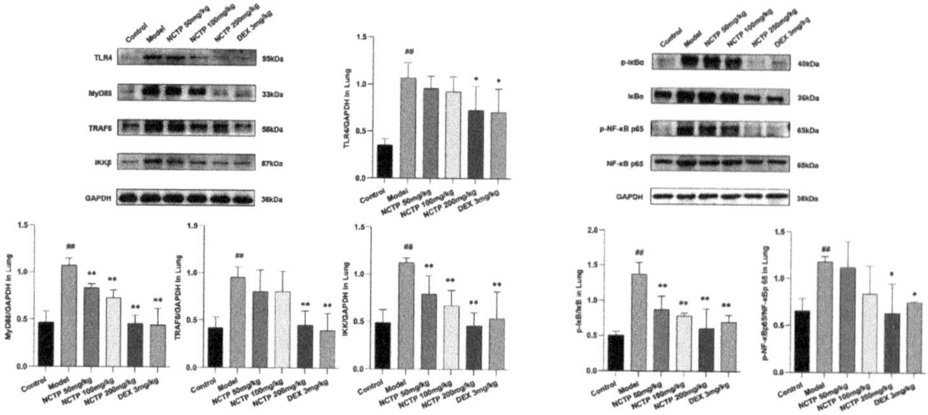

Figure 7. Effects of NCTP on proteins expression of lung tissues TLR-4/NF-κB in ALI mice. Data were presented as mean ± SD., $n = 6$. ## $p < 0.01$, versus control group; * $p < 0.05$, ** $p < 0.01$ versus model group.

As another important signal pathway for modulating the inflammatory process, the NLRP3 signaling pathway has been widely studied in recent years. Therefore, to further investigate the anti-inflammatory mechanism of NCTP in LPS-induced ALI, we also measured NLRP3 signaling pathway activation by NLRP3, ASC, and Caspase1 proteins 16 h after LPS treatment.

The results showed that NCTP could inhibit LPS induced by these three major proteins in the NLRP3 signaling pathway in a dose-dependent manner in a mouse model of ALI. As described in Figures 8 and S2 and Table S6, compared with the control group, the levels of NLRP3, ASC, and Caspase1 were significantly increased in the model group ($p < 0.01$), and the levels of NLRP3, ASC, and Caspase1 were strongly significantly inhibited by NCTP (200 mg/kg) ($p < 0.05$ or $p < 0.01$).

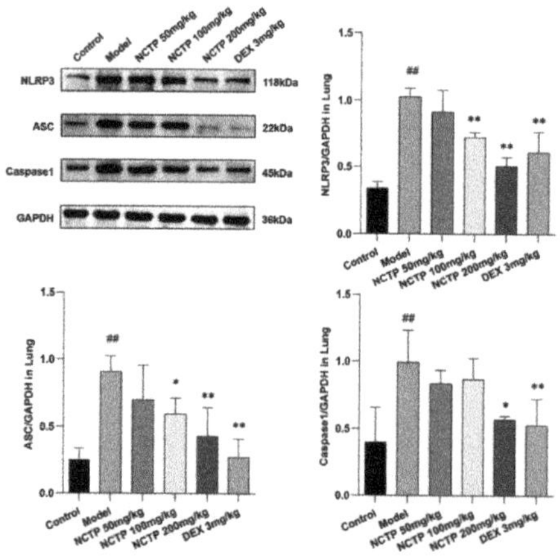

Figure 8. Effects of NCTP on proteins expression of lung tissues NLRP3 in ALI mice. Data were presented as mean ± SD, $n = 6$. ## $p < 0.01$, versus control group; * $p < 0.05$, ** $p < 0.01$ versus model group.

2.6. Protective Effect of NCTP on the Intestinal Mucosa

LPS administration altered the gut flora in mice, which resulted in slight swelling of the ileum, trifling edema, gut villus separation, and gut mucosal damage [25]. These changes were demonstrated in the model group but not in the control group. In contrast, the NCTP (200 mg/kg) group intervention ameliorated the intestinal injury caused by LPS administration when compared to the model group. Images of the pathological samples are shown in Figure 9A. The structures of crypts in the control group mice were obvious, showing a large number of goblet cells with mucin secreted in each crypt, and the colonic epithelium was covered with a thick mucus layer, which was continuous and intact (Figure 9B). The mice in the model group had a severe loss of mature goblet cells, a significant reduction in the number of goblet cells per crypt, and a destroyed mucus layer. Compared with the model group, the LPS-induced massive depletion of goblet cells and mucin was significantly ameliorated in the NCTP (200 mg/kg) groups.

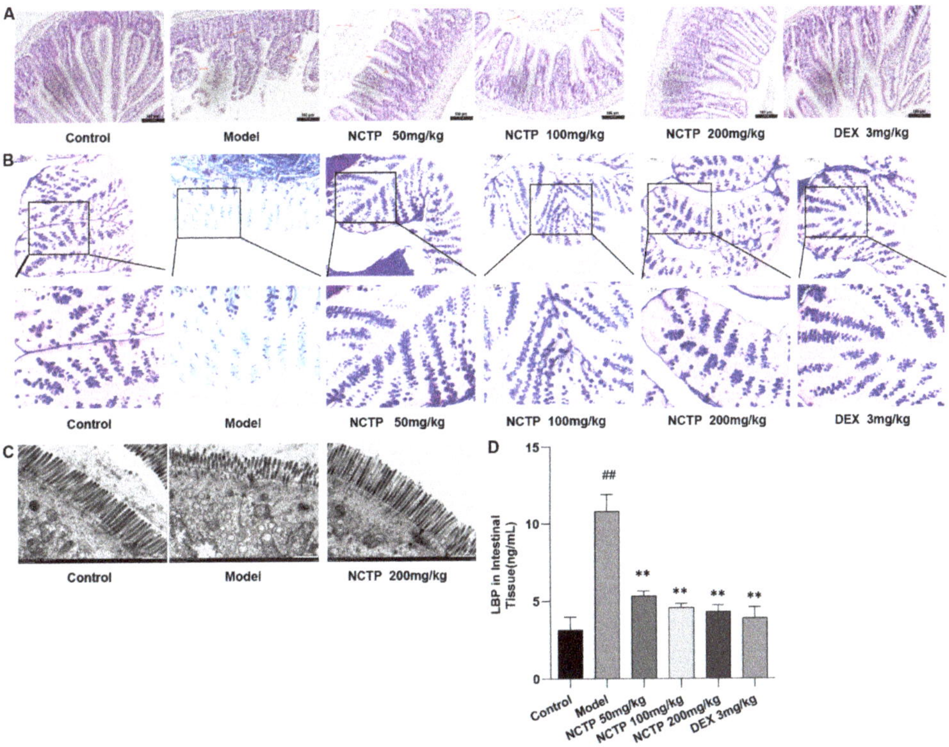

Figure 9. Effects of NCTP on the intestinal mucosa in ALI mice. (**A**) Effects of NCTP on histomorphology in ileum tissue of LPS-induced ALI mice. Scale bar, 100 μm. (**B**) Effect of NCTP in ileum tissue of LPS-induced ALI mice (AB-PAS stain) (at 10× magnification and 20× magnification). (**C**) Ultrastructural analysis in ileum tissue by transmission electron microscopy (at 1000× magnification). (**D**) Effect of NCTP on the levels of LBP in intestinal tissue. Data were presented as mean ± SD, $n = 6$. ## $p < 0.01$, versus control group; ** $p < 0.01$ versus model group.

As shown in the ileum tight junction structural morphology observed using an electron microscope (Figure 9C), regularly aligned microvilli, complete tight junctions, and intact organelles were observed in the intestinal epithelium in the control group. In the model group, the number of microvilli was reduced, lodging and shedding occurred, and their length and arrangement were irregular. At the same time, tight junctions were discontinuous, mitochondria were swollen, and vacuoles appeared between epithelial cells. In the NCTP (200 mg/kg) groups, the microvilli in the intestinal epithelium were regular, and the tight junctions tended to be intact with the mice in the model group. The LPS-binding proteins (LBPs) in intestinal tissue in the model group were highly expressed compared with the control group ($p < 0.01$). With NCTP (50,100 and 200 mg/kg) intervention, LBPs were decreased in comparison with the model group ($p < 0.05$ or $p < 0.01$) (Figure 9D and Table S7).

NCTP restored the ileum induced by LPS. As shown in Figures 10, S3 and S4 and Table S8, ZO-1 and occludin protein expressions were lower in the model group compared to the control group ($p < 0.01$). NCTP (50, 100, and 200 mg/kg) treatments had a greater expression of ZO-1 and occludin compared with the model group ($p < 0.01$).

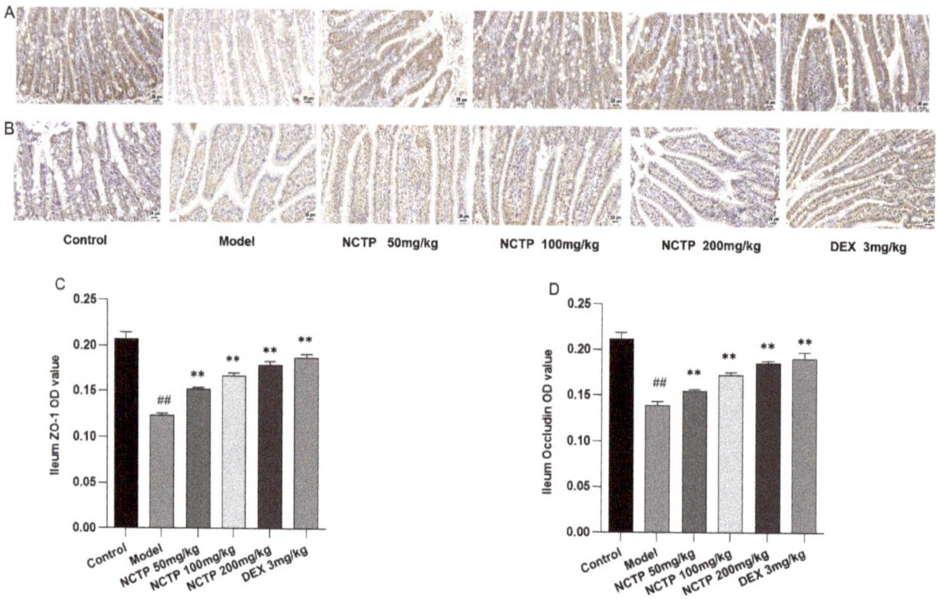

Figure 10. Effects of NCTP on ZO-1 and occludin protein expression in ALI mice. Immunohistochemical staining for ZO-1 (**A**,**C**) and occludin (**B**,**D**) in the ileum tissue (at 40× magnification). Data were presented as mean ± SD, $n = 6$. ## $p < 0.01$, versus control group; ** $p < 0.01$ versus model group.

2.7. NCTP Reversed the Dysbiosis of Microbiota Composition and the Reduction in SCFA Secretion in ALI Mice

The present results showed that the alpha diversity of the gut microbiome was significantly lower in the model group compared to the control group ($p < 0.01$) [26]. By contrast, the alpha diversity of the NCTP (200 mg/kg) group was significantly higher than the model group ($p < 0.01$) (Figure 11A–D and Table S9). The beta diversity of gut microbiota was analyzed to compare the similarity of different samples in species diversity [27] (Figure 11E,F). There was a significate difference in beta diversity among the groups. The control group and NCTP (200 mg/kg) group were clearly separated with the model groups, suggesting that the microbial community structure was different between the model group and control group. To clarify which bacteria were responsible for the differences in the gut microbiota structure among the groups, we explored the detailed differences in the relative abundance of bacterial communities at the phylum level.

At the phylum level, the intestinal flora of the organism is dominated by Bacteroidetes, Firmicutes, Proteobacteria, and Actinobacteria, which are beneficial bacteria, except Proteobacteria [28]. The major trend in the microbial community reaction to LPS-induced ALI was the loss of Firmicutes and the bloom of Proteobacteria [29]. Proteobacteria, Gram-negative aerobic bacteria, were considered a sign of gut microbiota dysbiosis and an important source of LPS, which could induce inflammation and be associated with many chronic diseases [30]. The relative abundance of Firmicutes (control, 0.34%; model, 0.18%) was decreased ($p < 0.01$), whereas the relative abundance of Proteobacteria (control, 0.08%; model 0.32%) was increased ($p < 0.01$) compared to the control group ($p < 0.01$). Moreover, we explored the microbiota relative abundance in the NCTP (200 mg/kg) groups. The abundance of Firmicutes (NCTP, 0.48%; model, 0.18%) increased significantly ($p < 0.01$), whereas Proteobacteria (NCTP, 0.13%; model, 0.32%) decreased in the NCTP (200 mg/kg) group compared with the model group ($p < 0.01$). The relative abundance of Bacteroidetes has no variances in all groups. As shown in the Venn diagram, the number of characteristic bacteria in the model group was decreased compared with the control group; the number of characteristic bacteria in the NCTP (200 mg/kg) group was restored compared with the model group. Furthermore, the differential analysis of LDA effect size (LEfSe) showed that Clostridiales and Lachnospiraceae were the representative bacterial order and family present in the NCTP (200 mg/kg) group (Figure 12 and Table S10). Lachnospiraceae belonged to Clostridiales in the order, which was the predominant species in NCTP.

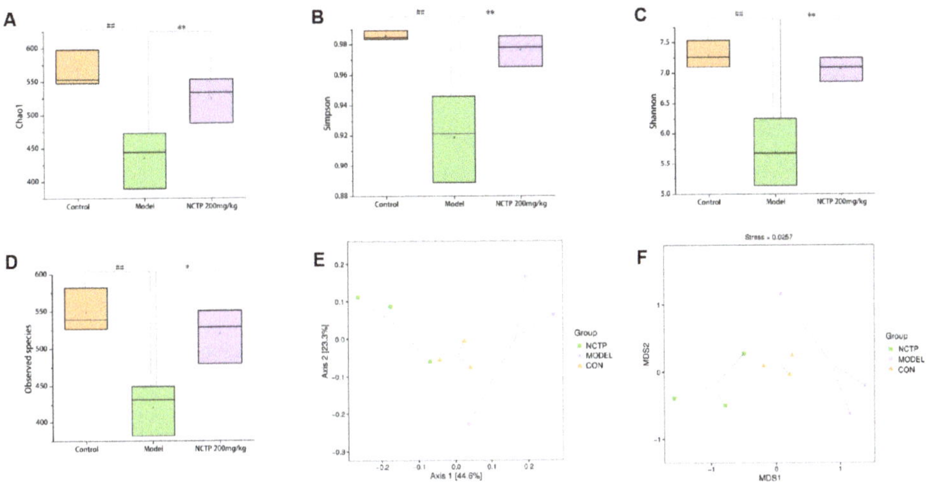

Figure 11. Effects of NCTP on the gut microbiota dysbiosis induced by ALI. The chao (**A**), Simpson (**B**), Shannon (**C**), and observed species (**D**) indices were used to assess the alpha diversity among groups. (**E**,**F**) The beta diversity among groups. Data were presented as mean ± SD, $n = 6$. ## $p < 0.01$, versus control group; * $p < 0.05$, ** $p < 0.01$ versus model group.

Compared with the control group, the fecal acetic acid, propionic acid, and butyric acid contents in the model group were obviously reduced ($p < 0.01$). Fecal acetic acid, propionic acid, and butyric acid levels were higher in NCTP (200 mg/kg) group mice compared with the model group [31], respectively ($p < 0.05$ or $p < 0.01$). Differences in other SCFAs were not significant (Figure 13 and Table S11).

In order to further investigate the association between mice intestinal flora and fecal SCFAs at the order level, the present study was based on the Spearman rank correlation test. We examined the correlation between the relative abundance of Clostridiales and Lachnospiraceae on the concentrations of three SCFAs (acetic acid, propionic acid, and butyric acid) (Figure 14). Spearman's correlation coefficients of $0.5 \leq r < 0.8$ were considered as moderate correlations, and $r \geq 0.8$ was considered as high correlations. Compared with the model group at the order level, acetic acid, propionic acid, and butyric acids were associated with Clostridiales, showing high positive correlation, as well as Lachnospiraceae at the family level.

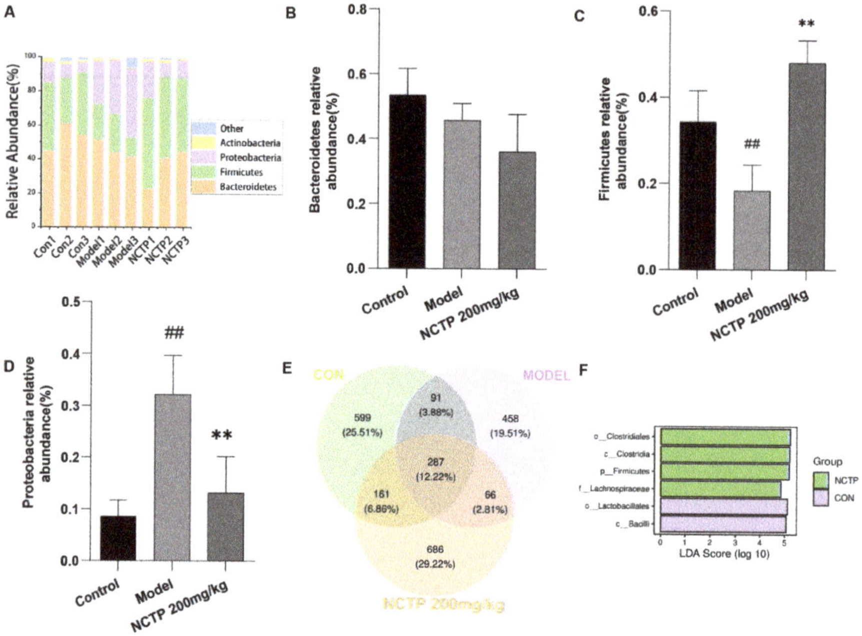

Figure 12. Effect of NCTP on phylum levels of gut microbiota. (**A**) The composition of gut microbiota of community bar plot analysis on phylum level. (**B**) The Bacteroidetes, (**C**) Firmicutes, (**D**) Proteobacteria on phylum level. (**E**) Venn diagram illustrated the overlap of OTUs in all tested mice. (**F**) Distribution histograms showing the results of LEfSe analysis. Data were presented as mean ± SD, $n = 6$. ## $p < 0.01$, versus control group; ** $p < 0.01$ versus model group.

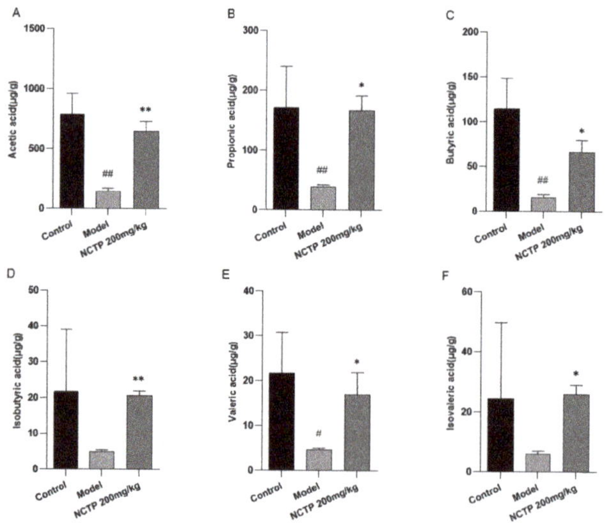

Figure 13. Effect of NCTP in the fecal content levels of short-chain fatty acids. (**A**) Acetic acid, (**B**) Propionic acid, (**C**) Butyric acid, (**D**) Isobutyric acid, (**E**) Valeric acid, (**F**) Isovaleric acid contents in fecal. Data were presented as mean ± SD., $n = 6$. # $p < 0.05$, ## $p < 0.01$, versus control group; * $p < 0.05$, ** $p < 0.01$ versus model group.

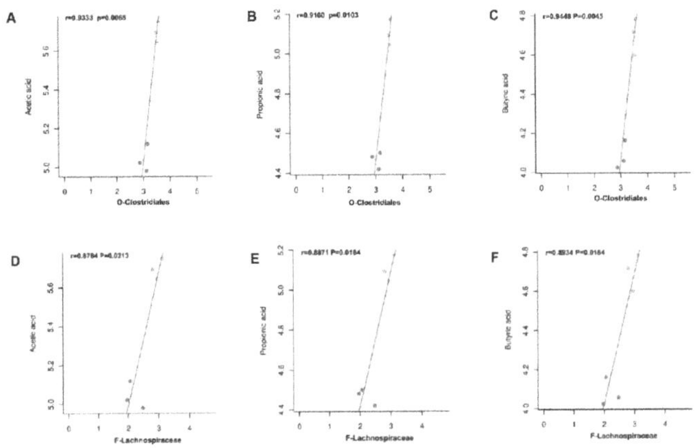

Figure 14. (**A–F**) Correlation analysis of the relative abundance of gut microbiota and fecal short-chain fatty acids. (The brown dots represent NCTP, 200 mg/kg, the blue dots represent model group).

3. Discussion

The pathogenesis of ALI involves several pathological processes, including immune cell overactivation, cytokine storm, and oxidative stress [32]. Intraperitoneal injection of LPS simulates the disease model of ALI caused by sepsis in intensive care units. ALI is a common complication of sepsis, and the most serious complication of sepsis is infectious shock, which has high morbidity and mortality. So, this experimental model has certain clinical significance [33,34]. As a severe pulmonary disease, ALI is closely related to the inflammatory/anti-inflammatory signaling pathways as well as the dysregulation of intestinal flora [35]. It has been reported that endothelial cell and alveolar epithelial injury, inflammatory cell influx into the alveolus, lung oedema, and increased inflammatory cytokines are related to the pathogenesis of ALI. Oedema is a very typical symptom of ALI, and we used the lung W/D ratio to quantify the direct effects of ALI. In this study, we found that NCTP could significantly reduce the lung coefficients of mice, attenuate pulmonary edema [36,37], and reduce the inflammatory cell infiltration of lung tissues and the pathological damage of lung tissues. We also found that NCTP can reduce the IL-1β, IL-6, and TNF-α expression levels in lung tissues and reduce the level of LPS in plasma [38], which suggests that NCTP can effectively ameliorate LPS-induced ALI.

Exogenous LPS enters the portal circulation, causing an imbalance in intestinal flora in vivo. Intestinal microflora disorders have been shown to induce G-bacteria to produce a large amount of LPS, which enters the circulation through gut leakage, increasing the immune response and activating the TLR4 signaling pathway in the lungs [39]. TLR4 has been shown to be the main signaling receptor for LPS-mediated injury in the body, where it has identified pathogenic bacteria that have translocated into the blood circulation, initiated downstream signal transduction pathways, activated the NF-κB pathway through the myeloid differentiation protein 88 (MyD88)-independent pathway and MyD88-dependent pathway, and upregulated the expression of various inflammatory factors [3]. In this study, NCTP suppressed the expression of TLR4, MyD88, TRAF6, IKKβ, p-IκBα, and p-NF-κB p65, which was upregulated by LPS [40]. These findings indicated that NCTP inhibited the activation of the TLR4/MyD88/NF-κB pathway in LPS-treated lung tissue, thereby suppressing NF-κB activation [41].

The production and secretion of the pro-inflammatory cytokine IL-1β is required not only by the NF-κB signaling pathway but also by the activation of the NLRP3 inflammasome. The NLRP3 inflammasome consists of an NLR family member, ASC, and Caspase-1 [42,43]. It can be activated by many factors, including bacteria and viruses, and evidence has suggested that the NLRP3 inflammasome is activated during ALI-induced LPS [44,45]. In the present study, we found that NCTP inhibited the proteins expression of

NLRP3, ASC, and Caspase-1 in LPS-induced ALI mice. These results suggested that the protective effects of NCTP may derive from its direct suppression of NLRP3 inflammasome activation [46].

The gut houses the largest collection of bacteria and endotoxins in the human body [47]. Substantial research has supported that the gut microbiota is closely related to the development and progression of ALI by transmitting gut–lung axis activation. The intestinal flora plays an important role in maintaining the normal immune function of the body, reducing bacterial translocation and improving lung injury [48]. Compared to the control group, the model group with an imbalance in the gut flora showed more obvious gut damage and decreased expression of ZO-1 and occludin [29,49]. In contrast, NCTP was verified to reconstruct the gut flora, upregulate ZO-1 and occludin expression, and reduce gut damage. Thus, the imbalance in the gut flora exacerbated bacterial translocation and stimulated inflammatory cascades. However, the abundance of intestinal flora in the NCTP group was increased compared with the model group, and the structure was relatively stable after NCTP intervention, indicating that NCTP improved the imbalance in the gut flora induced by ALI administration and regulated the abundance and diversity of gut flora.

Firmicutes is a major component of the intestinal flora and the main group of bacteria affected by NCTP. A variety of flora under Firmicutes are probiotics, which are involved in the fermentation of polysaccharide components, the production of short-chain fatty acids, lactic acid and other components, and the maintenance of intestinal flora homeostasis. SCFAs are the most important metabolites of the intestinal flora, which are absorbed through the intestines and enter the circulatory system to play a role in regulating immunity [50]. Lachnospiraceae is the dominant SCFA-producing species in the Firmicutes. Zhang [51] speculated that gut dysbiosis is a major factor, especially a decrease in the abundance of butyrate-producing bacteria such as Lachnospiraceae. Members of Lachnospiraceae are able to utilize lactate and acetate to produce butyrate via the butyryl-CoA or acetate CoA transferase pathways or the butyrate kinase pathway. Moreover, supplementation with butyrate-producing bacteria may be beneficial for alleviating gut dysbiosis. The experimental results showed that NCTP significantly increased the acetic acid, propionic acid, and butyric acids levels. NCTP inhibited the relative abundance of harmful bacteria by enriching the SCFAs in beneficial bacteria through the correlation analysis between SCFAs and intestinal flora.

In summary, NCTP could alleviate sepsis-induced acute lung injury by regulation of the NLRP3 and TLR-4/NF-κB pathways and the gut microbiota. Polyphenols are the main active ingredients of NCTP, such as isostrictiniin, ellagic acid, and nicotiflorin. Isostrictiniin has better anti-inflammatory and antioxidant activities [16]; ellagic acid has better anti-inflammatory, antioxidant, and immunomodulatory properties [52,53]; nicotiflorin possesses antioxidant, antibacterial, anti-inflammatory, and analgesic activities [54]. However, the preventive effects of these components on ALI need further study to reveal the pharmacodynamic material basis of NCTP.

4. Materials and Methods

4.1. Chemicals and Reagents

LPSs (lipopolysaccharides from *Escherichia coli* O55: B5, L2880-100 mg) were purchased from Shanghai Aladdin Biochemical Technology Co., Ltd. (Shanghai, China). Dexamethasone Acetate tablets were purchased from Chongqing Kerui Pharmaceutical Group Co., Ltd. (Chongqing, China). PBS (2246306-500 mL) was obtained from Shanghai Darthel Biotechnology Co., Ltd. (Shanghai, China). Further, 4% paraformaldehyde fixative solution (BL539A) was purchased from Beijing Labgic Technology Co., Ltd. (Beijing, China). Hematoxylin and eosin staining solution (BA4097-500 mL) was obtained from BaSO Biotech Inc. (Zhuhai, China). RIPA buffer (R0010) and PMSF (P0100) were purchased from Beijing Solarbio Science & Technology Co., Ltd. (Beijing, China). Phosphatase Inhibitor Mix (PR20015) was purchased from Wuhan Sanying Inc. (Wuhan, China). BCA Protein Concentration Assay Kit (PC0020) was purchased from Beijing Solarbio Science & Technology Co., Ltd.

(Beijing, China). LDH Kit (A020-1) was obtained from Nanjing Jiancheng Bioengineering Institute (Nanjing, China). ELISA kits for TNF-α (JL10484), IL-6 (JL20268), IL-1β (JL18442), LPS (JL18442), LBP (JL29644), and MPO (JL10367) were purchased from Shanghai Jianglai industrial Limited By Share Ltd. (Shanghai, China). NLRP3 antibody (ab270449), ASC antibody (ab309497), Caspase1 antibody (ab138483), MyD88 antibody (ab219413), TRAF6 antibody (ab33915), IKKβ antibody(ab124957), IκB antibody (ab32518), p-IκB antibody (ab133462), NF-κB p65 antibody (ab32536), p-NF-κB p65 antibody (ab76302), GAPDH antibody (ab181602), and occludin antibody (ab216327) were purchased from Abcam, Cambridge, UK. TLR4 antibody (#14358) was purchased from Cell Signaling Technology, Beverly, MA, USA. ZO-1 antibody (AF5145) was purchased from Affinity Biosciences (Changzhou, China).

4.2. Plant Materials and Preparation of NCTP

Nymphaea candida flower was purchased from Xinjiang Shengxiangcao Co. (Urumqi, China) and identified by Jiang He, a researcher at the Institute of Materia Medica of Xinjiang, and the voucher specimens (20210906N) were stored in the Institute of Materia Medica of Xinjiang. The extraction of NCTP was described in our previously published report [22]. This medicinal material was extracted with 70% ethanol by reflux three times for 1 h each time. We mixed the extract, concentrated under reduced pressure, and obtained the 70% ethanol extract. The extract was isolated by D101 resin, and 70% ethanol elutes were further purified using polyamide resins. The 70% ethanol elutes in enriched polyphenols were concentrated and dried to obtain NCTP.

4.3. Compound Analysis of NCTP

Compound analysis of NCTP was performed using a UHPLC-Q-Extractive mass spectrometry system consisting of an Ultimate 3000 UHPLC (Thermo Fisher Scientific, Bremen, Germany) combined with a Q Exactive Plus (Thermo Fisher Scientific, Germany). The chromatographic separation for NCTP was accomplished on an Agilent XDB C18 column (4.6 mm × 250 mm, 5 μm) (Agilent Technologies Inc., Santa Clara, CA, USA). The UHPLC flow rate was 1.0 mL/min, column temperature was 35 °C, and detection wavelength was 266 nm. The optimized mobile phases contained solvent A and solvent B, which were acetonitrile and 0.1% formic acid in water. The optimized mobile-phase gradient elution was as follows: 0~35 min, 5–15% A; 35–65 min, 15–18% A; 65~70 min, 18–20% A; 70–75 min, 20–5% A; 75–80 min, 5–5% A.

The detection parameters for the HRMS conditions were as follows: ESI spray voltage −2.8 kV (positive ions: 3.2 kV); sheath gas (sheath gas) 40 arb; auxiliary gas (auxiliary gas) 10 arb; curtain gas (CUR) 35; ion source temperature (HEAT TEMP) 350 °C; capillary temperature 300 °C; focusing voltage (FP) −350 V, DP2 −10 V. The whole process used nitrogen. Sample analysis was in DDA mode: Q-Orbitrap acquisition range was m/z 100–1500; fragment ion scanning range was set to m/z 50–1500; collision gas (NCE): 20%, 40%, 60%; MS resolution was 70,000 FWHM (m/z 200); MS^2 resolution was 17,600 FWHM (m/z 200). The UHPLC-Q-Orbitrap-HRMS system operation and data analysis were carried out with the Thermo Fisher Scientific Workstation software, which contains Data Acquisition (Xcalibur 4.0) software and Qualitative Analysis (Xcalibur 4.0) software.

4.4. Experimental Animals

Seven- to eight-week-old C57/BL6J male mice were purchased from Vitalriver (Beijing, China). The mice were housed in isolated cages under controlled environmental conditions at the Xinjiang Medical University (12 h light–dark cycle, 55 ± 5% humidity, 25 ± 1 °C), with free access to standard laboratory food and water. All animal studies were approved by the Experimental Animal Ethics Committee of Xinjiang Medical University (Approval No: IACUC-20220314-2) and complied with animal welfare regulations. The methods employed in this study were conducted in accordance with the approved guidelines.

4.5. Sepsis-Induced Acute Lung Injury (ALI) Model

After one week of adaptive feeding, mice were randomly divided into six groups, with 12 mice in each group: control group, model group, DEX (positive control, 3 mg·kg^{-1}) group, and dose NCTP (50, 100, 200 mg·kg^{-1}). On days 1–7, mice underwent intragastric administration of NCTP (50, 100, 200 mg·kg^{-1}) and DEX (3 mg·kg^{-1}), respectively, except for the control and model groups with the same amount of 0.5% CMC-Na solution. After 30 min of the last administration, LPS (10 mg·kg^{-1}) solution was intraperitoneal injection to induce ALI [23,55], except for the control group. Sixteen hours later, mice were euthanized, and bronchoalveolar lavage fluid (BALF), blood, lungs, ileum tissues, and fecal samples were performed, respectively.

4.6. Pathologic Analysis and Immunohistochemical Analysis

Histologic examination. The collected right upper lobe was fixed overnight in 4% paraformaldehyde, dehydrated, paraffin-embedded, cut into 5 μm thick sections, and stained with hematoxylin and eosin (H&E). Images were acquired using a light microscope (LEICA DM 3000, Wiesler, Hesse, Germany) at 20× magnification. The extent of lung damage was assessed according to previous studies [23]. The degree of lung injury was graded from 0 (normal) to 3 (severe) as follows: 0, no injury; I, mild injury; II, moderate injury; III, severe injury. Individual scores for each category and the sum of the scores of 6 mice in each category were calculated to determine the total lung injury score for histological assessment. The collected ileum tissue was fixed overnight in 4% paraformaldehyde, dehydrated, paraffin-embedded, cut into 5 μm thick sections, and stained with hematoxylin and eosin (H&E). AB-PAS staining. Mouse colon tissue was taken and fixed with Carnot's fixative, followed by paraffin embedding, sectioning, dewaxing, and staining with AB-PAS.

Ultrastructural analysis. Ileum tissue samples were immediately fixed with 3% glutaraldehyde immediately and stored at a temperature of 4 °C for 4 h. The samples were dehydrated with a graded series of ethanol and propylene oxide and then embedded in an epoxy resin. Ultrathin sections were cut, stained with uranyl acetate and lead citrate, and examined with transmission electron microscopy (Hitachi HT7800, Tokyo, Japan). The ultrastructure of the ileum tight junctions was evaluated by capturing high-magnification images.

Immunofluorescence and immunohistochemical analysis. Sections cut from paraffin-embedded ileum tissues were deparaffinized, washed with xylene and graded series of ethanol, subsequently recovered by antigen retrieval buffer with 10 mM citrate buffer, blocked with 10% goat serum, and incubated with primary antibody against ZO-1 (1/200), and occludin (1/200) at 4 °C overnight. Image J software (v1.8.0.112) was used for quantification.

4.7. BALF Collection, Processing, and Determination of Protein and LDH Concentration

The total protein content and LDH in bronchoalveolar lavage fluid (BALF) were examined. At the end of the experiment, mice were euthanized by cervical dislocation, 500 μL PBS was instilled twice through the trachea for 30 s, and BALF was collected. The collected BALF was used to determine the total protein content and LDH in the BALF supernatant using the BCA method and microplate spectrophotometer (Thermo Fisher Scientific, Multiskan GO, Waltham, MA, USA).

4.8. Measurement of Wet-to-Dry Ratio of the Lungs

Lung wet/dry weight ratio. The middle lobes of the right lung of mice were collected and rinsed with PBS. Excess surface water was removed by blotting with filter paper, and the lung tissue was then weighed using a sophisticated electronic balance to determine lung wet weight. Subsequently, the lung tissue was placed in a desiccator at a temperature of 80 °C and dried thoroughly for 72 h until a stable weight was obtained. The weight of the dried lung tissue was recorded as the lung dry weight. Finally, the lung wet/dry (W/D) weight ratio was calculated to assess the degree of pulmonary edema in mice.

4.9. Cell Count and Biochemical Assays

The white blood cells (WBCs), neutrophils (NEU), lymphocytes (LYM), monocytes (MON), eosinophils (EOS), and basophils (BAS) were detected by an Auto Hematology Analyzer (Mindray Healthcare International Inc. Mindray BC-5000, Shenzhen, China). The plasma LPS was determined by ELISA kits.

4.10. Determination of Lung Homogenate TNF-α, IL-1β, and IL-6

Lung homogenates were directly used to determine inflammatory mediators, including tumor necrosis factor α (TNF-α), interleukin-1β (IL-1β), and interleukin-6 (IL-6), by ELISA kits.

4.11. Analysis of Gut Microbiota

The cecum contents in different groups were collected, and total genomic DNA was extracted from each fecal sample with Stool DNA mini kits (Qiagen, Valencia, CA, USA). The purity and concentration of the extracted DNA were detected by gel electrophoresis; the DNA was stored at −80 °C for further analysis. The sample DNA was amplified by PCR amplification to enrich the V3-V4 region of 16S rDNA genes in bacteria. PCR products were quantified by a microplate reader (BioTek, FLx800, Winooski, VT, USA) and an Agilent Bioanalyzer 2100 system (Agilent Technologies Inc., Santa Clara, CA, USA). The whole genome sequencing was performed by double-end sequencing using a NovaSeq sequencer with NovaSeq 6000 SP Reagent Kit (500 cycles) (San Diego, CA, USA). The chao indices reflect the species richness, and the Shannon and Simpson indices reflect the species richness and uniformity in the sample. They were used to assess the alpha diversity among groups. The clustering characteristics of the gut microbiota were analyzed by beta diversity based on OTU abundances.

4.12. Fecal Short-Chain Fatty Acid (SCFA) Quantification Analysis

The concentrations of SCFAs were determined in the fecal supernatant using GC-MS (Thermo Trace 1300-Thermo ISQ 7000). Samples were homogenized for 1 min with 500 μL of water and 100 mg of glass beads, and then centrifuged at 4 °C for 10 min at 12,000 rpm. Then, 200 μL supernatant was extracted with 100 μL of 15% phosphoric acid and 20 μL of 375 μg/mL 4-methylvaleric acid solution as IS and 280 μL ether. Subsequently, the samples were centrifuged at 4 °C for 10 min at 12,000 rpm after vortexing for 1 min, and the supernatant was transferred into the vial prior to GC-MS analysis.

4.13. Western Blot

At the end of the experiment, 30 mg of the lower lobe of the right lung was weighed, and the total protein was extracted with RIPA lysis buffer. The protein was then quantified using the BCA method. Subsequently, 25 μg of protein was electrophoresed on an 8% denaturing polyacrylamide gel and transferred to a PVDF membrane. After blocking with the NcmBlot blocking buffer, the membrane was incubated overnight at 4 °C with anti-TLR4 (1:1000), MyD88 (1:1000), TRAF6 (1:5000), IKKβ (1:5000), IκB (1:2000), p-IκB (1:1000), NF-κB p65 (1:1000), p-NF-κBp 65 (1:1000), NLRP3 (1:1000), ASC (1:1000), Caspase-1 (1:1000), and GAPDH (1:1000). Then, it was incubated for another 1 h with horseradish peroxidase-conjugated rabbit anti-mouse IgG antibody (1:10,000). Finally, a chemiluminescence detection reagent and imaging system were used to visualize the membranes.

4.14. Statistical Analysis

Each experiment was conducted in triplicate, and the data are presented as mean ± standard deviation. Statistical comparisons between two groups were analyzed using the *t*-test, while differences among multiple groups were analyzed using one-way analysis of variance (ANOVA). GraphPad software version 9 (Prism, La Jolla, CA, USA) was utilized for statistical analysis. A p-value < 0.05 was considered statistically significant.

5. Conclusions

The experimental results suggest that NCTP could be considered as an anti-inflammatory agent for the treatment of ALI, and these findings can also confirm the traditional application of *N. candida* as well as its effects in classic formulas, including zukamu granules. This research could provide a reference for the development and application of the *N. candida* flower.

Supplementary Materials: The following supporting information can be downloaded at: https://www.mdpi.com/article/10.3390/ijms25084276/s1.

Author Contributions: C.L. and J.Z. designed the study; C.L., X.Q., L.X., Y.S. and Y.Y. performed the experiments; C.L. and X.Q. analyzed the data, and C.L. wrote the manuscript; L.X., Y.C. and J.Z. contributed to the interpretation of data and critically revised the manuscript. All authors have read and agreed to the published version of the manuscript.

Funding: This research was funded by the Key project of Natural Science Foundation of Xinjiang Uygur Autonomous Region in China, grant number 2021D01D14.

Institutional Review Board Statement: The animal study protocol was approved by the Experimental Animal Ethics Committee of Xinjiang Medical University (No: IACUC-20220314-2), Approval date: 14 March 2022.

Informed Consent Statement: Not applicable.

Data Availability Statement: Data is contained within the article and Supplementary Materials.

Conflicts of Interest: The authors declare no conflicts of interest.

References

1. Lu, Y.; Xu, D.; Liu, J.; Gu, L. Protective effect of sophocarpine on lipopolysaccharide-induced acute lung injury in mice. *Int. Immunopharmacol.* **2019**, *70*, 180–186. [CrossRef] [PubMed]
2. Mowery, N.T.; Terzian, W.T.H.; Nelson, A.C. Acute lung injury. *Curr. Probl. Surg.* **2020**, *57*, 100777. [CrossRef]
3. Wang, N.; Geng, C.; Sun, H.; Wang, X.; Li, F.; Liu, X. Hesperetin ameliorates lipopolysaccharide-induced acute lung injury in mice through regulating the TLR4–MyD88–NF-κB signaling pathway. *Arch. Pharmacal Res.* **2019**, *42*, 1063–1070. [CrossRef] [PubMed]
4. Li, L.; Huang, Q.; Wang, D.C.; Ingbar, D.H.; Wang, X. Acute lung injury in patients with COVID-19 infection. *Clin. Transl. Med.* **2020**, *10*, 20–27. [CrossRef]
5. Wang, Z.-F.; Yang, Y.-M.; Fan, H. Diagnostic value of miR-155 for acute lung injury/acute respiratory distress syndrome in patients with sepsis. *J. Int. Med. Res.* **2020**, *48*, 1–9. [CrossRef] [PubMed]
6. Fan, E.; Brodie, D.; Slutsky, A.S. Acute Respiratory Distress Syndrome. *JAMA* **2018**, *319*, 698–710. [CrossRef] [PubMed]
7. Tian, C.; Zhang, P.; Yang, J.; Zhang, Z.; Wang, H.; Guo, Y.; Liu, M. The protective effect of the flavonoid fraction of *Abutilon theophrasti* Medic. leaves on LPS-induced acute lung injury in mice via the NF-κB and MAPK signalling pathways. *Biomed. Pharmacother.* **2019**, *109*, 1024–1031. [CrossRef]
8. Huang, X.-T.; Liu, W.; Zhou, Y.; Hao, C.-X.; Zhou, Y.; Zhang, C.-Y.; Sun, C.-C.; Luo, Z.-Q.; Tang, S.-Y. Dihydroartemisinin attenuates lipopolysaccharide-induced acute lung injury in mice by suppressing NF-κB signaling in an Nrf2-dependent manner. *Int. J. Mol. Med.* **2019**, *44*, 2213–2222. [CrossRef] [PubMed]
9. Wu, X.; Lin, L.; Wu, H. Ferulic acid alleviates lipopolysaccharide-induced acute lung injury through inhibiting TLR4/NF-κB signaling pathway. *J. Biochem. Mol. Toxicol.* **2020**, *35*, e22664. [CrossRef] [PubMed]
10. Dang, A.T.; Marsland, B.J. Microbes, metabolites, and the gut–lung axis. *Mucosal Immunol.* **2019**, *12*, 843–850. [CrossRef]
11. Gao, L.-L.; Ma, J.-M.; Fan, Y.-N.; Zhang, Y.-N.; Ge, R.; Tao, X.-J.; Zhang, M.-W.; Gao, Q.-H.; Yang, J.-J. Lycium barbarum polysaccharide combined with aerobic exercise ameliorated nonalcoholic fatty liver disease through restoring gut microbiota, intestinal barrier and inhibiting hepatic inflammation. *Int. J. Biol. Macromol.* **2021**, *183*, 1379–1392. [CrossRef] [PubMed]
12. Tang, J.; Xu, L.; Zeng, Y.; Gong, F. Effect of gut microbiota on LPS-induced acute lung injury by regulating the TLR4/NF-kB signaling pathway. *Int. Immunopharmacol.* **2021**, *91*, 107272. [CrossRef] [PubMed]
13. Liu, J.; Chang, G.; Huang, J.; Wang, Y.; Ma, N.; Roy, A.-C.; Shen, X. Sodium Butyrate Inhibits the Inflammation of Lipopolysaccharide-Induced Acute Lung Injury in Mice by Regulating the Toll-Like Receptor 4/Nuclear Factor κB Signaling Pathway. *J. Agric. Food Chem.* **2019**, *67*, 1674–1682. [CrossRef] [PubMed]
14. Liu, S.; Chen, Q.; Liu, J.; Yang, X.; Zhang, Y.; Huang, F. Sinomenine protects against *E. coli*-induced acute lung injury in mice through Nrf2-NF-κB pathway. *Biomed. Pharmacother.* **2018**, *107*, 696–702. [CrossRef] [PubMed]
15. Hu, X.M. *Chinese Materia Medica—Uyghur Medicine Volume*; Shanghai Scientific & Technical Publishers: Shanghai, China, 1999.
16. Zhao, J.; Tuersunmaimaiti, M.; Ji, T.; Liu, T.; Xu, F. Hepatoprotective activity of isostrictiniin from *Nymphaea candida* on Con A-induced acute liver injury in mice. *Nat. Prod. Res.* **2019**, *35*, 1662–1666. [CrossRef] [PubMed]

17. Yin, Q.; Hou, S.; Yin, H.; Mu, D.; Jiang, D.; Tian, F.; Li, J.J.; Huang, F. The analgesic and anti-inflammatory effects of zukamu granules, a traditional Chinese medical formulation. *Pharm. Biol.* **2019**, *57*, 729–735. [CrossRef]
18. Yu, C.; Li, Y.; Li, Y.; Li, S.; Zeng, F.; Yu, J.; Ji, Z.; Li, K.; Zhai, H. A novel mechanism for regulating lung immune homeostasis: Zukamu granules alleviated acute lung injury in mice by inhibiting NLRP3 inflammasome activation and regulating Th17/Treg cytokine balance. *J. Ethnopharmacol.* **2024**, *324*, 117831. [CrossRef]
19. Li, S.; Liu, G.; Gu, M.; Li, Y.; Li, Y.; Ji, Z.; Li, K.; Wang, Y.; Zhai, H.; Wang, Y. A novel therapeutic approach for IPF: Based on the "Autophagy-Apoptosis" balance regulation of Zukamu Granules in alveolar macrophages. *J. Ethnopharmacol.* **2022**, *297*, 115568. [CrossRef]
20. Zhao, J.; Zhang, S.; You, S.; Liu, T.; Xu, F.; Ji, T.; Gu, Z. Hepatoprotective Effects of Nicotiflorin from *Nymphaea candida* against Concanavalin A-Induced and D-Galactosamine-Induced Liver Injury in Mice. *Int. J. Mol. Sci.* **2017**, *18*, 587. [CrossRef]
21. Dong, H.j.; Guo, Y.; Liu, T.; Zhao, J. Preventive effect of isostrictiniin from *Nymphaea candida* on carbon tetrachloride-induced hepatic fibrosis in mice. *Nat. Prod. Res.* **2022**, *37*, 628–632. [CrossRef]
22. Sun, Y.; Qi, Z.; Xu, Y.; Li, C.; Zhao, J.; Liu, T. Anti-inflammatory, analgesic, antitussive and antipyretic activities of polyphenol-enriched fraction from *Nymphaea candida*. *J. Ethnopharmacol.* **2024**, *324*, 117789. [CrossRef] [PubMed]
23. Li, L.; Fu, W.-W.; Wu, R.-T.; Song, Y.-H.; Wu, W.-Y.; Yin, S.-H.; Li, W.-J.; Xie, M.-Y. Protective effect of Ganoderma atrum polysaccharides in acute lung injury rats and its metabolomics. *Int. J. Biol. Macromol.* **2020**, *142*, 693–704. [CrossRef] [PubMed]
24. Serra, G.; Incani, A.; Serreli, G.; Porru, L.; Melis, M.P.; Tuberoso, C.I.G.; Rossin, D.; Biasi, F.; Deiana, M. Olive oil polyphenols reduce oxysterols -induced redox imbalance and pro-inflammatory response in intestinal cells. *Redox Biol.* **2018**, *17*, 348–354. [CrossRef]
25. He, Y.; Wen, Q.; Yao, F.; Xu, D.; Huang, Y.; Wang, J. Gut–lung axis: The microbial contributions and clinical implications. *Crit. Rev. Microbiol.* **2016**, *43*, 81–95. [CrossRef] [PubMed]
26. Ying, M.; Yu, Q.; Zheng, B.; Wang, H.; Wang, J.; Chen, S.; Nie, S.; Xie, M. Cultured Cordyceps sinensis polysaccharides modulate intestinal mucosal immunity and gut microbiota in cyclophosphamide-treated mice. *Carbohydr. Polym.* **2020**, *235*, 115957. [CrossRef]
27. Jiang, T.; Li, P.; Zhao, J.; Dai, L.; Sun, D.; Liu, M.; An, L.; Jia, L.; Jing, X.; Wang, H.; et al. Long-chain polyunsaturated fatty acids improve airway pathological features and gut microbial imbalances in BALB/c mice with ovalbumin-induced asthma. *J. Funct. Foods* **2021**, *81*, 104465. [CrossRef]
28. Li, N.; Dai, Z.; Wang, Z.; Deng, Z.; Zhang, J.; Pu, J.; Cao, W.; Pan, T.; Zhou, Y.; Yang, Z.; et al. Gut microbiota dysbiosis contributes to the development of chronic obstructive pulmonary disease. *Respir. Res.* **2021**, *22*, 274. [CrossRef] [PubMed]
29. Zhou, X.; Liao, Y. Gut-Lung Crosstalk in Sepsis-Induced Acute Lung Injury. *Front. Microbiol.* **2021**, *12*, 779620. [CrossRef]
30. Lyu, M.; Wang, Y.-F.; Fan, G.-W.; Wang, X.-Y.; Xu, S.-Y.; Zhu, Y. Balancing Herbal Medicine and Functional Food for Prevention and Treatment of Cardiometabolic Diseases through Modulating Gut Microbiota. *Front. Microbiol.* **2017**, *8*, 270196. [CrossRef]
31. Guo, C.-E.; Cui, Q.; Cheng, J.; Chen, J.; Zhao, Z.; Guo, R.; Dai, X.; Wei, Z.; Li, W. Probiotic-fermented Chinese dwarf cherry [Cerasus humilis (Bge.) Sok.] juice modulates the intestinal mucosal barrier and increases the abundance of Akkermansia in the gut in association with polyphenols. *J. Funct. Foods* **2021**, *80*, 104424. [CrossRef]
32. Huppert, L.; Matthay, M.; Ware, L. Pathogenesis of Acute Respiratory Distress Syndrome. *Semin. Respir. Crit. Care Med.* **2019**, *40*, 31–39. [CrossRef] [PubMed]
33. Tuerdi, B.; Zuo, L.; Ma, Y.; Wang, K. Downregulation of miR-155 attenuates sepsis-induced acute lung injury by targeting SIRT1. *Int. J. Clin. Exp. Pathol.* **2018**, *11*, 4483–4492. [PubMed]
34. Wang, X.; Xu, T.; Jin, J.; Ting Gao, M.M.; Wan, B.; Gong, M.; Bai, L.; Lv, T.; Song, Y. Topotecan reduces sepsis-induced acute lung injury and decreases the inflammatory response via the inhibition of the NF-kappaB signaling pathway. *Pulm. Circ.* **2022**, *12*, e12070. [CrossRef] [PubMed]
35. Ding, Q.; Feng, S.W.; Xu, G.H.; Chen, Y.Y.; Shi, Y.Y. Effects of total ginsenosides from Panax ginseng stems and leaves on gut microbiota and short-chain fatty acids metabolism in acute lung injury mice. *Zhongguo Zhong Yao Za Zhi* **2023**, *48*, 1319–1329. [CrossRef] [PubMed]
36. Li, W.; Li, W.; Zang, L.; Liu, F.; Yao, Q.; Zhao, J.; Zhi, W.; Niu, X. Fraxin ameliorates lipopolysaccharide-induced acute lung injury in mice by inhibiting the NF-kappaB and NLRP3 signalling pathways. *Int. Immunopharmacol.* **2019**, *67*, 1–12. [CrossRef] [PubMed]
37. Tao, W.; Su, Q.; Wang, H.; Guo, S.; Chen, Y.; Duan, J.; Wang, S. Platycodin D attenuates acute lung injury by suppressing apoptosis and inflammation in vivo and in vitro. *Int. Immunopharmacol.* **2015**, *27*, 138–147. [CrossRef] [PubMed]
38. Hu, L.; Wang, Y.; Sun, H.; Xiong, Y.; Zhong, L.; Wu, Z.; Yang, M. An untargeted metabolomics approach to investigate the wine-processed mechanism of Scutellariae radix in acute lung injury. *J. Ethnopharmacol.* **2020**, *253*, 112665. [CrossRef]
39. Jacobs, M.C.; Lankelma, J.M.; Wolff, N.S.; Hugenholtz, F.; de Vos, A.F.; van der Poll, T.; Wiersinga, W.J. Effect of antibiotic gut microbiota disruption on LPS-induced acute lung inflammation. *PLoS ONE* **2020**, *15*, e0241748. [CrossRef]
40. Ju, M.; Liu, B.; He, H.; Gu, Z.; Liu, Y.; Su, Y.; Zhu, D.; Cang, J.; Luo, Z. MicroRNA-27a alleviates LPS-induced acute lung injury in mice via inhibiting in fl ammation and apoptosis through modulating TLR4/MyD88/NF-kappaB pathway. *Cell Cycle* **2018**, *17*, 2001–2018. [CrossRef]
41. Gao, H.; Xiao, D.; Gao, L.; Li, X. MicroRNA-93 contributes to the suppression of lung inflammatory responses in LPS-induced acute lung injury in mice via the TLR4/MyD88/NF-kappaB signaling pathway. *Int. J. Mol. Med.* **2020**, *46*, 561–570. [CrossRef]

42. Wang, X.; Deng, R.; Dong, J.; Huang, L.; Li, J.; Zhang, B. Eriodictyol ameliorates lipopolysaccharide-induced acute lung injury by suppressing the inflammatory COX-2/NLRP3/NF-kappaB pathway in mice. *J. Biochem. Mol. Toxicol.* **2020**, *34*, e22434. [CrossRef] [PubMed]
43. Liu, Q.; Ci, X.; Wen, Z.; Peng, L. Diosmetin Alleviates Lipopolysaccharide-Induced Acute Lung Injury through Activating the Nrf2 Pathway and Inhibiting the NLRP3 Inflammasome. *Biomol. Ther.* **2018**, *26*, 157–166. [CrossRef] [PubMed]
44. Jiang, Y.; Yang, W.; Gui, S. Procyanidin B2 protects rats from paraquat-induced acute lung injury by inhibiting NLRP3 inflammasome activation. *Immunobiology* **2018**, *223*, 555–561. [CrossRef] [PubMed]
45. Zhang, Y.; Wang, X.; Liu, Z.; Yu, L. Dexmedetomidine attenuates lipopolysaccharide induced acute lung injury by targeting NLRP3 via miR-381. *J. Biochem. Mol. Toxicol.* **2018**, *32*, e22211. [CrossRef] [PubMed]
46. Yu, J.; Ni, L.; Zhang, X.; Zhang, J.; Abdel-Razek, O.; Wang, G. Surfactant Protein D Dampens Lung Injury by Suppressing NLRP3 Inflammasome Activation and NF-kappaB Signaling in Acute Pancreatitis. *Shock* **2019**, *51*, 557–568. [CrossRef]
47. Chioma, O.S.; Mallott, E.K.; Chapman, A.; Van Amburg, J.C.; Wu, H.; Shah-Gandhi, B.; Dey, N.; Kirkland, M.E.; Blanca Piazuelo, M.; Johnson, J.; et al. Gut microbiota modulates lung fibrosis severity following acute lung injury in mice. *Commun. Biol.* **2022**, *5*, 1401. [CrossRef] [PubMed]
48. Wang, Y.; Liu, J.; Li, F.; Ma, S.; Zhao, L.; Ge, P.; Wen, H.; Zhang, Y.; Liu, X.; Luo, L.; et al. Mechanisms of Qingyi Decoction in Severe Acute Pancreatitis Associated Acute Lung Injury via Gut Microbiota: Targeting the Short-Chain Fatty Acids-Mediated AMPK/NF-kB/NLRP3 Pathway. *Microbiol. Spectr.* **2023**, *11*, e0366422. [CrossRef]
49. Bai, R.-X.; Xu, Y.-Y.; Qin, G.; Chen, Y.-M.; Wang, H.-F.; Wang, M.; Du, S.-Y. Repression of TXNIP-NLRP3 axis restores intestinal barrier function via inhibition of myeloperoxidase activity and oxidative stress in nonalcoholic steatohepatitis. *J. Cell Physiol.* **2019**, *234*, 7524–7538. [CrossRef]
50. Wang, Z.; Li, F.; Liu, J.; Luo, Y.; Guo, H.; Yang, Q.; Xu, C.; Ma, S.; Chen, H. Intestinal Microbiota—An Unmissable Bridge to Severe Acute Pancreatitis-Associated Acute Lung Injury. *Front. Immunol.* **2022**, *13*, 913178. [CrossRef]
51. Zhang, J.; Song, L.; Wang, Y.; Liu, C.; Zhang, L.; Zhu, S.; Liu, S.; Duan, L. Beneficial effect of butyrate-producing Lachnospiraceae on stress-induced visceral hypersensitivity in rats. *J. Gastroenterol. Hepatol.* **2019**, *34*, 1368–1376. [CrossRef]
52. Zhang, X.; Wang, S.; Wu, Y.; Liu, X.; Wang, J.; Han, D. Ellagic Acid Alleviates Diquat-Induced Jejunum Oxidative Stress in C57BL/6 Mice through Activating Nrf2 Mediated Signaling Pathway. *Nutrients* **2022**, *14*, 1103. [CrossRef] [PubMed]
53. Rønning, S.B.; Voldvik, V.; Bergum, S.K.; Aaby, K.; Borge, G.I.A. Ellagic acid and urolithin A modulate the immune response in LPS-stimulated U937 monocytic cells and THP-1 differentiated macrophages. *Food Funct.* **2020**, *11*, 7946–7959. [CrossRef] [PubMed]
54. Wang, Y.; Chen, P.; Tang, C.; Wang, Y.; Li, Y.; Zhang, H. Antinociceptive and anti-inflammatory activities of extract and two isolated flavonoids of *Carthamus tinctorius* L. *J. Ethnopharmacol.* **2014**, *151*, 944–950. [CrossRef] [PubMed]
55. Ding, Z.; Zhong, R.; Yang, Y.; Xia, T.; Wang, W.; Wang, Y.; Xing, N.; Luo, Y.; Li, S.; Shang, L.; et al. Systems pharmacology reveals the mechanism of activity of Ge-Gen-Qin-Lian decoction against LPS-induced acute lung injury: A novel strategy for exploring active components and effective mechanism of TCM formulae. *Pharmacol. Res.* **2020**, *156*, 104759. [CrossRef] [PubMed]

International Journal of
Molecular Sciences

MDPI

Review

Effects of Common Fig (*Ficus carica* L.) and Its Extracts on Certain Cancer Types: Focusing on the Mechanism of Action

Elif Nisa Gökçen [1], Sevgi Gezici [2], Bence L. Raposa [3], Dávid Szép [4], Ferenc Budán [4,*,†] and Duygu Ağagündüz [5,*,†]

1 Department of Nutrition and Dietetics, Faculty of Health Science, Kilis 7 Aralık University, 79000 Kilis, Turkey; elifnisa.pak@kilis.edu.tr
2 Department of Medical Biology and Genetics, Faculty of Medicine, Gaziantep University, 27000 Gaziantep, Turkey
3 Institute of Basics of Health Sciences, Midwifery and Health Visiting, Faculty of Health Science, University of Pécs, 7621 Pécs, Hungary; raposa.laszlo@pte.hu
4 Institute of Physiology, Medical School, University of Pécs, 7624 Pécs, Hungary; szep.david@pte.hu
5 Department of Nutrition and Dietetics, Faculty of Health Science, Gazi University, 06000 Ankara, Turkey
* Correspondence: budan.ferenc@pte.hu (F.B.); duyguturkozu@gazi.edu.tr (D.A.)
† These authors contributed equally to this work.

Abstract

Cancer continues to be a global health problem due to high mortality rates and resistance to treatment. Since conventional chemotherapies cause serious side effects, interest in natural complementary therapies has increased. In this context, common fig (*Ficus carica* L.) (*F. carica*), which stands out with its rich phytochemical content, has been used in traditional medicine for a long time and attracts attention with its anticancer potential. The purpose of this review is to evaluate the biological effects of extracts obtained from different parts of the *F. carica* plant on cancer cells. Recent in vitro studies have shown that *F. carica* extracts suppress proliferation, induce apoptosis and reduce oxidative stress in various cancer cell lines. However, factors such as the plant part used, extraction method, dose and application time have caused differences in the results. In vivo studies are limited and there is no clinical study. Some studies report that high doses, especially latex, may cause toxic effects. *F. carica* extracts are promising against cancer. However, comprehensive in vivo and clinical studies with standardized extracts are needed to transfer this potential to clinical practice.

Keywords: *Ficus carica*; anticancer activity; bioactive compounds; plant extracts; phytochemicals

1. Introduction

Cancer is characterized by unregulated cell proliferation and remains the second leading cause of death worldwide. Globally, an estimated 20 million new cancer cases and 9.7 million deaths from cancer were reported in 2022 [1]. The cancer burden is projected to increase by approximately 77% by 2050, further straining healthcare systems, individuals, and communities [1,2]. One of the main goal in the field of cancer treatment is to formulate innovative approaches that specifically target malignant cells while sparing healthy tissues and reducing the development of resistance to chemotherapeutic compounds. Effective anticancer drugs are expected to exert their effects through mechanisms such as induction of apoptosis, inhibition of cell proliferation, promotion of cellular differentiation, suppression of DNA topoisomerase function, and modulation of intracellular signaling pathways.

 https://doi.org/10.3390/ijms27010056

However, despite significant medical and technological advances, drugs used in traditional cancer treatments still face, in many cases, serious limitations, such as unacceptable toxicity, development of resistance, and the risk of targeting healthy cells and causing adverse effects [3,4]. Therefore, studies investigating the anticancer potential of naturally occurring compounds are gaining increasing importance. These limitations of conventional cancer therapies have intensified interest in exploring naturally occurring compounds as complementary or alternative strategies. Many plant-derived molecules interact with key oncogenic pathways—such as proliferation control, apoptosis induction, and oxidative stress regulation—suggesting that phytochemicals may help bridge current therapeutic gaps and support the development of safer, more targeted anticancer approaches [5,6]. In this context, plants that have been used in traditional medicine for a long time and have a rich pharmacological profile are attracting attention as alternative therapeutic agents. One of these plants, the Ficus genus, is an important group that stands out with its medicinal properties, and one of the most well-known species of this genus is the common fig (*Ficus carica L.)*, which is especially valued for its nutritious fruit [7,8]. *Ficus carica* L. (the common fig) has a long history of use in traditional medicine in the Mediterranean basin, the Middle East, North Africa, and parts of South Asia and East Africa. Different plant parts are used depending on tradition. The fruits and seeds are eaten for their nutritional and digestive benefits, while leaf decoctions are used for glycemic control and gastrointestinal complaints. The latex is applied topically for warts and other skin conditions. Ethnobotanical studies and pharmacological research document the use of the leaves, fruit, latex, bark, and roots for conditions such as digestive disorders, diabetes, skin conditions, infections, and inflammation. Examples include the traditional topical use of fig latex for verruca vulgaris and the use of fig leaf decoctions in experimental and clinical glycemic control studies [9–11]. This species, which naturally spreads from the intersection of Asia and Europe to Northern India, is commercially cultivated in many regions, especially in the Mediterranean countries, thanks to its high adaptability to hot climate conditions. However, its production is limited to certain regions due to its special pollination and drying requirements [12]. The fig is considered a quality fruit due to its high nutritional value. Beyond its nutritional benefits, *F. carica* serves as a rich source of polyphenols, flavonoids, and other bioactive compounds [13]. Minerals, vitamins, carbohydrates, phenolic compounds, organic acids and various sugars have been determined to be present in high amounts in dried fruits of *F. carica*. Fiber and polyphenol contents are remarkable in both fresh and dried fruits. Phytochemical analyses show that *F. carica* is a rich source not only of polyphenols and flavonoids, but also of many biologically active compounds such as β-amyrins, arabinose, β-carotene, β-sitosterols and xanthoxol, as well as various flavonoids and phenolic glycosides (including rutin and quercetin derivatives) [14,15]. Bioactive substances isolated from different parts of the plant include phenolic compounds, organic acids, phytosterols, coumarins, triterpenoids, aliphatic alcohols, and hydrocarbons. In addition to these groups, a number of volatile compounds and other secondary metabolites have been reported in *F. carica* cultivars. This richness of content is an important indicator supporting the pharmacological potential of the plant. As summarized in the review by Hajam and Saleem (2022) [16], several studies have reported that phenolic compounds, coumarins, flavonoids, and terpenoids isolated from *F. carica* may exert anticancer effects by inhibiting cancer cell proliferation, inducing apoptosis through mitochondrial pathways, and modulating oxidative stress. These conclusions are based on in vitro findings from human cancer cell lines, including breast, colorectal, and hepatocellular carcinoma models, in which isolated components such as psoralen, bergapten, and quercetin derivatives demonstrated cytotoxic or pro-apoptotic activity [16]. However, the current scientific evidence on these potential biological effects of *F. carica* is still limited. Comprehensive and systematic studies are

https://doi.org/10.3390/ijms27010056

needed to more clearly reveal the protective or therapeutic roles of different components of the plant on cancer development.

The aim of this review is to evaluate current scientific studies investigating the anticancer potential of extracts obtained from various parts of the *F. carica* plant. In line with the studies conducted in recent years, it has been reported that bioactive compounds obtained from different parts of the plant, such as fruit, leaf, latex, bark, and seed, have various biological effects on cancer cells through mechanisms such as inhibiting cell proliferation, inducing apoptosis, reducing oxidative stress and modulating tumor cell-specific signaling pathways [17–19]. In this context, in addition to variables such as plant part, extraction method, solvent, dose, and application time used in the review, the effects of the cancer cell lines used on the results were evaluated comparatively.

2. Bioactive Compounds of *F. Carica*

Phytochemical studies on *F. carica* have revealed that many biologically active compounds are present in various parts of the plant (latex, leaves, fruits, and roots). These compounds include phytosterols, flavonoids (including anthocyanins), organic acids, aliphatic alcohols and hydrocarbons. When examined in terms of organic acids, malic, citric, oxalic, shikimic and fumaric acids were detected, especially in the leaf extract, along with high amounts of quinic acid. In terms of flavonoid content, luteolin stands out as the main compound in the leaves, followed by quercetin and biochanin A. The antioxidant, anti-inflammatory and anticarcinogenic effects of these compounds have been reported in the literature. Phenolic compounds are found in all parts of the fruit; phenols such as caffeoylquinic acid isomers, quercetin derivatives, ferulic acid, psoralen, and bergapten were detected especially intensely in the leaves. Among phytosterols, β-sitosterol was found in the highest amount, while other sterols such as lupeol, α-amyrin, and betulol were also identified [14,16,20,21] (Tables 1 and 2). In addition, the content of five different polyphenols (-epicatechin, chlorogenic acid, syringic acid, psoralen, and rutin) was determined in *F. carica* seeds [22].

Table 1. Phytochemical content of *F. carica* according to its parts [14,16,23].

Compound Type	Fruits	Leaves	Latex
Flavanols	Quercetin, catechin, rutin, epicatechin, gallocatechin, kaempferol	Epicatechin, catechin, taxifolin, kaempferol, rutin, quercetin	-
Flavones	Chrysin, apigenin, galangin	Isoorientin, orientin, apigenin, luteolin, vitexin, biochanin A/B,	-
Flavanones	Pinostrobin, pineocembrin	Hesperidin, naringenin	-
Anthocyanins	(Epi)catechin-(4 → 8) cyanidin-3,5-diglucoside, cyanidin-3 glucoside/rutinoside	-	-
Coumarins	Scopoletin, bergapten, psoralen, umbelliferone	Psoralen, umbelliferone	-

https://doi.org/10.3390/ijms27010056

Table 1. *Cont.*

Compound Type	Fruits	Leaves	Latex
Monoterpenes	Menthol, α-pinene, β-pinene, limonene, inasole	Menthol, limonene	Limonene, α-thujene, α-pinene, β-pinene
Sesquiterpenes	α-Caryophyllene, β-copaene, α-zingiberene, β-caryophyllene, (E)-β-bergamotene, β-thujone	α-Cubebene, α-zingiberene, β-caryophyllene, α-humulene, β-eudesmol, α-bourbonene	-
Triterpenes/Tetraterpenes	β-carotene, zeaxanthin, α-carotene, β-cryptoxanthin, lutein	α-Taraxasterol acetate, bauerenol, lupeol acetate	Lupeol, β-amyrin, α-amyrin, lanosterol
Hydroxybenzoic Acids	Gallic acid	Gallic acid, syringic acid, protocatechuic acid	-
Hydroxycinnamic Acids	p-Coumaric, ferulic, chlorogenic, cinnamic	p-Coumaric, ferulic, chlorogenic	-

Table 2. Phytosterol and organic acid content of *F. carica* according to its parts [14,23].

Compound Type	Fruits	Leaves	Latex
Phytosterols	β-Sitosterol, stigmasterol, campesterol	β-Sitosterol	β-Sitosterol
Organic Acids	-	Citric, oxalic, fumaric, malic, and quinic acid	-

3. Effects of *F. carica* on Cancer and Its Mechanism of Action

Studies investigating the anticancer effects of *F. carica* have revealed the potential therapeutic value of various extracts of this plant against different types of cancer. These studies, which were mainly conducted in in vitro and in vivo model systems, have revealed that extracts obtained from different parts of *F. carica*, such as fruit, leaves, latex and bark, exhibit various biological effects such as cytotoxicity, induction of apoptosis, cell cycle arrest, reduction in cell viability, suppression of migration and invasion [24–27]. It has been observed that factors such as the extraction methods used, type of solvent, dose intervals, and the exposure times are determinants of the biological effect. In addition, these studies report that the effects of *F. carica* mostly occur through mechanisms such as ROS production, disruption of mitochondrial membrane potential, caspase activation, p53 regulation and modulation of intracellular signaling pathways. However, some studies have also reported dose-dependent or cell line-specific ineffectiveness. Current evidence suggests that *F. carica* extracts exhibit significant cytotoxic and apoptotic effects, particularly on breast, liver, colon, and cervix cancer cell lines. Tables 3 and 4 summarizes the findings from studies investigating the anticancer effects of *F. carica*.

https://doi.org/10.3390/ijms27010056

Table 3. The potential anticancer effects and underlying mechanisms of *F. carica* based on in vitro research.

Type of Cancer	Cell Line	Part of *F. carica*	Dosage and Exposure Time	(+) Effects	No Effect/ (−) Effects	Mechanisms	References
Breast	AMJ13	Fruit extract (n-hexane) 2-benzhydrylsulfinyl-N-hydroxyacetamide-Na isolation (methanol solvent)	100, 50, 25, 12,5, 6,25 μg/mL 72 h.	↓ cell proliferation, ↓ cell viability, ↑ apoptosis	Not specified	↓ MMP, ↓ BCL-2, ↑ BAX, ↑ P53, ↑ caspase-8 ↑ caspase-9	[24]
Breast	MCF7	Leaf extract (ethanol solvent)	5000, 2500, 1250, 625, 312.5, and 156 μg/mL 48 h.	↓ cell viability, ↑ peripheral blood mononuclear cells	Not specified	↑ ROS	[25]
Breast	MDA-MB-231	Latex extract (no specified)	%0.1, %0.25, %0.5 ve %1 24, 48, and 72 h	↑ apoptosis, ↓ cell proliferation, ↓ cell viability, ↓ cell migration, ↑ cytotoxicity and genotoxicity	Not specified	↓ ERK, ↓ CREB, ↓ AKT, ↓ GSK-3α/β, ↓ AMPK ↑ DNA damage	[28]
Breast	MCF7	Latex extract (ethanol solvent)	5 -400 μg/mL 4 h.	↑ cytotoxicity ↓ cell viability	healthy CCD45-SK cells cytotoxicity ↑	Not specified	[19]
Breast	MDA-MB-231	Leaf extract nanoparticles	50 μ M 24 h.	↑ cytotoxicity ↑ anti-tumor activity	Not specified	Not specified	[29]
Breast	MCF7	Fruit extract (methanol solvent)	500 μg mL^{-1} 24–72 h.	↑ cytotoxicity ↑ apoptosis	Not specified	Not specified	[30]
Breast	MCF7	Leaf extract (ethanol solvent)	8 mg/mL Not specified	↑ cytotoxicity	Not specified	Not specified	[31]
Breast	MCF7 MDA-MB231 MDA-MB436	Leaf extract (ethanol solvent)	From 6.25 to 300 μg/mL 72 h.	↑ cytotoxicity ↓ cell viability	Not specified	↑ Caspase-3	[32]
Breast	MCF7	Wood bark extract (asetone solvent)	From 5 to 100 μg/mL 4 h.	↑ cytotoxicity ↓ cell viability	>100 μg/mL healthy CCD45-SK cells viability ↓	Not specified	[33]

https://doi.org/10.3390/ijms27010056

Table 3. *Cont.*

Type of Cancer	Cell Line	Part of *F. carica*	Dosage and Exposure Time	(+) Effects	No Effect/ (−) Effects	Mechanisms	References
Breast	MCF7 MDA-MB231	Latex extract (Not specified)	1/1500–1/12 48 h.	↑ cytotoxicity ↓ cell viability	Not specified	Not specified	[34]
Colon	CaCo2	Leaf extract (ethanol solvent)	5000, 2500, 1250, 625, 312.5, and 156 μg/mL 48 h.	↓ cell viability ↑ peripheral blood mononuclear cells	Not specified	↑ ROS	[25]
Colon	HT29	Latex extract (pure water phosphate-buffered saline solvent)	10, 20, 25, 30, 40, 50, 75, 100 μg/mL 24, 48, and 72 h	↑ cytotoxicity ↑ apoptosis	↔ cytotoxicity (10 μg/mL) ↔ apoptosis (40 μg/mL 24 h.)	↑ Caspase-3/7	[35]
Colon	HCT116	Latex extract (ethanol solvent)	5 -400 μg/mL 4 h.	↑ cytotoxicity ↓ cell viability	healthy CCD45-SK cells cytotoxicity ↑	Not specified	[19]
Colon	C26	Leaf extract (petroleum ether solvent) Bergapten isolation Cubic nanoparticles	10 μM and 100 μM 48 snd 72 h. 22 days.	↑ cytotoxicity ↓ cell viability ↓ tumor growth	Not specified	Inhibit P-glycoprotein	[36]
Colon	CaCo2	Leaf extract (ethanol solvent)	8 mg/mL Not specified	Not found	↔ cytotoxicity	Not specified	[31]
Colon	HT29	Fruit extract (n-hexane, dichloromethane, methanol, ethanol, Water)	1/10–1/100,000 (3330 mg/mL stock solution) 24–72 h.	↑ cytotoxicity ↑ apoptosis ↑ Cell Cycle Arrest ↓ cell viability ↓ cell proliferation	Not specified	Not specified	[37]
Colon	CaCo2	Latex (Cellulose acetate/polyethylene oxide nanofiber)	1, 5, 10, 15, 30, 60 ve 120 μg/mL 24 h.	↑ cytotoxicity ↑ apoptosis	↓ TP53, ↓ TNF-α, ↑ Bcl2 ↓ P21(10 μg/mL)	↑ P21	[17]
Colon	HT116	Wood bark extract (asetone solvent)	From 5 to 100 μg/mL 4 h.	↑ cytotoxicity ↓ cell viability	>100 μg/mL healthy CCD45-SK cells viability ↓	Not specified	[33]

https://doi.org/10.3390/ijms27010056

Table 3. *Cont.*

Type of Cancer	Cell Line	Part of *F. carica*	Dosage and Exposure Time	(+) Effects	No Effect/ (−) Effects	Mechanisms	References
Colon	HT29	Latex extract (methanol solvent)	0,5–2,5 mg/mL 48 h.	↓ cell viability ↑ anti-colonization	Not specified	Not specified	[38]
Liver	HepG2 NIH3T3	Latex extract (ethyl acetate,n-hexane, chloroform, and methanol solvents)	0.6, 1.25, 2.5, 5 and 10 mg/mL 24, 48, and 72 h.	↑ cytotoxic effect ((n-hexane, chloroform, ethyl acetate form) ↑ apoptosis (chloroform)	↑ growth stimulation activity(methanol form) ↔ cell death	Not specified	[26]
Liver	HepG2	Leaf extract (ethanol solvent)	5000, 2500, 1250, 625, 312.5, and 156 μg/mL 48 h.	↓ cell viability ↑ peripheral blood mononuclear cells	↔ cell viability (156 μg/mL)	↑ ROS	[25]
Liver	HepG2	Leaf extract (asetone solvent)	0.125, 0.25, 0.5, 1, 2, and 4 mg/mL 24, 48, and 72 h	↑ cytotoxicity ↑ apoptosis ↓ cell proliferation ↑ cell cycle arrest ↑ anti-colonization	↑ CDK-10, ↓ TP53	↑ ROS, ↓ MMP, ↓ Bcl-2, ↓ CDK1/5/9, ↓ DNA synthesis	[39]
Liver	HepG2	Latex extract (ethanol solvent)	5 –400 μg/mL 4 h.	↑ cytotoxicity ↓ cell viability	healthy CCD45-SK cells cytotoxicity ↑	Not specified	[19]
Liver	HepG2	Fruit extract essential oils isolation Water and n-hexane solvent)	Not specified	↑ Cell Cycle Arrest (water form) ↑ apoptosis ↓ cell viability	↔Cell Cycle Arrest (n-hexane form) ↓ ROS	Not specified	[40]
Liver	HepG2	Fruit extract (methanol solvent)	500 μg mL^{-1} 24–72 h.	↑ cytotoxicity ↑ apoptosis	Not specified	Not specified	[30]
Liver	HepG2	Latex (Cellulose ac-etate/polyethylene oxide nanofiber)	1, 5, 10, 15, 30, 60 ve 120 μg/mL 24 h.	↑ cytotoxicity ↑ apoptosis	↓ TP53, ↓ TNF-α, ↑ Bcl2 ↓ P21(10 μg/mL)	↑ P21	[17]

https://doi.org/10.3390/ijms27010056

Table 3. *Cont.*

Type of Cancer	Cell Line	Part of *F. carica*	Dosage and Exposure Time	(+) Effects	No Effect/ (−) Effects	Mechanisms	References
Liver	HepG2	Wood bark extract (asetone solvent)	From 5 to 100 μg/mL 4 h.	↔ cell viability	>100 μg/mL healthy CCD45-SK cells viability ↓	Not specified	[33]
Cervical	HeLa	Fruit extract (ethanol solvent)	Not specified	Not found	↔ cytotoxic effect	Not specified	[27]
Cervical	HeLa HPV18+ CaSki HPV16+	Latex extract (DMSO solvent)	5, 10, 50, 100, and 200 μg/mL 72 h.	↑ apoptosis ↑ cytotoxicity ↓ Cell growth ↑ Cell Cycle Arrest	Not specified	Nonsense-Mediated Decay pathway (↑ RPS27A, RNF111, and RPS6) Cell Cycle pathway (↓ PCNA, POLD3, PRIM1, and ORC2) ↑ TP53, ↓ E6 and E7 oncoproteins ↓ CDK4, CDK1 and MAPK1	[41]
Cervical	HeLa(HPV18+) CaSki(HPV16+) C33A(HPV−)	Latex extract (DMSO solvent)	5, 10, 50, 100, and 200 μg/mL 72 h.	↑ cytotoxicity ↑ Cell Cycle Arrest ↓ Cell growth ↑ apoptosis	Not specified	Degradation and Antigen Presentation via the Ubiquitin-Proteasome and MHC-I Pathways (↑ RPS27A, RNF111, CUL5, FBXO4, FBXL4, and CALR) ↓ E5 oncoprotein	[42]
Cervical	HeLa	Latex extract (methanol solvent)	0.5–2.5 mg/mL 48 h.	↓ cell viability ↑ anti-colonization	Not specified	Not specified	[38]

https://doi.org/10.3390/ijms27010056

Table 3. *Cont.*

Type of Cancer	Cell Line	Part of *F. carica*	Dosage and Exposure Time	(+) Effects	No Effect/ (−) Effects	Mechanisms	References
Pancreas	Panc1 QGP1	Fruit extract (ethanol solvent)	150, 300, 600 µg/mL and 1.2 mg/mL	↑ cytotoxicity ↑ apoptosis ↑ cell death ↓ cell viability ↓ cell migration, ↓ metastasis and invasion ↑ anti-colonization ↑ cell senescence and autophagy	↔ Caspase 3, 7, 8 and 9	↓ PARP-1, ↓ N –kadherin, ↑ E-kadherin, ↑ ROS, ↓ Atg-5, ↓ Beclin-1 ve ↓ p62, ↑ LC3A/B, ↓ MMP, ↑ lipid peroxidation	[43]
Pancreas	Panc1 AsPC1	Fruit extract (ethyl acetate solvent)	0.1, 0.5, and 1 mg/mL 24 h.	↑ cytotoxicity	↔ cell proliferation ↑ cell migration	Not specified	[44]
Ovarian	SKOV3	Fruit extract (n-hexane) 2-benzhydrylsulfinyl-N-hydroxyacetamide-Na isolation (methanol solvent)	100, 50, 25, 12.5, 6.25 µg/mL 72 h.	↓ cell proliferation ↓ cell viability ↑ apoptosis	Not specified	↓ MMP, ↓ BCL-2, ↑ BAX, ↑ P53, ↑ caspase-8 ↑ caspase-9	[24]
Ovarian	SKOV3	Fruit extract (n-hexane) 2-(Benzhydryl sulfinyl)-N-sec-butylacetamide) isolation (methanol solvent)	25 µg/mL 60 min. 500 µg/kg	↑ bone marrow-derived macrophages phagocytic activity ↑ anti-tumor activity	Not specified	↑ Fcγ receptor expression ↑ ROS	[45]
Gastric	AGS	Fruit extract (methanol solvent)	20, 40, 60, 80, 100 mg/mL 24 h.	↑ cytotoxicity ↓ cell viability ↑ apoptosis	Not specified	↑ ROS, ↓ Glutathione, ↓ MMP, ↑ DNA damage	[18]

https://doi.org/10.3390/ijms27010056

Table 3. *Cont.*

Type of Cancer	Cell Line	Part of *F. carica*	Dosage and Exposure Time	(+) Effects	No Effect/ (−) Effects	Mechanisms	References
Gastric	AGS	Leaf extract (ethanol solvent)	8 mg/mL Not specified	↑ cytotoxicity	Not specified	Not specified	[31]
Osteosarcoma	MG63	Fruit extract (n-hexane, dichloromethane, methanol, ethanol, Water)	1/10–1/100,000 (3330 mg/mL stock solution) 24–72 h.	↑ cytotoxicity ↑ apoptosis ↓ cell viability ↓ cell proliferation	↔ Cell Cycle Arrest	Not specified	[37]
Lung	A549	Latex extract (Not specified)	1/1500–1/12 48 h.	↑ cytotoxicity ↓ cell viability	Not specified	Not specified	[34]
Prostate	PC3	Latex extract (pure water phosphate-buffered saline solvent)	10, 20, 25, 30, 40, 50, 75, 100 µg/mL 24, 48, and 72 h	↑ cytotoxicity	Not specified	↑ Caspase-3/7	[35]
Skin	B16F10	Leaf extract (methanol solvent)	150, 250, 350, 450, 550, 650, 750 ve 850 µg / mL 24 and 48 h.	↑ cytotoxicity ↑ apoptosis	Not specified	↑ p53, ↑ Bax, ↑ Caspase 3 and 9	[46]
Laryngeal	Hep2	Leaf extract (ethanol solvent)	5000, 2500, 1250, 625, 312.5, and 156 µg/mL 48 h.	↓ cell viability ↑ peripheral blood mononuclear cells	↔ cell viability (156 µg/mL)	↑ ROS	[25]

AGS: Adenocarcinoma Gastric Cell Line, **AKT**: Protein Kinase B, **AMPK**: AMP-activated Protein Kinase, **Atg-5**: Autophagy-Related 5, **Bax**: Bcl-2-associated X Protein, **Bcl-2**: B-cell Lymphoma 2, **CaCo2**: Human Colorectal Adenocarcinoma Cell Line, **CALR**: Calreticulin, **CaSki**: Cervical Cancer Cell Line (HPV16+), **CDK**: Cyclin-Dependent Kinase, **CREB**: cAMP Response Element-Binding Protein, **CUL5**: Cullin 5, **DNA**: Deoxyribonucleic Acid, **DMSO**: Dimethyl Sulfoxide, **E-cadherin**: Epithelial Cadherin, **E5/E6/E7**: Human Papillomavirus (HPV) Oncoproteins, **ERK**: Extracellular Signal-Regulated Kinase, **FBXL4**: F-box and Leucine-Rich Repeat Protein 4, **FBXO4**: F-box Only Protein 4, **Fcγ receptor**: Fc Gamma Receptor, **GSK-3α/β**: Glycogen Synthase Kinase 3 Alpha/Beta, **HeLa**: Human Cervical Cancer Cell Line, **Hep2**: Human Laryngeal Carcinoma Cell Line, **HepG2**: Human Hepatocellular Carcinoma Cell Line, **HPV**: Human Papillomavirus, **HT29**: Human Colorectal Adenocarcinoma Cell Line, **LC3A/B**: Microtubule-associated Protein 1A/1B-light Chain 3A/B, **MAPK1**: Mitogen-Activated Protein Kinase 1, **MCF7**: Human Breast Adenocarcinoma Cell Line, **MDA-MB-231/MDA-MB-436**: Triple-negative Breast Cancer Cell Lines, **MHC-I**: Major Histocompatibility Complex Class I, **MG63**: Human Osteosarcoma Cell Line, **MMP**: Mitochondrial Membrane Potential, **NIH3T3**: Mouse Embryonic Fibroblast Cell Line, **N-cadherin**: Neural Cadherin, **ORC2**: Origin Recognition Complex Subunit 2, **P21**: Cyclin-Dependent Kinase Inhibitor 1A, **PARP-1**: Poly(ADP-ribose) Polymerase 1, **PC3**: Human Prostate Cancer Cell Line, **PCNA**: Proliferating Cell Nuclear Antigen, **POLD3**: DNA Polymerase Delta 3, **PRIM1**: DNA Primase Subunit 1, **QGP1**: Human Pancreatic Neuroendocrine Tumor Cell Line, **RNF111**: Ring Finger Protein 111, **ROS**: Reactive Oxygen Species, **RPS27A**: Ribosomal Protein S27a, **SEC**: Secundum, **SKOV3**: Human Ovarian Cancer Cell Line, **TP53**: Tumor Protein p53, **TNF-α**: Tumor Necrosis Factor Alpha, **Ubiquitin-Proteasome**: Protein Degradation Pathway.

https://doi.org/10.3390/ijms27010056

Table 4. The potential anticancer effects and underlying mechanisms of F. carica based on in vivo research.

Animal Model	Plant Part/Extract	Dose & Duration	Evaluated Toxicity Parameters	Main Findings	References
Balb/c mice	Fruit extract (n-hexane) 2-benzhydrylsulfi-nyl-N-hydroxyacetamide-Na isolation	1 mg/kg (intraperitoneal) 4 weeks	The liver, kidney, lung, spleen histopathology and liver enzymes GOT, GPT, and ALP.	No significant changes and toxicity.	[24]
Syrian mice	Latex extract (chloroform, and methanol solvents)	1, 2 and 3 g/kg (intraperitoneal) 14 days	abdominal organs and liver enzymes AST, ALT, and ALP	3 deaths in 3 g/kg; no death in 1 and 2 g/kg. No significant changes and toxicity in 1 and 2 g/kg; significant changes and toxicity in 3 g/kg.	[26]
BALB/c mice	Fruit extract (ethanol sol-vent)	200 mg/kg) once a day (gavage) 7 days.	tumor growth and cytotoxicity on normal pancreatic cells	Tumor volume and tumor weight significantly lower than in the control group. No significant toxicity in normal pancreatic cells.	[43]
Mice	Leaf extract (Bergapten isolation Cubic nanoparticles)	10 µM–100 µM (intravenous) 22 days.	Anti-tumor activity	Inhibited tumor growth.	[36]
C57/BL6 mice	Fruit extract (n-hexane) 2-(Benzhydryl sulfi-nyl)-N-sec-butylacetamide) isolation	500 µg/kg (intraperitoneal)	Tumoricidal activity	Enhanced the functional activities of bone marrow-derived macrophages against tumor cells.	[45]
Wistar rats	Latex extract	0.5 mL (intratumorally injected) 4 weeks	tumor growth	Significant reduction in the tumor volume.	[47]
Albino Wistar rats	Morin	50 mg/kg/day (intragastric intubation) 30 days.	tumor multiplicity BAX/BCL2 ratio anti-inflammatory effects	Significantly reduced tumor multiplicity. Modulating BAX/BCL2 ratio. Regulating cytokines and eicosanoid. Attenuated NF-κB signaling in colon tumors	[48]

GOT: Glutamic Oxaloacetic Transaminase, GPT: Glutamic Pyruvic Transaminase, ALP: Alkaline Phosphatase, AST: Aspartat Aminotransferaz, ALT: Alanin Aminotransferaz.

https://doi.org/10.3390/ijms27010056

3.1. Breast Cancer

The activity of *F. carica* against breast cancer has been investigated using extracts obtained from its latex, fruit, leaves and bark (Table 3).

A recent study has shown that exposure to varying concentrations (1/1500, 1/900, 1/600, 1/300, 1/180, 1/60, 1/36 and 1/12) of fig latex resulted in a significant decrease in the survival of breast cancer cells, exhibiting a dose-dependent response. The latex showed remarkable cytotoxic activity against MCF7 and MDA-MB-231 breast cancer models, especially at the highest concentration tested (1/12). Cell viability after treatment was observed as 20.63% and 16.66%, respectively, with half-maximal inhibitory concentrations (IC50) measured at 1/40 and 1/45 [34]. In a different study, researchers characterized the phenolic composition of the latex extract, identifying the presence of luteolin, rutin, quercetin, chlorogenic acid, and caffeic acid. Treatment of MCF-7 cells with this extract resulted in a significant decrease in cell viability. The strongest cytotoxic effect was observed at a concentration of 400 μg/mL, which reduced cell viability to 14 ± 2% (IC50: 134.60 ± 13.52 μg/mL). The researchers attributed this effect to the hydrophilic and hydrophobic interactions between phenolic compounds and proteins expressed by cancer cells. However, the study also reported that latex extract exhibited cytotoxic effects on healthy human skin fibroblasts cells (CCD-45 SK) [19]. Another study reported that *F. carica* latex exhibited antiproliferative and antimetastatic effects in MDA-MB-231 breast cancer cells. The treatment dose-dependently reduced cell proliferation and migration, induced apoptosis and led to morphological changes. Cytotoxic and genotoxic effects were confirmed by DNA damage analyses, and it was determined that these effects occurred at the molecular level by downregulation of AMPKα, ERK2, CREB, and GSK-3α/β signaling pathways [28].

In a recent study, it was determined that *F. carica* leaf extract showed cytotoxic effects comparable to cisplatin in MCF-7, MDA-MB-231 and MDA-MB-436 breast cancer cells. This effect was reported to be related to caspase-3 activation by rutin, naringin and catechin compounds found in the extract. It was also reported that the extract showed lower toxicity against healthy peripheral blood mononuclear cells and was more selective against cancer cells [32]. A different study found that the leaf extract showed a significant inhibitory effect on cell viability in MCF-7 cells, while also exhibiting an increase in the percentage of cellular viability of peripheral blood mononuclear cells [25]. These differences from the findings reported by Yahiaoui et al. [19] are likely due to the distinct phytochemical compositions of latex and leaf extracts, which commonly lead to variations in biological activity rather than representing a true contradiction. In a study conducted by Farooq et al. [29], it was reported that nanoparticles obtained from *F. carica* leaf extract reduced cell viability in MDA-MB-231 breast cancer cells and increased the effectiveness of the chemotherapeutic drug integrated into these nanoparticles. Shiraishi et al. [31], reported that the leaf extract showed moderate cytotoxic effects in MCF-7 cells. In another study, it was observed that *F. carica* fruit extract promoted apoptosis in MCF-7 cells, but only had a mild cytotoxic effect [30]. In one study, 2-benzhydrylsulfinyl-N-hydroxyacetamide-Na compound found in *F. carica* fruit extract was identified and shown to trigger mitochondrial apoptosis in AMJ-13 breast cancer cells by increasing Bax, suppressing Bcl-2, as well as activating p53, caspase-8 and -9 via increasing ROS. The compound effectively inhibited cell proliferation and induced mitochondrial pathway-dependent apoptosis. Furthermore, no changes were observed in liver enzyme levels and lung and spleen tissues, and no significant toxicity was reported [24]. Finally, A recent study revealed that *F. carica* bark extract did not exhibit cytotoxic effects on healthy macrophage (RAW 264.7) and human skin fibroblast (CCD45-SK) cell lines, even at a high concentration of 400 μg/mL. However, the extract showed significant cytotoxic effect on MCF-7 breast cancer cells at lower concentration

https://doi.org/10.3390/ijms27010056

(IC50: 143.30 μg/mL), indicating its selective anticancer potential [33]. Ghandehari and Fatemi, in an in vivo experiment, demonstrated the anticancer effect of *F. carica* latex extract, decreasing tumor size and growth as well as inhibiting metastasis in breast tumor-bearing Wistar rats [47]. The mechanism of action of *F. carica* on breast cancer is summarized in Figure 1.

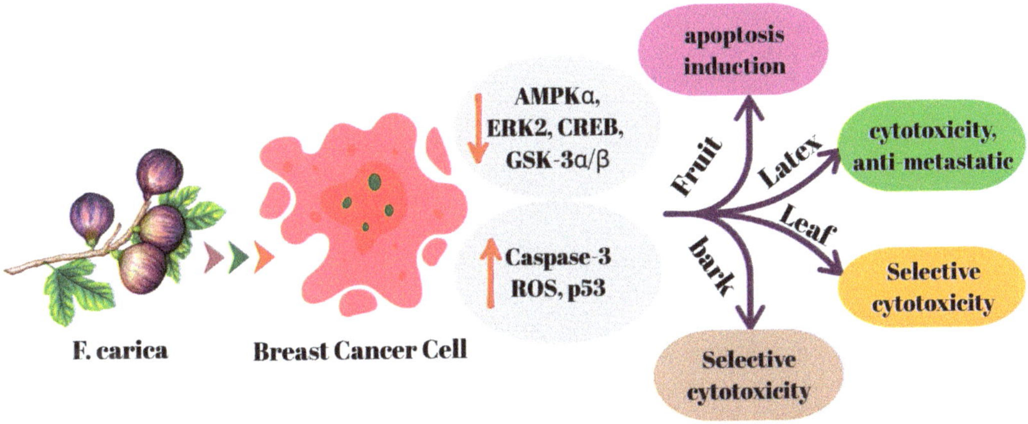

Figure 1. The mechanism of action of *F. carica* on breast cancer.

The evidence obtained indicates that *F. carica* has remarkable anticancer potential against breast cancer cells. Latex, leaf, fruit, and bark extracts have been reported to exhibit cytotoxic, antiproliferative, and apoptosis-inducing effects on different breast cancer cell lines. The latex modulates signaling pathways related to energy and proliferation, while leaf extracts stand out with caspase-3 activation and phytochemical selectivity. Compounds isolated from fruit extracts potently activate mitochondrial apoptosis, while the peel extract is notable for its low toxicity to healthy cells. Although the obtained findings are generally promising, it is understood that phytochemical differences between different types of extracts are decisive for the results. Therefore, before *F. carica* can be evaluated as a complementary agent in breast cancer treatment, extract standardization, toxicity profile, and mechanisms of action need to be investigated more comprehensively with advanced in vitro and in vivo studies.

3.2. Colon Cancer

The effects of *F. carica* against colon cancer have been evaluated in various studies with extracts from latex, fruit, leaves and bark of the plant.

A recent study investigated the impact of *F. carica* latex extracts on HT-29 colon cancer cells, revealing a dose- and time-dependent reduction in cell viability relative to the control group, both before and after digestion. Furthermore, both forms of the extract inhibited colony formation in HT-29 cells, though the undigested extracts exhibited lower IC50 values compared to their digested counterparts. In this study, the digested and undigested forms of F. carica latex extracts were compared using an in vitro gastrointestinal digestion model based on the standardized INFOGEST protocol [49]. Digestion was initiated by subjecting the extracts to sequential oral, gastric, and intestinal phases [38]. A study was conducted on HC-116 colon cancer cells to evaluate the cytotoxic effects of latex extracts from three different varieties of *F. carica*. The results indicated that the Aberkane and Aghanime varieties exhibited cytotoxic effects, while the El-bakour variety did not show any toxicity. Moreover, it was reported that all varieties exhibited elevated concentrations of cytotoxic

https://doi.org/10.3390/ijms27010056

activity against the healthy fibroblast cell line (CCD-45SK) [19]. Similarly, another study found that latex obtained from the Yellow Lop variety did not show cytotoxic effects at 10 μg/mL, but reduced HT-29 cell viability by 28.91% after 72 h at 20 μg/mL. Alternatively, latex from the Aydın Siyah variety at the same concentration caused 85.33% cell death. Furthermore, 40 μg/mL Aydın Black latex increased apoptosis by 10-fold, while Yellow Lop latex increased it by 2-fold, and this effect was shown to be related to caspase-3 and -7 activation [35]. The findings of these two studies reveal that *F. carica* variety is an important factor in terms of biological activity. The difference in activity between varieties may be due to differences in their phytochemical compositions. A recent study reported that nanofibers loaded with *F. carica* latex showed lower toxicity compared to 5-fluorouracil on normal cells (WI-38) and had significant cytotoxic effects on colon cancer cells (Caco-2) (IC50: 23.96 μg/mL). Molecular analysis revealed upregulation of *BCL-2* and downregulation of *P53* and *TNF*, indicating that this treatment could potentially weaken tumor suppressive mechanisms. However, increased *P21* gene expression suggests that the antiproliferative effect may be supported by cell cycle arrest [17].

In a study using the Caco-2 cell line, a significant inhibition of cell proliferation was observed when *F. carica* leaf extract was applied in the concentration range of 5000–156 μg/mL [25]. However, in another study using a similar cell line, no significant cytotoxic effect was detected at an extract concentration of 8 mg/mL [31]. Despite using a comparable cell line and applying concentrations within a similar range, the discrepancies in results may stem from variations in the specific *F. carica* cultivar used, differences in incubation periods, or unavoidable phytochemical degradation occurring between leaf collection and extract preparation. Abdou et al. [36] evaluated the in vitro and in vivo cytotoxic effects of bergapten isolated from *F. carica* leaves on C-26 colon cancer cells by loading it into cubosomes. Bergapten-loaded cubosomes exhibited stronger cytotoxicity than free bergapten (IC50 values of 0.015 ± 0.001 μM and 0.062 ± 0.003 μM, respectively, for 72 h incubation), while empty cubosomes did not show toxicity. Additionally, significant reductions in tumor volume and weight were observed in mice treated with bergapten cubosomes in in vivo experiments [36]. A significant cytotoxic effect was observed in HT-29 cells treated with *F. carica* fruit extract, depending on the dose and time. The increase in lactate dehydrogenase (LDH) levels in the culture medium suggested that this cytotoxic effect may be due to the disruption of cell membrane integrity. Furthermore, the researchers reported that the extract arrested HT-29 cells in S phase ($p < 0.007$) and that apoptosis rates increased 12-fold in the early phase and 1.5-fold in total after treatment [37]. In a study in which HT-116 cell lines were treated with *F. carica* bark extract, only a negligible cytotoxic effect was observed even at the highest concentration (400 μg/mL), but this varied among different cultivars [33]. According to Sharma et al. the morin flavonoid content of *F. carica* inhibited the 1,2-dimethylhydrazine-induced colon cancer in Wistar rats by downregulating the inflammatory and cell proliferation-stimulating NF-κB pathway, including interleukin 6 (IL-6), TNF-α, prostaglandin (PGE-2), and cyclooxygenase 2 (COX-2). Furthermore, it increased the BAX/BCL2 ratio, favoring the apoptotic pathway [48]. Hence, morin downregulates both Pumilio-1 (PUM1) and CD133, leading to inhibition of Cancer Stem Cells (CSCs) that may remain after colon cancer treatment [50]. The potential effects of *F. carica* on colon cancer are summarized in Figure 2.

Various studies on colon cancer cells have demonstrated that extracts from different parts of *F. carica* exhibit significant effects, which are dependent on both dose and plant variety. In particular, it appears that the digestive process can weaken the biological activity of extracts, while differences in plant variety can significantly alter toxicity levels. Furthermore, the use of nanotechnological formulations such as cubosomes and nanofibers has been reported to increase the efficacy of extracts while potentially reducing toxicity to

https://doi.org/10.3390/ijms27010056

healthy cells. However, the inconsistencies in the levels of efficacy observed in different studies are due to differences in the *F. carica* varieties used, extract preparation methods, and experimental conditions. This suggests that the potential of *F. carica* in the treatment of colon cancer depends on the development of standardized extract preparation and dosing protocols. Overall, the data support the existence of anticancer effects of the plant, but its efficacy needs to be optimized and its mechanisms investigated in more detail.

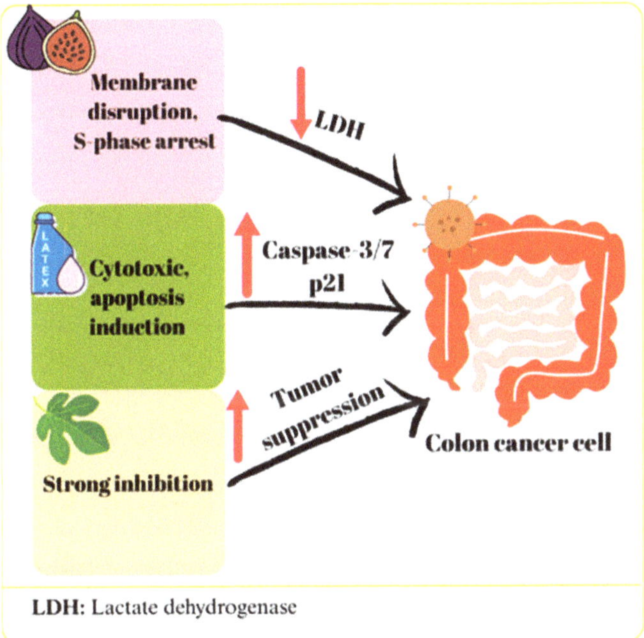

Figure 2. The mechanism of action of *F. carica* on colon cancer.

3.3. Liver Cancer

The therapeutic capacity of *F. carica* against liver cancer has been addressed in studies evaluating extracts from various tissue parts, including latex, fruit, leaves, and bark.

In a study, *F. carica* latex-loaded nanofiber was determined to have cytotoxic effects on liver cancer cells (HepG2), and the IC50 value was determined as 23.97 μg/mL [17]. Similarly, in HepG2 hepatocellular carcinoma cells, exposure to the highest tested concentration (100 μg/mL) of latex from the three *F. carica* cultivars (Aberkanei, Aghanime, and El-Bakour) resulted in a marked decrease in cell viability, with the highest IC50 value reported as 19.40 ± 2.69 μg/mL [19]. In another study, *F. carica* latex was tested with four different solvents in HepG2 cells. The methanol fraction exhibited no cytotoxic activity and promoted cell growth, while the other fractions showed toxic effects. The strongest effect was observed in the chloroform fraction with an IC50 value of 0.2 ± 0.02 mg/mL and was reported to induce apoptosis. This effect was attributed to lupeol acetate and lupeol palmitate identified in the fraction [26]. Mustafa et al. [39] reported that *F. carica* leaf extract produced dose-dependent cytotoxic effects on HepG2 cells (IC50: 0.081 mg/mL), whereas it did not show significant toxicity up to 2 mg/mL in healthy HUVEC cells. While significant morphological deteriorations were observed in treated HepG2 cells, HUVECs maintained their normal appearance. These differences suggested that the extract exhibited selective apoptotic effects. The researchers stated that this effect was associated with increased ROS production, DNA damage, disruption of mitochondrial membrane potential, suppression

https://doi.org/10.3390/ijms27010056

of Bcl-2, and downregulation of CDK-1, CDK-5, and CDK-9 [39]. Similarly, Abdel-Rahman et al. [25] reported that leaf extract showed a significant effect in HepG cells with a 66.9% inhibition rate [25]. A study reported that treatment of HepG2 cells with *F. carica* dried fruit extract provided moderate cytotoxic effects [30]. In one study, the anticancer effects of *F. carica* fruit essential oils extracted using a hot distillation procedure with water (FCFW) and hexane (FCFH) were investigated in HepG2 cells. It was determined that both extracts showed moderate cytotoxicity; FCFW arrested the cells in S phase, while FCFH did not affect the cell cycle. In the apoptosis analysis, FCFH increased early apoptosis (43.1%), while FCFW increased late apoptosis (30.2%) [40]. In addition, it was observed that both extracts reduced ROS production, which contradicts what was reported by Mustafa et al. (2021) [39]. This discrepancy indicates differences in the anticancer mechanisms of fruit and leaf extracts and suggests that ROS reduction may be a different pathway from classical chemotherapeutic mechanisms [40]. Finally, *F. carica* wood bark extract showed a weak cytotoxic effect against HepG-2 cells (IC50: ≥100 µg/mL) [33]. The potential effects of *F. carica* on liver cancer are summarized in Figure 3.

In general, different tissue extracts of *F. carica* show anticancer potential in HepG2 cells. Notably, noteworthy that nanofiber formulations increase efficacy while reducing toxicity to healthy cells. The findings vary depending on the tissue used, extraction method, and test conditions. Therefore, standardization of extracts, isolation of active components, and more comprehensive studies at the mechanistic level are required to fully demonstrate the therapeutic potential of F. carica against liver cancer.

3.4. *Cervical Cancer*

The effects of *F. carica* on cervical cancer were evaluated through studies performed with latex and fruit-derived extracts of the plant.

A study investigated the anticancer activity of digested (in vitro) and undigested *F. carica* latex (white and black varieties) in the HeLa cell line. Although in vitro gastrointestinal digestion was observed to cause a decrease in anticancer properties, all calculated % viability values were reported to be lower compared to the control groups [38]. In two similar studies, the effects of *F. carica* latex on HPV16+ CaSki, HPV18+ HeLa, and HPV-negative C33A cervical cancer cell lines were evaluated. IC50 values were determined as 106 µg/mL for HeLa, 110 µg/mL for CaSki, and 108 µg/mL for C33A, and no toxicity was observed in healthy HCKT1 cells. Latex application led to upregulation of genes related to immune response, RNA processing and protein degradation, especially in HeLa and CaSki cells, while it activated apoptosis pathways in C33A cells. In addition, it was determined that MHC-I expression and TP53-mediated pathways increased, and the immune escape effect of E5 protein decreased. It has been reported that latex affects cell cycle and immune responses through regulatory molecules such as TAF1, CDK1/4, and MAPK1 [41,42]. In a different study, it was suggested that *F. carica* fruit extract did not exhibit any cytotoxic effect on HeLa cancer cells (IC50: >1000 µg/mL). Furthermore, when combined with olive oil, it diminished the cytotoxic effect of olive oil [27]. The potential effects of *F. carica* on cervical cancer are summarized in Figure 4.

Current studies on cervical cancer cells have shown that latex extract the latex extract has a unique activity profile, specifically targeting genetic and immune response mechanisms. In addition, its selective effect is supported by the fact that it does not cause toxicity in healthy cells. However, no cytotoxic effect was observed with the fruit extract. These findings show that the potential of *F. carica* varies depending on the part of the plant and the method of application. Further in vivo and clinical studies are needed to confirm the efficacy and clarify the mechanisms.

 https://doi.org/10.3390/ijms27010056

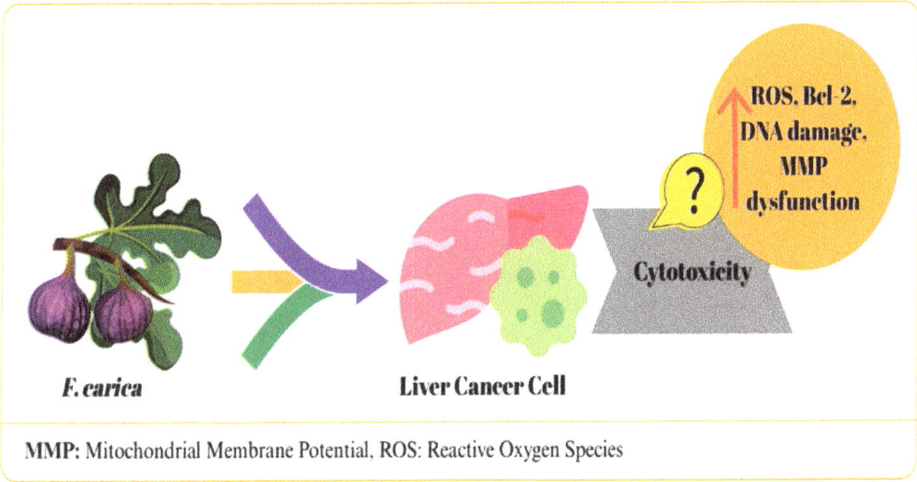

Figure 3. The mechanism of action of *F. carica* on liver cancer.

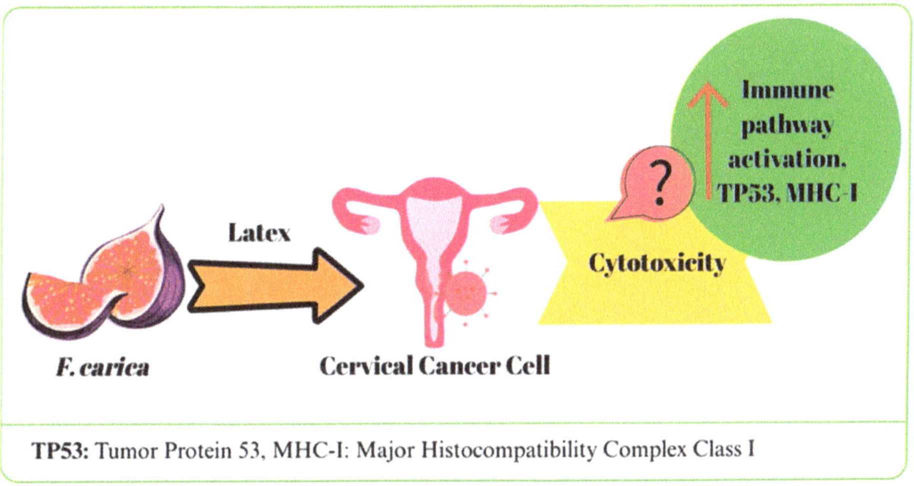

Figure 4. The mechanism of action of *F. carica* on cervical cancer.

3.5. Other Types of Cancer

The fruit extract of *F. carica* has been the subject of investigation in the context of pancreatic, ovarian, gastric, and osteosarcoma cancer cell lines.

In one study, the extract (0.1, 0.5, 1.0 mg/mL for 24 h) exhibited low toxicity in Panc-1 and AsPC-1 pancreatic cancer cells, with no significant antiproliferative effect. While the extract slightly reduced cancer cell viability, it did not substantially impact cell proliferation [44]. Conversely, another study reported that *F. carica* fruit extract exhibited dose- and time-dependent cytotoxic effects on pancreatic cancer cells. The 600 μg/mL extract caused a 60% and 70% decrease in viability in Panc-1 and QGP-1 cells, respectively. Apoptosis was confirmed by increased PARP-1 activation, and cell migration and invasion were suppressed. The extract induced cell death by increasing ROS production, while simultaneously balancing this effect by stimulating autophagy. The decrease in mitochondrial membrane potential and the increase in lipid peroxidation suggested the potential

 https://doi.org/10.3390/ijms27010056

for ferroptosis. In vivo experiments, oral administration of 200 mg/kg extract significantly suppressed tumor growth in mice with QGP-1-induced tumors; increased PARP-1 and LC3A/B expression indicated a collaborative role for apoptosis and autophagy. No toxicity was observed in normal pancreatic cells [43]. The observed variations in results between the two studies may be due to differences in extract composition, solvent choice, preparation techniques, cell models, administered concentrations, exposure duration, and overall experimental preparation. Notably, the extract's impact on cancer cells seems influenced by both concentration and treatment duration.

Ali et al. [45] reported that the 2-(benzhydrylsulfinyl)-N-sec-butylacetamide compound isolated from *F. carica* fruit promoted macrophage activation in SKOV-3 ovarian cancer cells by increasing Fcγ receptor expression. When co-administered with trastuzumab, the compound enhanced phagocytic activity and exerted in vivo tumoricidal effects through the induction of reactive oxygen species (ROS) generation and mitochondrial activation. Similarly, 2-benzhydrylsulfinyl-N-hydroxyacetamide-Na compound derived from the same plant showed dose-dependent suppression of proliferation and induction of apoptosis in SKOV-3 cells. This effect was associated with downregulation of BCL-2, activation of pro-apoptotic factors, and cytochrome c release due to a decrease in mitochondrial membrane potential [24].

A recent study reported that *F. carica* fruit extract exhibited cytotoxic effects in gastric cancer cells (AGS) in a dose-dependent manner by both increasing ROS production and via apoptosis and DNA damage [18]. Another study reported that the fruit extract promoted cell death in osteosarcoma cancer cells (MG-63) in proportion to dose and time by disrupting the cell membrane structure, reducing mitochondrial activity and DNA damage, although no significant changes were observed in any cell cycle stage [37].

F. carica latex extract was investigated for its cytotoxic effects on prostate and lung cancer cell lines. Aysin et al. [34] reported that latex extract dose-dependently reduced cell viability in A549 lung cancer cells, and the IC50 value was 1/26 dilution [34]. The same observation was documented by Baohong et al., namely that the alcoholic extract of fig fruit latex in A549 xenograft mice significantly inhibited the tumor growth without inducing obvious damage to the liver or kidney normal mouse tissues [51]. Similarly, in another study, the effects of latex obtained from different fig varieties on PC-3 prostate cancer cell line were evaluated; while viability decreased from 84.7% to 19.1% in Sarı Lop variety, this rate decreased from 80.6% to 2.0% in Black variety. These findings indicate that *F. carica* latex has significant cytotoxic potential against both cancer types [35].

The leaf extract of *F. carica* has been the subject of investigation in the context of skin and laryngeal cancer cell lines. A study demonstrated that *F. carica* leaf extract exhibits cytotoxic effects on melanoma cancer cells (B16F10). Notably, the extract decreased cell viability by 50% at 750 μg/mL after 24 h of incubation and 650 μg/mL after 48 h. Nonetheless, these concentration levels did not result in notable toxicity in normal HEK-293 epithelial cells. Furthermore, findings from the study suggested that the extract induced programmed cell death in melanoma cells via the mitochondrial pathway, leading to an upregulation of *caspase-3*, *caspase-9*, *P53*, and *BAX* gene expression. Notably, at elevated doses (650–750 μg/mL), the expression of apoptosis-associated genes exhibited a substantial increase compared to the control reference gene [46]. Additionally, research by Abdel-Rahman et al. [25] provided evidence of the cytotoxic impact of *F. carica* leaf extract on laryngeal cancer cell lines [25].

In pancreatic cancer models, *F. carica* extract-induced autophagy appears to be regulated via the mTOR/AMPK axis, as suggested by concurrent PARP-1 and LC3A/B upregulation. However, upstream regulators such as Nrf2 and MAPK remain to be investigated [6].

https://doi.org/10.3390/ijms27010056

3.6. Comparative Summary Across Cancer Types

Across different cancer types, *F. carica* extracts commonly induce mitochondrial-mediated apoptosis through ROS generation, P53 activation, and caspase signaling. Latex and leaf extracts show similar molecular patterns, whereas fruit extracts exert milder and more variable responses. When comparing the extracts, latex generally demonstrates the strongest cytotoxic and pro-apoptotic activity but also presents a greater risk of toxicity in normal cells. Leaf extracts exhibit strong anticancer potential with better selectivity, likely due to their high polyphenolic content (quercetin, luteolin, rutin). Fruit extracts tend to show moderate effects, possibly due to lower concentrations of triterpenes and coumarins. These findings suggest that leaf-based extracts may offer the most favorable therapeutic profile, while latex requires further safety validation.

3.7. Contradictions and Gaps

Several studies have reported opposing trends regarding ROS modulation, indicating both antioxidant and pro-oxidant behavior depending on extract composition and dose. This duality highlights the need for standardized extraction methods and comparative mechanistic assays to reconcile the discrepancies observed across studies.

A concise summary of the comparative properties of different *F. carica* extracts is presented in Table 5.

Table 5. Summary of the comparative properties of different *F. carica* extracts.

Extract Type	Typical Mechanisms	Efficacy	Toxicity Profile	References
Latex	ROS ↑, Caspase ↑, p53 ↑	Strong	Moderate–High	[19,26,28,41]
Leaf	ROS ↑/↓, Bcl-2 ↓, Caspase ↑	Strong	Low	[25,32,39]
Fruit	Moderate apoptosis, variable ROS	Moderate	Low–Moderate	[18,24,43]
Bark	Limited data	Weak	Low	[33]

4. Safety

The safety of *F. carica* and its extracts remains an important issue in the evaluation of their potential therapeutic applications. In popular medicine for several hundred years, the fruits of *F. carica* have been used for laxative, nose bleeding, or skin problems, e.g., leprosy, or as an emollient, demulcent. Also, it was utilized as an antipyretic, aphrodisiac, lithontriptic, or painkiller. The latex of *F. carica* has been employed as an expectorant, diuretic, and anthelmintic. The leaves have been utilized to treat diabetes, among others [52]. Remarkably, *F. carica* exhibits a diverse array of advantageous effects, including, but not limited to, antioxidant, anti-neurodegenerative, antimicrobial, antiviral, anti-inflammatory, anti-arthritic, antiepileptic, anticonvulsant, anti-hyperlipidemic, anti-angiogenic, antidiabetic, anticancer, and antimutagenic properties [53]. In summary, the multifarious utilization could be underpinned by the anti-inflammatory effects of different plant parts, as demonstrated in this study. Hence, the mentioned indications probably determined the categories of utilization, excluding inadequate use, for example, the consumption of the fruit with diarrhea.

Most in vitro studies have reported significant anticancer activity in different cancer cell lines; nevertheless, in vivo evidence regarding toxicity and safety is rather limited and insufficient. Studies have shown that *F. carica* extracts have a selective affinity for cancer cells while showing minimal toxicity to normal cells. However, this selectivity is largely dependent on factors such as the type of extract, the solvent used, the dosage, and the duration of treatment [52].

https://doi.org/10.3390/ijms27010056

The systemic effects of *F. carica* extracts at higher doses in in vivo models have not been adequately investigated, and some reports indicate that certain plant parts—particularly the latex—may be toxic to normal cells at elevated concentrations. Given the scarcity of in vivo data and these toxicity concerns, the review cannot provide definitive information regarding the therapeutic applicability of *F. carica* extracts. Therefore, current findings remain insufficient to guide clinical or systemic use, highlighting the need for well-designed in vivo studies, toxicity assessments, and extract standardization before translation into therapeutic practice can be considered. For example, fig latex extracts exhibited cytotoxic effects on normal fibroblast cells (CCD-45SK) at concentrations above 100 μg/mL [19], whereas leaf extracts showed no significant toxicity up to 2 mg/mL in HUVEC cells [39]. These data suggest a relatively wider therapeutic window for leaf extracts.

Furthermore, while several studies describe antioxidant properties that may contribute to its anticancer effects, others indicate potential pro-oxidant activity that could be detrimental under specific conditions. Toxic manifestations reported at high doses include irritation, oxidative damage, or mitochondrial dysfunction, particularly with latex or essential oil fractions rich in thujones and coumarins. To date, there is no standardized toxicological assessment, and the safe-use limits have not been clearly established [54].

On the other hand, in one patient anaphylactic reaction occurred after ingestion of a fresh fig fruit. In that case, the IgE-dependent allergic mechanism was demonstrated based on the positivity of the prick test with fresh *Ficus 5* extract. However, presumably, weeping fig (*Ficus benjamina*) initiated sensitization in that patient [55]. Furthermore, rarely acute contact urticaria to *F. carica* may occur [56]. Hence, ficin is the major allergen component of *F. carica*, belonging to the cysteine protease family, such as Der p 1 [57].

Furthermore, ficin can activate the coagulation factor X, a key step in the blood clotting cascade [58]. Also, the vitamin K in *F. carica* can interfere with the effectiveness of anticoagulant drugs (blood thinners) [53,59].

Another critical concern is the complete lack of clinical studies evaluating the long-term safety of *F. carica* in humans. Without comprehensive pharmacokinetic and toxicological evaluations, it is difficult to determine the safe dosage range for therapeutic use. In addition, interactions with conventional chemotherapy or other drugs have not yet been investigated, raising concerns about possible side effects [53]. Given these uncertainties, comprehensive in vivo and longitudinal randomized controlled trials are imperative to evaluate *F. carica* extracts' safety. Standardized toxicity assessments, dose optimization studies, and long-term safety analyses will be critical to determining whether *F. carica* can be safely incorporated into anticancer treatment strategies. Until evidence is available, caution should be exercised when evaluating its use beyond experimental settings.

5. Discussion

F. carica is a plant with a long history of use in traditional medicine and shows remarkable potential in modern oncological research. Current literature suggests that extracts derived from *F. carica* fruit, leaves, and latex can exhibit anticancer activity in various cancer cell lines, such as pancreas, ovary, stomach, osteosarcoma, lung, prostate, skin, and larynx cancer models. These effects have been associated with mitochondrial-mediated apoptosis, increased ROS production, activation of anti-proliferative mechanisms, induction of autophagy, and, in some cases, initiation of alternative cell death pathways such as ferroptosis. In addition, some studies have shown that compounds isolated from *F. carica* may also play a role in tumor suppression via the immune system. Nonetheless, the mechanisms explaining the pharmacological effects of *F. carica* and the specific bioactive compounds that cause these effects remain largely undefined. A significant limitation of the current findings is that the anticancer effect is largely based on in vitro data. A critical parameter in

https://doi.org/10.3390/ijms27010056

assessing the clinical applicability of herbal extracts is whether the concentrations used are pharmacologically relevant. Although no universally accepted standard defines precise potency thresholds for crude extracts, the literature commonly refers to approximate ranges to guide interpretation. Generally accepted efficacy thresholds for extracts in the literature are reported as IC50 < 50 μg/mL for strong cytotoxicity, 50–100 μg/mL for moderate activity, and > 200 μg/mL for poor pharmacological applicability. These ranges, however, are not definitive and may vary depending on experimental design and methodological differences among studies [60,61]. While some of the studies included in this review demonstrated efficacy below this range, some examples, particularly latex and polar extracts, appear to produce cytotoxic effects only at very high concentrations [25,33]. This suggests that the observed in vitro effect may have limited translational relevance in terms of in vivo bioavailability and tissue distribution. Significant inconsistencies among studies are due to methodological differences, such as the plant parts from which the extracts were prepared, the solvents used, doses, application times, and the variety of cell lines used in the studies. This heterogeneity complicates the comparative assessment of the anticancer potential of *F. carica*. Also, the use of crude extracts in many studies leaves the true biological source of the anticancer effect unclear. Given the known low oral bioavailability, rapid metabolic transformation, and poor systemic stability of phenolic compounds, the use of extracts as direct therapeutic agents is limited from a pharmacokinetic perspective. Therefore, the use of biodirected fractionation methods to isolate individual bioactive compounds and define the pharmacodynamic mechanisms specific to these molecules appears to be a more rational approach. Indeed, some isolated compounds appear to have significantly lower IC50 values compared to crude extracts. Regarding safety, although many studies indicate that *F. carica* extracts selectively target cancer cells and cause minimal toxicity to normal cells, these results are generally short-term and confined to cell culture. It has been reported that some plant parts, especially latex, can show toxicity in normal cells at high concentrations. Also, contrary to some findings, limited data suggest that F. carica may promote tumor development under certain conditions, highlighting the condition-dependent effects of the plant. Therefore, the safety profile should be examined from a broad perspective, not only through toxicity assessments but also by identifying potential tumor-promoting effects. Moreover, in addition to antioxidant effects, the potential for pro-oxidant activity may also pose a risk under certain conditions. Currently, there is insufficient pharmacokinetic, toxicological, or clinical evidence to evaluate the long-term safety of *F. carica* in humans.

6. Conclusions

In conclusion, although *F. carica* demonstrates promising anticancer potential against various types of cancer, the bioactive compounds mediating these effects and their underlying mechanism remain inadequately characterized. Before clinical application, comprehensive in vivo studies and well-designed controlled clinical trials utilizing standardized extracts are essential to rigorously evaluate both the efficacy and safety profile of *F. carica*. For example, in preclinical small animal studies, a careful choice of parts (excluding leaves) or purification of components, or an absence genetic version of *F. carica* will be needed to avert anticoagulant effects of coumarin as well as liver- and neurotoxic effects of thujones. Such studies are critical to determine whether *F. carica* can be incorporated into anticancer treatment strategies on a scientific basis. Future research should elucidate the upstream signaling regulators, such as Nrf2, MAPK, and mTOR, that may integrate oxidative stress responses and autophagy pathways induced by *F. carica* compounds.

Here we need to highlight that potential allergic and anticoagulant effects, as well as blood sugar level-decreasing effects of *F. carica*, indicate precautions before treatment of potentially sensitive patients with its extracts or even the first consumption of the fruits.

https://doi.org/10.3390/ijms27010056

Author Contributions: E.N.G.: Conceptualization, Formal Analysis, Investigation, Writing—Original Draft, Writing—Review and Editing. S.G.: Formal Analysis, Writing, Review and Editing. B.L.R.: Supervision, Review and Editing. D.S.: Supervision, Review and Editing. F.B.: Supervision, Review and Editing. D.A.: Supervision, Review and Editing, Project Administration. All authors have read and agreed to the published version of the manuscript.

Funding: This research received no external funding.

Institutional Review Board Statement: Not applicable.

Informed Consent Statement: Not applicable.

Data Availability Statement: No new data were created or analyzed in this study. Data sharing is not applicable to this article.

Conflicts of Interest: The authors declare no conflicts of interest.

References

1. WHO. Cancer Fact Sheet. 2022. Available online: https://www.who.int/news-room/fact-sheets/detail/cancer (accessed on 4 December 2025).
2. WHO. World Cancer Day. 2024. Available online: https://www.emro.who.int/media/news/world-cancer-day-2024.html (accessed on 4 December 2025).
3. Zhou, R.; Tang, X.; Wang, Y. Emerging strategies to investigate the biology of early cancer. *Nat. Rev. Cancer* **2024**, *24*, 850–866. [CrossRef] [PubMed]
4. Kiri, S.; Ryba, T. Cancer, metastasis, and the epigenome. *Mol. Cancer* **2024**, *23*, 154. [CrossRef] [PubMed]
5. Ranjan, A.; Ramachandran, S.; Gupta, N.; Kaushik, I.; Wright, S.; Srivastava, S.; Das, H.; Srivastava, S.; Prasad, S.; Srivastava, S.K. Role of Phytochemicals in Cancer Prevention. *Int. J. Mol. Sci.* **2019**, *20*, 4981. [CrossRef]
6. Liu, S.; Li, L.; Ren, D. Anti-Cancer Potential of Phytochemicals: The Regulation of the Epithelial-Mesenchymal Transition. *Molecules* **2023**, *28*, 5069. [CrossRef]
7. Alzahrani, M.Y.; Alshaikhi, A.I.; Hazzazi, J.S.; Kurdi, J.R.; Ramadan, M.F. Recent insight on nutritional value, active phytochemicals, and health-enhancing characteristics of fig (*Ficus craica*). *Food Saf. Health* **2024**, *2*, 179–195. [CrossRef]
8. Holic, R.; Xu, Y.; Caldo, K.M.P.; Singer, S.D.; Field, C.J.; Weselake, R.J.; Chen, G. Bioactivity and biotechnological production of punicic acid. *Appl. Microbiol. Biotechnol.* **2018**, *102*, 3537–3549. [CrossRef]
9. Bohlooli, S.; Mohebipoor, A.; Mohammadi, S.; Kouhnavard, M.; Pashapoor, S. Comparative study of fig tree efficacy in the treatment of common warts (*Verruca vulgaris*) vs. cryotherapy. *Int. J. Dermatol.* **2007**, *46*, 524–526. [CrossRef]
10. Stephen Irudayaraj, S.; Christudas, S.; Antony, S.; Duraipandiyan, V.; Naif Abdullah, A.D.; Ignacimuthu, S. Protective effects of *Ficus carica* leaves on glucose and lipids levels, carbohydrate metabolism enzymes and β-cells in type 2 diabetic rats. *Pharm. Biol.* **2017**, *55*, 1074–1081. [CrossRef]
11. Mon, A.M.; Shi, Y.; Yang, X.; Hein, P.P.; Oo, T.N.; Whitney, C.W.; Yang, Y. The uses of fig (Ficus) by five ethnic minority communities in Southern Shan State, Myanmar. *J. Ethnobiol. Ethnomed.* **2020**, *16*, 55. [CrossRef]
12. Ayuso, M.; Carpena, M.; Taofiq, O.; Albuquerque, T.G.; Simal-Gandara, J.; Oliveira, M.B.P.; Prieto, M.A.; Ferreira, I.C.; Barros, L. Fig "*Ficus carica* L." and its by-products: A decade evidence of their health-promoting benefits towards the development of novel food formulations. *Trends Food Sci. Technol.* **2022**, *127*, 1–13. [CrossRef]
13. Bayrak, Ç.; Birinci, C.; Kemal, M.; Kolayli, S. The Phenolic Composition and Antioxidant Properties of Figs (*Ficus carica* L.) Grown in the Black Sea Region. *Plant Foods Hum. Nutr.* **2023**, *78*, 539–545. [CrossRef]
14. Madrigal-Santillán, E.; Portillo-Reyes, J.; Morales-González, J.A.; Sánchez-Gutiérrez, M.; Izquierdo-Vega, J.A.; Valadez-Vega, C.; Álvarez-González, I.; Chamorro-Cevallos, G.; Morales-González, Á.; Garcia-Melo, L.F.; et al. A review of *Ficus* L. genus (Moraceae): A source of bioactive compounds for health and disease. Part 1. *Am. J. Transl. Res.* **2024**, *16*, 6236–6273. [CrossRef]
15. Zhao, C.; Li, S.; Li, C.; Wang, T.; Tian, Y.; Li, X. Flavonoids from Fig (*Ficus carica Linn.*) Leaves: The Development of a New Extraction Method and Identification by UPLC-QTOF-MS/MS. *Appl. Sci.* **2021**, *11*, 7718. [CrossRef]
16. Hajam, T.A.; Saleem, H. Phytochemistry, biological activities, industrial and traditional uses of fig (*Ficus carica*): A review. *Chem.-Biol. Interact.* **2022**, *368*, 110237. [CrossRef]
17. El Fawal, G.; Sobhy, S.E.; Hafez, E.E. Biological activities of fig latex -loaded cellulose acetate/poly(ethylene oxide) nanofiber for potential therapeutics: Anticancer and antioxidant material. *Int. J. Biol. Macromol.* **2024**, *270*, 132176. [CrossRef] [PubMed]
18. Metin Guler, E.; Nur Ozkan, B.; Bozali, K.; Dundar, T.; Kocyigit, A. *Ficus carica* L. extract: A glimmer of hope on AGS human gastric adenocarcinoma. *World Cancer Res. J.* **2024**, *11*, e2746. [CrossRef]

https://doi.org/10.3390/ijms27010056

19. Yahiaoui, S.; Kati, D.E.; Ali, L.M.A.; El Cheikh, K.; Morére, A.; Menut, C.; Bachir-bey, M.; Bettache, N. Assessment of antioxidant, antiproliferative, anti-inflammatory, and enzyme inhibition activities and UPLC-MS phenolic determination of *Ficus carica* latex. *Ind. Crops Prod.* **2022**, *178*, 114629. [CrossRef]
20. Zhang, X.Y.; Song, X.Y.; Hu, X.S.; Chen, F.; Ma, C. Health benefits of proanthocyanidins linking with gastrointestinal modulation: An updated review. *Food Chem.* **2023**, *404*, 134596. [CrossRef] [PubMed]
21. Vivarelli, S.; Costa, C.; Teodoro, M.; Giambo, F.; Tsatsakis, A.M.; Fenga, C. Polyphenols: A route from bioavailability to bioactivity addressing potential health benefits to tackle human chronic diseases. *Arch. Toxicol.* **2023**, *97*, 3–38. [CrossRef]
22. Nakilcioglu-Tas, E.; Otles, S. Influence of extraction solvents on the polyphenol contents, compositions, and antioxidant capacities of fig (*Ficus carica* L.) seeds. *An. Acad. Bras. Cienc.* **2021**, *93*, e20190526. [CrossRef]
23. Badgujar, S.B.; Patel, V.V.; Bandivdekar, A.H.; Mahajan, R.T. Traditional uses, phytochemistry and pharmacology of *Ficus carica*: A review. *Pharm. Biol.* **2014**, *52*, 1487–1503. [CrossRef]
24. Al-Salman, H.N.K.; Ali, E.T.; Jabir, M.; Sulaiman, G.M.; Al-Jadaan, S.A.S. 2-Benzhydrylsulfinyl-N-hydroxyacetamide-Na extracted from fig as a novel cytotoxic and apoptosis inducer in SKOV-3 and AMJ-13 cell lines via P53 and caspase-8 pathway. *Eur. Food Res. Technol.* **2020**, *246*, 1591–1608. [CrossRef]
25. Abdel-Rahman, R.; Ghoneimy, E.; Abdel-Wahab, A.; Eldeeb, N.; Salem, M.; Salama, E.; Ahmed, T. The therapeutic effects of *Ficus carica* extract as antioxidant and anticancer agent. *South Afr. J. Bot.* **2021**, *141*, 273–277. [CrossRef]
26. Jeivad, F.; Yassa, N.; Ostad, S.N.; Hassannejad, Z.; Gheshlaghi, G.H.; Sabzevari, O. *Ficus carica* l. Latex: Possible chemo-preventive, apoptotic activity and safety assessment. *Iran. J. Pharm. Res.* **2020**, *19*, 231–240. [CrossRef]
27. Widyaningrum, N.; Hussaana, A.; Puspitasari, N.A. Invitro Antioxidant and Cytotoxic Potential of *Ficus carica* L. and *Olea europeae* L. Against Cervical Cancer. *Sains Med.* **2020**, *11*, 6. [CrossRef]
28. AlGhalban, F.M.; Khan, A.A.; Khattak, M.N.K. Comparative anticancer activities of *Ficus carica* and Ficus salicifolia latex in MDA-MB-231 cells. *Saudi J. Biol. Sci.* **2021**, *28*, 3225–3234. [CrossRef]
29. Farooq, I.; Noor, S.; Sajjad, S.; Ashraf, S.; Sadaf, S. Functionalization of Biosynthesized CQDs Capped AgO/ZnO Ternary Composite Against Microbes and Targeted Drug Delivery to Tumor Cells. *J. Clust. Sci.* **2023**, *34*, 2681–2693. [CrossRef]
30. Sangsopha, J.; Moongngarm, A.; Pornchaiyaphum, S.; Bandorn, A.; Khaejonlarp, R.; Pluemjia, B. Chemical compositions, phytochemicals, and biological activity of dried edible wild fruits. *Pak. J. Agric. Sci.* **2023**, *60*, 299–306. [CrossRef]
31. Shiraishi, C.S.H.; Zbiss, Y.; Roriz, C.L.; Dias, M.I.; Prieto, M.A.; Calhelha, R.C.; Alves, M.J.; Heleno, S.A.; da Cunha Mendes, V.; Carocho, M.; et al. Fig Leaves (*Ficus carica* L.): Source of Bioactive Ingredients for Industrial Valorization. *Processes* **2023**, *11*, 1179. [CrossRef]
32. Tikent, A.; Laaraj, S.; Bouddine, T.; Chebaibi, M.; Bouhrim, M.; Elfazazi, K.; Alqahtani, A.S.; Noman, O.M.; Hajji, L.; Rhazi, L.; et al. Antioxidant potential, antimicrobial activity, polyphenol profile analysis, and cytotoxicity against breast cancer cell lines of hydro-ethanolic extracts of leaves of (*Ficus carica* L.) from Eastern Morocco. *Front. Chem.* **2024**, *12*, 1505473. [CrossRef]
33. Yahiaoui, S.; Kati, D.E.; Chaalal, M.; Ali, L.M.A.; El Cheikh, K.; Depaepe, G.; Morère, A.; Menut, C.; Bettache, N.; Bachir-Bey, M. Antioxidant, antiproliferative, anti-inflammatory, and enzyme inhibition potentials of *Ficus carica* wood bark and related bioactive phenolic metabolites. *Wood Sci. Technol.* **2024**, *58*, 1051–1075. [CrossRef]
34. Aysin, F.; Şimşek Özek, N.; Acet, N.; Koc, K. The Anti-Proliferative Effects of *Ficus carica* Latex on Cancer and Normal Cells [*Ficus carica* Sütünün Kanser ve Normal Hücre Üzerindeki Anti-proliferatif Etkileri]. *J. Anatol. Environ. Anim. Sci.* **2024**, *9*, 145–151. [CrossRef]
35. Boyacıoğlu, O.; Kara, B.; Tecimen, S.; Kılıç, M.; Delibaş, M.; Erdoğan, R.; Özdemir, E.; Bahadır, A.; Örenay-Boyacıoğlu, S. Antiproliferative effect of *Ficus carica* latex on cancer cell lines is not solely linked to peroxidase-like activity of ficin. *Eur. J. Integr. Med.* **2021**, *45*, 101348. [CrossRef]
36. Abdou, R.; Mojally, M.; Attia, H.G.; Dawoud, M. Cubic nanoparticles as potential carriers for a natural anticancer drug: Development, in vitro and in vivo characterization. *Drug Deliv. Transl. Res.* **2023**, *13*, 2463–2474. [CrossRef]
37. Yalçınkaya, T.; Kozacı, L.D.; Çarhan, A. *Ficus carica* extract causes cell cycle arrest and induces apoptosis in MG-63 and HT-29 cancer cell lines. *Turk. Hij. Den. Biyol. Derg.* **2023**, *80*, 73–88. [CrossRef]
38. Yeniocak, S.; Karaduman-Yesildal, T.; Arslan, M.E.; Toraman, G.C.; Yücetepe, A. Effect of In Vitro Digestion on Anticancer and Antioxidant Activity of Phenolic Extracts From Latex of Fig Fruit (*Ficus carica* L.). *Chem. Biodivers.* **2024**, *22*, e202401624. [CrossRef]
39. Mustafa, K.; Yu, S.X.; Zhang, W.; Mohamed, H.; Naz, T.; Xiao, H.F.; Liu, Y.X.; Nazir, Y.; Fazili, A.A.; Nosheen, S.; et al. Screening, characterization, and in vitro-ROS dependent cytotoxic potential of extract from *Ficus carica* against hepatocellular (HepG2) carcinoma cells. *South. Afr. J. Bot.* **2021**, *138*, 217–226. [CrossRef]
40. Kalefa, N.K.; Al-Shawi, A.A.A. The essential oils isolated from the fruit of *Ficus carica* induced reactive oxygen species and apoptosis of human liver cancer cells. *Egypt. J. Chem.* **2023**, *66*, 257–265. [CrossRef]
41. Cakir, M.O.; Bilge, U.; Ghanbari, A.; Ashrafi, G.H. Regulatory Effect of *Ficus carica* Latex on Cell Cycle Progression in Human Papillomavirus-Positive Cervical Cancer Cell Lines: Insights from Gene Expression Analysis. *Pharmaceuticals* **2023**, *16*, 1723. [CrossRef] [PubMed]

 https://doi.org/10.3390/ijms27010056

42. Cakir, M.O.; Bilge, U.; Naughton, D.; Ashrafi, G.H. *Ficus carica* Latex Modulates Immunity-Linked Gene Expression in Human Papillomavirus Positive Cervical Cancer Cell Lines: Evidence from RNA Seq Transcriptome Analysis. *Int. J. Mol. Sci.* **2023**, *24*, 13646. [CrossRef]
43. Ou, A.X.; Zhao, X.X.; Lu, Z.M. Autophagy is involved in *Ficus carica* fruit extract-induced anti-tumor effects on pancreatic cancer. *Biomed. Pharmacother.* **2022**, *150*, 112966. [CrossRef]
44. Dekdouk, N.; Ameddah, S.; Bensouici, C.; Ibtissem, M.; Ahmed, M.; Martel, F. Antioxidant properties and anticancer activity of *Olea europaea* L. olive and *Ficus carica* fruit extracts against pancreatic cancer cell lines. *Trends Phytochem. Res.* **2024**, *8*, 80–92. [CrossRef]
45. Ali, E.T.; Al-Salman, H.N.K.; Rasool, K.H.; Jabir, M.S.; Ghimire, T.R.; Shari, F.H.; Hussein, H.H.; Al-Fregi, A.A.; Sulaiman, G.M.; Khalil, K.A.A.; et al. 2-(Benzhydryl sulfinyl)-N-sec-butylacetamide) isolated from fig augmented trastuzumab-triggered phagocytic killing of cancer cells through interface with Fcγ receptor. *Nat. Prod. Res.* **2023**, *37*, 4112–4120. [CrossRef]
46. Sadeghizade, M.; Baharara, J.; Salek, F.; Amini, E. Assessment the Effect of *Ficus carica* Leaf Extract on $B_{16}F_{10}$ Melanoma Cancer Cells: The Roles of p53, Caspase-3 & Caspase-9 on Induction of Intrinsic Apoptosis Cascade. *Jundishapur J. Nat. Pharm. Prod.* **2022**, *17*, e117429. [CrossRef]
47. Ghandehari, F.; Fatemi, M. The effect of *Ficus carica* latex on 7 2018, 12-dimethylbenz (a) anthracene-induced breast cancer in rats. *Avicenna J. Phytomed* **2018**, *8*, 286–295.
48. Sharma, S.H.; Kumar, J.S.; Chellappan, D.R.; Nagarajan, S. Molecular chemoprevention by morin—A plant flavonoid that targets nuclear factor kappa B in experimental colon cancer. *Biomed. Pharmacother.* **2018**, *100*, 367–373. [CrossRef] [PubMed]
49. Minekus, M.; Alminger, M.; Alvito, P.; Ballance, S.; Bohn, T.; Bourlieu, C.; Carrière, F.; Boutrou, R.; Corredig, M.; Dupont, D.; et al. A standardised static in vitro digestion method suitable for food—An international consensus. *Food Funct.* **2014**, *5*, 1113–1124. [CrossRef]
50. Gor, R.; Saha, L.; Agarwal, S.; Karri, U.; Sohani, A.; Madhavan, T.; Pachaiappan, R.; Ramalingam, S. Morin inhibits colon cancer stem cells by inhibiting PUM1 expression in vitro. *Med. Oncol.* **2022**, *39*, 251. [CrossRef] [PubMed]
51. Baohong, L.; Zhongyuan, L.; Ying, T.; Beibei, Y.; Wenting, N.; Yiming, Y.; Qinghua, C.; Qingjun, Z. Latex derived from *Ficus carica* L. inhibited the growth of NSCLC by regulating the caspase/gasdermin/AKT signaling pathway. *Food Funct.* **2023**, *14*, 2239–2248. [CrossRef] [PubMed]
52. Morovati, M.R.; Ghanbari-Movahed, M.; Barton, E.M.; Farzaei, M.H.; Bishayee, A. A systematic review on potential anticancer activities of *Ficus carica* L. with focus on cellular and molecular mechanisms. *Phytomedicine* **2022**, *105*, 154333. [CrossRef]
53. Fazel, M.F.; Abu, I.F.; Mohamad, M.H.N.; Mat Daud, N.A.; Hasan, A.N.; Aboo Bakkar, Z.; Md Khir, M.A.N.; Juliana, N.; Das, S.; Mohd Razali, M.R.; et al. Physicochemistry, Nutritional, and Therapeutic Potential of *Ficus carica*—A Promising Nutraceutical. *Drug Des. Dev. Ther.* **2024**, *18*, 1947–1968. [CrossRef]
54. Gaur, R.; Chauhan, A.; Kanta, C. A critical review of antioxidant potential and pharmacological applications of important Ficus species. *J. HerbMed Pharmacol.* **2024**, *13*, 537–549. [CrossRef]
55. Dechamp, C.; Bessot, J.C.; Pauli, G.; Deviller, P. First report of anaphylactic reaction after fig (*Ficus carica*) ingestion. *Allergy* **1995**, *50*, 514–516. [CrossRef] [PubMed]
56. Cruz, N.V.; Bahna, S.L.; Knight, A.K. Fig Allergy: Not Just Oral Allergy Syndrome (OAS). *J. Allergy Clin. Immunol.* **2007**, *119*, S110. [CrossRef]
57. Urbani, S.; Aruanno, A.; Nucera, E. Adverse reaction to *Ficus Carica*: Reported case of a possible cross-reactivity with Der p1. *Clin. Mol. Allergy* **2020**, *18*, 9. [CrossRef] [PubMed]
58. Richter, G.; Schwarz, H.P.; Dorner, F.; Turecek, P.L. Activation and inactivation of human factor X by proteases derived from *Ficus carica*. *Br. J. Haematol.* **2002**, *119*, 1042–1051. [CrossRef]
59. Hakima, B.; Sara, H. Anticoagulant Activity of Some *Ficus carica* Varieties Extracts Grown in Algeria. *Acta Sci. Nat.* **2019**, *6*, 73–78. [CrossRef]
60. Canga, I.; Vita, P.; Oliveira, A.I.; Castro, M.; Pinho, C. In Vitro Cytotoxic Activity of African Plants: A Review. *Molecules* **2022**, *27*, 4989. [CrossRef]
61. Belete, T.M.; Beyna, A.T. Review on the Ethnopharmacological Use of Medicinal Plants and Their Anticancer Activity from Preclinical to Clinical Trial. *Nat. Prod. Commun.* **2025**, *20*, 1934578X251333027. [CrossRef]

 https://doi.org/10.3390/ijms27010056

International Journal of
Molecular Sciences

MDPI

Review

Genetic Divergence and Functional Significance of Bioactive Compounds in Rice and Barley: Implications for Biofortification and Human Health

Essam ElShamey [1,2,*,†], Jiazhen Yang [1,†], Xiaomeng Yang [1], Md. Mahmudul Hasan [1], Tao Yang [1] and Yawen Zeng [1,*]

[1] Biotechnology and Germplasm Resources Institute, Yunnan Academy of Agricultural Sciences, Kunming 650205, China; yangjiazhen415@163.com (J.Y.); yxm89ccf@126.com (X.Y.); mmhasanum@gmail.com (M.M.H.); yt52279076@163.com (T.Y.)

[2] Rice Research Department, Field Crops Research Institute, Agricultural Research Center, Cairo 12619, Egypt

* Correspondence: essamelshamey@gmail.com (E.E.); zengyw1967@126.com (Y.Z.); Tel.: +86-15752883410 (E.E.)

† These authors contributed equally to this work.

Abstract: The functional components in cereals (rice and barley), such as gamma-aminobutyric acid (GABA), resistant starch (RS), and alkaloids, play crucial roles in human health, offering benefits such as improved cardiovascular function, enhanced gut microbiota, and potential anticancer properties. Rice (*Oryza sativa*) and barley (*Hordeum vulgare*) are key dietary staples with distinct genetic architectures influencing the biosynthesis and accumulation of these bioactive compounds. In this study, we explore the interaction and divergence of gene loci associated with GABA, RS, and alkaloid pathways in rice and barley, leveraging comparative genomics to identify conserved and species-specific regulatory mechanisms. We highlight key quantitative trait loci (QTLs) and candidate genes, such as GAD (glutamate decarboxylase) for GABA synthesis, *SSIIa* and *GBSS* for RS formation, and alkaloid biosynthesis genes including *CYP80G2*. Additionally, we discuss the health implications of these functional components, including their roles in reducing hypertension, managing diabetes, and exhibiting neuroprotective effects. Understanding the genetic differences between rice and barley in accumulating these compounds can guide biofortification strategies to enhance nutritional quality in cereal crops, ultimately benefiting human health and dietary outcomes.

Keywords: GABA; resistant starch; alkaloids; rice; barley; QTL mapping; biofortification; health benefits

1. Introduction

Rice (*Oryza sativa*) and barley (*Hordeum vulgare*) are among the most widely consumed cereals, serving as staple foods for billions of people worldwide [1,2]. Beyond their roles as primary energy sources, these grains contain bioactive compounds that contribute to human health, including gamma-aminobutyric acid (GABA), resistant starch (RS), and alkaloids [3]. The interactions among these components influence the nutritional and functional properties of rice and barley, making them valuable in the development of functional foods aimed at preventing chronic diseases and improving metabolic and neurological health [4–6]. The human chronic disease outbreak is associated with the change in staple food from brown rice and barley for ancient people to white rice and barley flour for modern people, which divided the six major dietary structures. In fact, sustaining the consumption of major foods plus barley grass powder is the healthiest dietary guideline for modern humans. More than 30 functional ingredients in barley grass exert potent preventive effects on over 20 chronic diseases, and 15 functional ingredients in barley grains prevent 11 chronic diseases [7,8]. In this article, we review the key functional components in both rice and barley, as well as their potential health benefits. GABA, a major inhibitory neurotransmitter in the central nervous system, has been linked to anti-hypertensive, anti-stress, and neuroprotective effects [9,10]. In rice and barley, GABA levels increase during germination, fermentation, and other stress-induced processes, enhancing their functional food potential [11]. Meanwhile, resistant starch, a type of dietary fiber that escapes digestion in the small intestine, acts as a prebiotic, promoting gut microbiota diversity and improving insulin sensitivity [12]. The fermentation of RS by gut bacteria may also influence GABA production, suggesting a potential gut–brain axis interaction [13]. Rice is a rich source of starch [12], which provides energy and acts as a functional ingredient in food processing. RS in rice has prebiotic effects, promoting gut health by feeding beneficial gut bacteria [14]. Brown rice and whole-grain rice contain significant amounts of dietary fiber, which aids digestion, reduces cholesterol levels [15,16], and helps regulate blood sugars [17]. Additionally, rice contains B vitamins (e.g., thiamine and niacin) and minerals such as magnesium, phosphorus, and selenium [18]. These nutrients support metabolism, bone health, and immune function.

Although they are less studied in cereals compared to medicinal plants, alkaloids exhibit bioactive properties, including antioxidant, anti-inflammatory, and neuromodulatory effects [19,20]. Certain alkaloids may interact with GABA receptors or affect starch digestion, further complicating the metabolic interplay among these compounds. The presence and bioactivity of these molecules depend on genetic factors, environmental conditions, and post-harvest processing techniques, such as soaking, germination, and fermentation [21,22]. Understanding the synergistic and antagonistic interactions among GABA, resistant starch, and alkaloids in rice and barley is crucial for optimizing their health benefits. For instance, GABA-enriched germinated brown rice is marketed for its anti-anxiety and cardiovascular benefits, while high-RS barley products are used for diabetes management and gut health promotion. However, the potential modulation of these effects by alkaloids remains underexplored.

Barley is considered one of the richest sources of beta-glucans, which are functional components and a type of soluble fiber. These are known to lower cholesterol, improve heart health, and regulate blood glucose levels [23]. Additionally, barley contains both soluble and insoluble fiber, which aid digestion, promote satiety, and support gut health [24]. Similarly to rice, barley contains resistant starch, which acts as a prebiotic and supports gut microbiota [25]. Alkaloids are a diverse group of naturally occurring nitrogen-containing compounds that have garnered significant attention due to their pharmacological, ecological, and nutritional properties. Regarding which alkaloids are commonly associated with medicinal plants, recent research has highlighted their presence and functional roles in stable crops such as barley (*Hordeum vulgare*) [26]. Functional foods are designed to provide health benefits beyond basic nutrition, often through bioactive compounds such as polyphenols, prebiotics, and probiotics. A key factor in their efficacy is the synergistic interaction between different components, which can enhance bioavailability, bioactivity, and overall health effects compared to isolated compounds. This study on rice and barley highlights how GABA, RS, and alkaloids interact to improve metabolic, neurological, and gastrointestinal health. Their synergistic effects are crucial in functional food development because some bioactive compounds improve the absorption of others; for example, resistant starch acts as a prebiotic, promoting gut microbiota and metabolizing GABA, potentially increasing its bioavailability. In addition, GABA (a neuroactive compound) and RS (a prebiotic) may work together to influence gut microbiota, which in turn affects neurotransmitter production and mental health. In barley, β-glucans and RS collectively improve insulin sensitivity and cholesterol metabolism, offering better outcomes than does either compound alone.

One of the key challenges in cereal genomics is bridging the genetic gap between rice (a model monocot with a well-annotated genome) and barley (a more complex, diploid Triticeae crop with a large genome). Comparative genomics allows researchers to identify conserved and divergent quantitative trait loci (QTLs) associated with critical agronomic traits. Among these, glutamate decarboxylase (*GAD*), starch synthase IIa (*SSIIa*), and cytochrome *P450* (*CYPs*) genes play crucial roles in stress response, grain quality, and metabolic pathways [27,28]. This in-depth analysis explores the genetic and functional differences between rice and barley by comparing QTLs and allelic variations in *GAD*, *SSIIa*, and *CYPs*, highlighting their implications for crop improvement. For rice genome structure and evolution, a small genome (~430 Mb), which is highly sequenced and annotated, belongs to the BEP clade (Bambusoideae, Ehrhartoideae, Pooideae) but diverged early from Triticeae [29]. Also, it has high gene density and synteny with other grasses, making it a model for cereal genomics. Meanwhile, the barley genome structure and evolution involves a large genome (~5.1 Gb), with ~80% repetitive elements, and is a member of the Triticeae tribe (along with wheat and rye) [30], which underwent recent polyploidization events, more complex regulatory mechanisms, and structural variations compared to rice.

Challenges in comparative genomics include synteny breakdown and the existence of colinear regions, extensive rearrangements, and transposable element expansions in barley reducing direct gene-to-gene comparisons. Certain gene families (e.g., CYPs) have undergone differential expansions in barley and rice. Divergent selection pressures for rice (tropical/subtropical) and barley (temperate/cold-adapted) have led to the evolution of distinct stress-responsive pathways. Rice *GADs* have been more extensively studied in terms of heat/drought responses, whereas barley *GADs* have been linked to cold tolerance. Barley may lack some regulatory elements found in rice, affecting GABA accumulation under stress. Starch Synthase *IIa* (*SSIIa*) catalyzes amylopectin elongation, affecting grain texture and cooking quality. Rice *SSIIa* has a stronger effect on cooking quality, while barley *SSIIa* is crucial for malting. Barley shows more allelic diversity, likely due to human selection for brewing traits. Cytochrome *P450* (*CYPs*) is involved in detoxification, hormone synthesis, and secondary metabolism. Rice *CYPs* are more studied in herbicide resistance, whereas barley *CYPs* are linked to Fusarium resistance (DON detoxification). Barley has expanded *CYP* clans (e.g., *CYP71*) for harsher temperate climate adaptations. The genetic gap between rice and barley is shaped by divergent evolution, genome complexity, and selection pressures. While rice serves as a genomic model, barley's larger genome and adaptive traits present unique challenges and opportunities. Comparative analysis of *GAD*, *SSIIa*, and *CYP* QTLs reveals both conserved and divergent mechanisms underlying stress tolerance, grain quality, and metabolism. Bridging this gap through advanced genomics and biotechnology will enhance precision breeding for improved cereal crops. This comparative framework provides a foundation for future studies aiming to leverage rice genomic resources for barley improvement and vice versa.

In this review, we analyze the occurrence, biosynthesis, and functional significance of alkaloids in rice and barley in detail, emphasizing their potential as bioactive compounds significant to human health, plant defense, and agricultural applications. We examine the biochemical pathways, mutual interactions, and health implications of GABA, resistant starch, and alkaloids in rice and barley, emphasizing their roles in functional food development. By analyzing current research, we highlight the gaps in knowledge and future directions for harnessing these interactions to design tailored functional foods for enhancing metabolic, neurological, and gastrointestinal health.

2. Interaction of and Differences in Gene Loci Among Rice Functional Components

Rice (*Oryza sativa*) contains several bioactive compounds; these components are governed by distinct genetic loci and metabolic pathways, and their interactions and differences in gene regulation contribute to rice's functional properties [31,32]. GABA is a non-protein amino acid that functions as a major inhibitory neurotransmitter in the central nervous system of mammals [11]. It also plays significant roles in plants, including in stress responses, signaling, and growth regulation. GABA is found in various foods, including rice and barley, and its concentration can be influenced by processing methods, such as fermentation or germination. The compound can be measured according to [33], and Figure 1 explains its metabolism process in rice and barley [34].

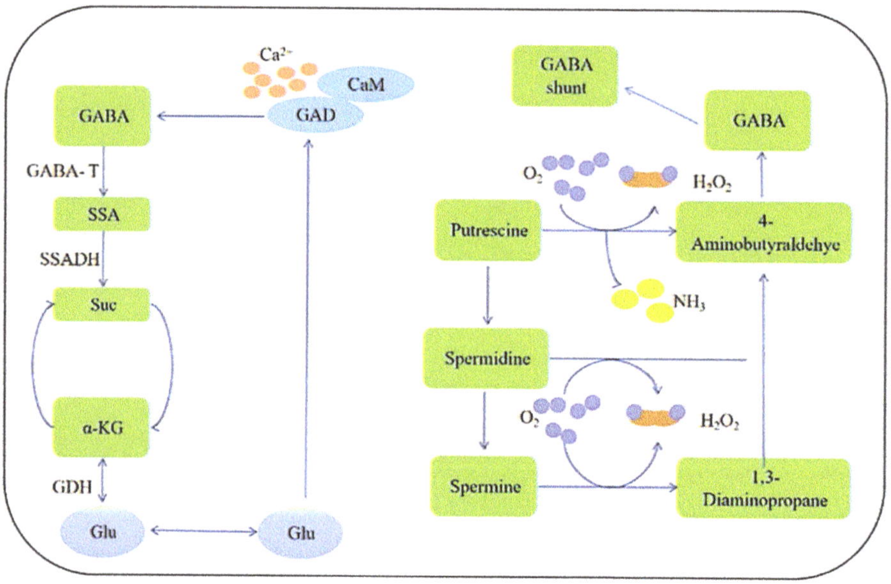

Figure 1. Metabolism process of GABA in plants.

A kind of starch known as RS can function as a prebiotic in the colon and support gut health by avoiding digestion in the small intestine [35,36]. It is found in various foods, including rice and barley. Below is the detailed information about resistant starch in both rice and barley, and its types are shown in Figure 2 [37]. RS content can be measured according to the modified Goñi method [38].

Resistant starch (RS) is a fraction of starch that escapes digestion in the small intestine and functions similarly to dietary fiber, providing numerous health benefits. Among the five types of resistant starch, Type 5 (RS5) refers to the amylose–lipid complex (ALC), where amylose forms a helical inclusion complex with lipids, reducing enzymatic digestibility. RS5 is naturally present in cereals such as rice and barley and can be enhanced through processing techniques such as heat-moisture treatment (HMT) and annealing. The molecular basis of amylose–lipid complexes (ALCz) are amylose helix formation through amylose chains adopting a single helical V-type crystalline structure when complexed with endogenous lipids (e.g., free fatty acids, monoacylglycerols). Lipid binding through hydrophobic interactions between the lipid tail and the amylose helix core stabilizes the complex, reducing accessibility to amylolytic enzymes (α-amylase, amyloglucosidase). RS5 (amylose–lipid complex) in rice and barley represents a promising dietary component for improving metabolic health, gut microbiota balance, and glycemic control. While rice RS5 is highly influenced by processing, barley's natural β-glucan content provides additional health synergies. Future research should focus on optimizing RS5 levels through agronomic practices, food processing, and genetic engineering to harness its full nutraceutical potential [39].

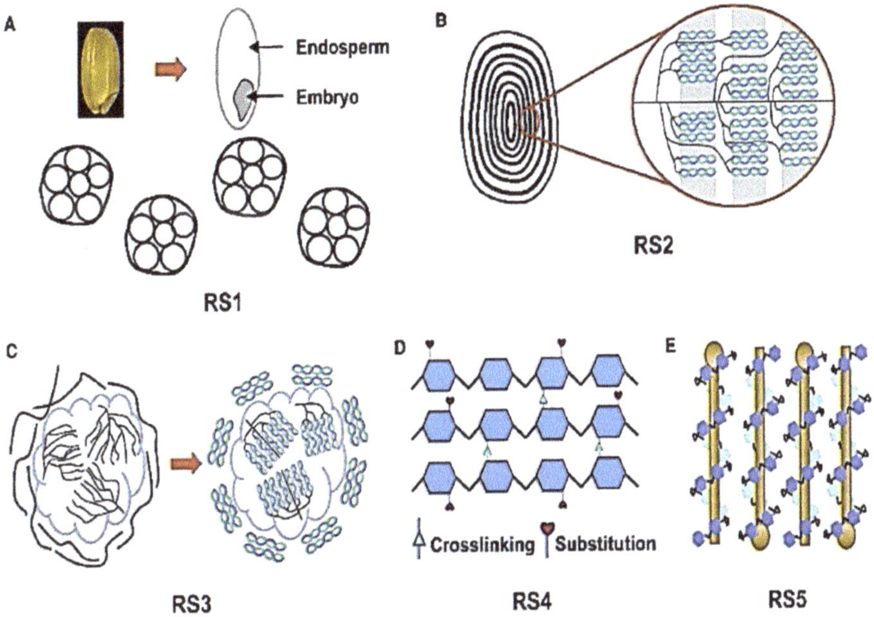

Figure 2. Types of resistant starch in plants. (**A**) RS1, a type of RS that is physically inaccessible to digestion. The starch granules are enclosed by cell walls or other tissues that prevent contact and reaction with amylase. (**B**) RS2, native starch granules with a compact crystalline structure that are relatively dehydrated. RS2 is tightly packed in a radial pattern in raw starch granules. (**C**) RS3, retrograded starches that are primarily formed via the gelatinization and retrogradation process during food processing and manufacturing. (**D**) RS4, chemically modified starch that generates new chemical bonds by substitution, esterification, or crosslinking. Through chemical modification, the molecular structure of starch is altered, thereby increasing its resistance to amylase. (**E**) RS5, a new type of RS that forms a complex of amylose and lipid. RS5 is usually generated in high-amylose starch cereals [37].

Secondary metabolites called alkaloids are essential to plant physiology, defense systems, and interactions with the environment. Although they were traditionally studied in medicinal plants such as opium poppy and cinchona, alkaloids are increasingly being identified in cereal crops, including rice and barley [40,41]. These compounds contribute to the nutritional and functional properties of these grains, making them important for both human health and agricultural sustainability. This review explores the types of alkaloids found in rice and barley, their biosynthetic pathways, and their functional roles in plants and human diets (Figure 3) [42].

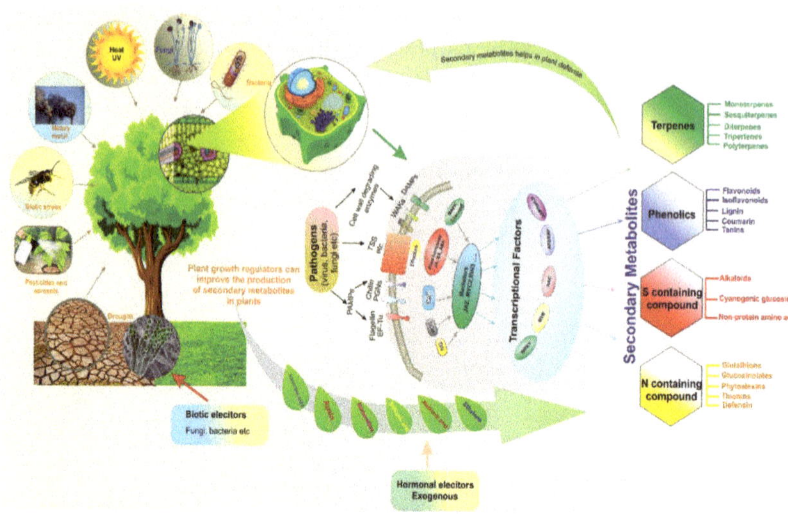

Figure 3. Plant growth and development are impacted by a variety of biotic and abiotic stressors [42].

2.1. *GABA in Rice*

2.1.1. Natural Occurrence

GABA is naturally present in rice, particularly in the germ and bran layers of brown rice. White rice contains significantly less of this compound compared to brown rice because the polishing process removes the germ and bran [43]. Brown rice has GABA contents ranging from 1 to 10 mg per 100 g [44], depending on the variety and germination conditions. In germinated brown rice (GBR), this content is significantly increased [45]. GBR can contain anything from 10 to 50 mg of GABA per 100 g or even more in some cases [46]. In white rice, GABA content is generally lower, typically less than 1 mg per 100 g, due to the removal of the bran layer during milling [47].

2.1.2. Enhancement Through Germination

GABA is a non-protein amino acid and important for human health and plant stress responses. Soaking, germination, and fermentation are common methods to enhance the GABA content in rice, and fermentation using lactic acid bacteria or yeast can further increase the levels of this compound [48]. Its accumulation is influenced by environmental conditions, processing methods, and genetic factors [49]. Enhancing the GABA content in rice through germination, fermentation, or bio-fortification can improve its nutritional value, offering potential benefits such as anti-hypertensive, anti-diabetic, and neuroprotective effects [50,51]. Further research into GABA biosynthesis regulation and optimized processing techniques could lead to the development of functional rice varieties with enhanced health-promoting properties, contributing to both food security and human well-being. Germinated brown rice (GBR) or GABA rice is a popular functional food product. During germination, the activity of the enzyme glutamate decarboxylase (GAD) increases, converting glutamate into GABA. GABA levels in germinated brown rice can be 5–10 times higher than those in non-germinated brown rice [52]. Concentrations of GABA in different types of rice are shown in Table 1.

Table 1. GABA concentration in rice (mg/100 g dry weight).

Grain Type	Processing Method	GABA Concentration (mg/100 g)	Key Factors Influencing GABA Levels	Refs.
White Rice	Raw (unprocessed)	1.5–5.0	Milling makes levels of endogenous GABA low	[53–57]
Brown Rice	Raw (unprocessed)	5.0–15.0	Bran layer contains more GABA	
Germinated Brown Rice (GBR)	Soaking + germination (24–72 h)	25–350	Activation of glutamate decarboxylase (GAD)	
Fermented Rice	Lactic acid bacteria (LAB) fermentation	50–400	Microbial activity enhances GABA synthesis	
Black Rice	Germination	100–500	Anthocyanin-rich varieties contain more GABA	

2.1.3. GABA (γ-Aminobutyric Acid) Biosynthesis and Gene Loci

In rice (*Oryza sativa*), GABA (γ-aminobutyric acid) biosynthesis is prompted by L-glutamate through the action of the enzyme glutamate decarboxylase (GAD), which catalyzes the removal of carboxyl groups [58]. This reaction requires pyridoxal-5′-phosphate (PLP, a vitamin B6 derivative) as a cofactor [59–61]. GABA can be further metabolized in the GABA shunt (mitochondrial pathway) via GABA transaminase (GABA-T) catalysis, the conversion of GABA to succinic semialdehyde (SSA), succinic semialdehyde dehydrogenase (SSADH) catalysis, and the conversion of succinic semialdehyde (SSA) to succinate, thereby integrating into the tricarboxylic acid (TCA) cycle. GABA biosynthesis is regulated in rice through stress responses: the compound accumulates under hypoxia (flooding), heat stress, salinity, mechanical damage, and calcium signaling; Ca^{2+}/calmodulin activates OsGAD1 under stress and developmental regulation; and GABA levels vary during seed germination and seedling growth.

Key Genes Involved in GABA Biosynthesis in Rice

GABA biosynthesis primarily occurs via the glutamate-decarboxylase-(GAD)-mediated decarboxylation of glutamate, followed by conversion via GABA transaminase (GABA-T). The key genes involved in this pathway include GAD family members (e.g., *OsGAD1*, *OsGAD2*), which regulate GABA production, and *OsGABA-T*, which modulates GABA catabolism. Additionally, the calcium/calmodulin (CaM)-binding domain in GAD enzymes links GABA synthesis to stress signaling. Several genes encode enzymes involved in GABA metabolism in rice (Table 2). The most critical ones include the following:

- Glutamate decarboxylase (GAD) genes: *OsGAD1* (LOC_Os03g15220) is found at chromosome number three; it converts glutamate to GABA and contains a calmodulin-binding domain (CaMBD), allowing calcium/calmodulin-mediated activation under stress (e.g., hypoxia, heat, salinity). *OsGAD2* (LOC_Os08g33690) is found at chromosome number eight; it functions similarly to OsGAD1 but may have different expression patterns.
- GABA transaminase (GABA-T) genes: *OsGABA-T1* (LOC_Os03g20310) is found at chromosome number three; it converts GABA to succinic semialdehyde (SSA). *OsGABA-T2* (LOC_Os01g48960) is found at chromosome number one; it functions as an alternative to GABA transaminase isoforms.
- Succinic semialdehyde dehydrogenase (SSADH) gene: *OsSSADH* (LOC_Os01g10940) is found at chromosome number one; it converts SSA to succinate, linking GABA metabolism to the TCA cycle.

Table 2. Summary of key GABA-related genes in rice.

Gene	Locus ID	Chromosome	Function	Refs.
OsGAD1	LOC_Os03g15220	3	Glutamate decarboxylase	[62–65]
OsGAD2	LOC_Os08g33690	8	Alternative GAD isoform	
OsGABA-T1	LOC_Os03g20310	3	GABA transaminase	
OsGABA-T2	LOC_Os01g48960	1	Alternative GABA-T isoform	
OsSSADH	LOC_Os01g10940	1	Succinic semialdehyde dehydrogenase	

The biosynthesis of GABA in rice is primarily mediated by *OsGAD1/2*, while GABA-T and SSADH regulate its catabolism. These genes are crucial for stress responses and nutritional enhancement in rice. The genetic manipulation of these loci can lead to high-GABA rice varieties with improved stress tolerance and health benefits.

Applications in Rice Breeding

Increasing GABA content in rice through breeding methods presents a promising approach to enhancing its nutritional value. This review explores various breeding strategies, such as conventional selection, marker-assisted breeding, and genomic approaches, in developing high-GABA rice varieties. Key genetic factors influencing GABA accumulation, including glutamate decarboxylase (GAD) activity and polyamine degradation pathways, are discussed. Additionally, the potential of mutation breeding and CRISPR-Cas9 gene editing to modulate GABA biosynthesis is highlighted. Successful case studies of biofortified rice cultivars with elevated GABA levels are examined, demonstrating the feasibility of breeding-based enhancement. Challenges, including yield trade-offs and regulatory considerations, are also addressed. In high-GABA rice varieties (e.g., germinated brown rice (GBR) used in functional foods), mutations in GABA-T or the overexpression of GAD can increase GABA content. Additionally, abiotic stress tolerance can enhance GABA synthesis and improve flood/drought resistance.

In plants, GABA accumulates in response to stress and plays a role in nitrogen metabolism, pH regulation, and defense mechanisms. Pigmented rice varieties, such as Venere rice (an Italian black rice), Thai jasmine black rice, and Chinese purple rice, have been found to contain significantly higher GABA levels compared to conventional white rice. This is attributed to their unique genetic makeup, anthocyanin-rich bran layers, and stress-induced metabolic pathways that enhance GABA biosynthesis. GABA is synthesized primarily via the glutamate decarboxylase (GAD) pathway, where the GAD decarboxylates L-glutamate into GABA. Pigmented rice varieties exhibit higher GAD enzyme activity due to genetic polymorphisms in GAD genes, enhanced stress responses (e.g., anaerobic germination, UV exposure, and cold storage), which upregulate GABA accumulation, and interaction with phenolic compounds, where anthocyanins (e.g., cyanidin-3-glucoside) may stabilize GABA or modulate its metabolic flux. Studies report GABA concentrations in pigmented rice ranging from 15–50 mg/100 g, compared to 2–10 mg/100 g in polished white rice. Germination (sprouting) and fermentation further amplify GABA levels, making germinated black rice a potent functional food.

2.1.4. Health Benefits

GABA functions as a major inhibitory neurotransmitter in the central nervous system and has been linked to numerous health benefits. In rice (*Oryza sativa* L.), this compound accumulates through the decarboxylation of glutamate via glutamate decarboxylase (GAD), particularly under anaerobic conditions such as germination (sprouting) or fermentation. Recent studies highlight that GABA-rich rice, including germinated brown rice (GBR), exhibits enhanced nutritional and therapeutic properties. GABA has been associated with neuroprotective effects, including anxiety and stress reduction, improved sleep quality, and the potential alleviation of depression via the modulation of GABA receptors in the brain. Additionally, it demonstrates antihypertensive activity by inhibiting angiotensin-converting enzyme (ACE), thus aiding in blood pressure regulation. GABA-enriched rice also shows antidiabetic potential, improving insulin sensitivity and reducing oxidative stress in diabetic models. Furthermore, it enhances hepatic and renal function by mitigating oxidative damage and inflammation. GABA's antioxidant properties contribute to reducing lipid peroxidation and enhancing cellular defense mechanisms, while its role in immune modulation may support metabolic and cardiovascular health. Fermented or germinated rice with elevated GABA levels also promotes gut health by influencing beneficial microbiota. Given these benefits, GABA-rich rice represents a functional food with promising applications in preventing and managing metabolic, neurological, and cardiovascular disorders [57,66]. Further research is needed to optimize GABA bioaccumulation in rice grains and validate its long-term health impacts in human clinical trials.

Human clinical benefits of GABA from pigmented rice include neuroprotective and antistress effects. Anxiolytic and sedative properties in dietary GABA from pigmented rice have been shown to reduce cortisol levels and improve sleep quality in clinical trials. Cognitive enhancement is achieved through GABA crossing the blood–brain barrier (albeit limitedly) and may enhance alpha-wave activity, improving relaxation and focus. Also, in terms of cardiovascular health, for blood pressure regulation, GABA acts as an ACE (angiotensin-converting enzyme) inhibitor, reducing hypertension in hypertensive patients. Antioxidant synergy through anthocyanins and GABA mitigates oxidative stress, lowering LDL oxidation and improving endothelial function. Pigmented rice varieties, particularly those with high anthocyanin content such as Venere rice, represent a natural reservoir of GABA, with demonstrated neuroprotective, cardiometabolic, and anti-inflammatory benefits. Further clinical research is needed to optimize bioavailability and establish dietary guidelines for maximal therapeutic effects. The integration of GABA-rich pigmented rice into functional diets could serve as a sustainable strategy for preventive healthcare.

2.1.5. Differences in GABA Between Rice and Barley

Rice (*Oryza sativa*) and barley (*Hordeum vulgare*), despite being evolutionarily related, exhibit significant differences in GABA accumulation, regulation, and functional roles due to their distinct genetic backgrounds, environmental adaptations, and domestication histories. This in-depth comparative analysis explores the genetic, enzymatic, and metabolic mechanisms governing GABA biosynthesis, degradation, and physiological functions in rice and barley, highlighting key similarities, divergences, and potential biotechnological applications for crop improvement. The genetic and enzymatic basis of GABA metabolism for GABA is primarily synthesized via the glutamate decarboxylase (GAD) pathway, where GAD enzymes convert glutamate to GABA, followed by further metabolism through GABA transaminase (GABA-T) and succinic semialdehyde dehydrogenase (*SSADH*). While both rice and barley possess conserved GAD gene families, their structural and regulatory features differ. Rice contains five GAD genes (*OsGAD1-5*), with *OsGAD1* and *OsGAD2* being the most active in GABA synthesis under stress. These genes possess calmodulin-binding domains (CaMBDs), allowing calcium-mediated activation during abiotic stress. Meanwhile, barley has fewer GAD homologs (*HvGAD1-4*), with HvGAD1 showing strong induction under cold stress. Unlike rice, some barley GAD isoforms may lack CaMBD, suggesting alternative regulatory mechanisms. GABA levels in rice and barley are dynamically regulated under different environmental conditions. Rice GABA rapidly accumulates under heat, drought, and hypoxia, functioning as a signaling molecule to mitigate oxidative damage. High GABA rice varieties (e.g., Koshihikari) are prized for their nutritional benefits. On the other hand, barley GABA plays a critical role in cold acclimation and salinity tolerance, with elevated levels observed in response to low temperatures. Barley malting processes also induce GABA, influencing beer quality. GABA-rich rice (e.g., germinated brown rice) is promoted for its anti-hypertensive and neuroprotective effects. High-GABA barley is linked to improved malt fermentability and potential functional food applications. While rice and barley share core GABA metabolic pathways, their genetic regulation, stress adaptation strategies, and end-use qualities differ significantly. Understanding these distinctions provides a foundation for tailoring GABA-enriched cereals through precision breeding and genetic engineering, enhancing both agricultural resilience and nutritional value in a changing climate. Future research should focus on transgenic validation, multi-omics integration and field trials to optimize GABA traits in both crops. Later, we discuss GABA as a functional component in barley in detail.

2.2. Resistant Starch in Rice

2.2.1. Types of Resistant Starch in Rice

Rice (*Oryza sativa*), a staple food for over half of the global population, contains varying levels of resistant starch depending on genetic, environmental, and processing factors. Resistant starch in rice is classified into five main types: RS1 (physically inaccessible starch) is found in whole or partially milled rice grains and trapped within intact cell walls, limiting enzymatic digestion; RS2 (resistant granules) is present in high-amylose rice varieties, where the native crystalline structure of starch (B-type polymorphs) resists digestion; RS3 (retrograded starch) is formed when cooked and cooled rice undergoes retrogradation, increasing amylose realignment and resistance to digestion; RS4 (chemically modified starch) is artificially produced through chemical treatments (e.g., cross-linking, esterification) to enhance resistance and is less common in natural rice; and RS5 (amylose–lipid complexes), often found in rice with higher lipid content, is produced by starch–lipid interactions that reduce digestibility [36,67]. The proportion of these RS types in rice depends on varietal differences (e.g., high-amylose vs. waxy rice), processing methods (e.g., parboiling, cooling), and cooking techniques. High-amylose rice cultivars, such as those with *ALK* gene mutations, tend to have higher RS2 and RS3 contents. Additionally, post-harvest treatments such as fermentation and hydrothermal processing can further modify RS levels. Understanding these variations is crucial for developing functional rice products with enhanced health benefits. This review explores the mechanisms, influencing factors, and potential applications of resistant starch in rice for improved human nutrition [68].

2.2.2. Factors Affecting Resistant Starch Content

There are many factors affecting resistant starch content [69,70], including rice variety; some varieties, such as high-amylose rice (e.g., basmati or parboiled rice), have higher resistant starch contents compared to low-amylose rice (e.g., jasmine or sticky rice). Cooking rice and then cooling it increases resistant starch content due to retrogradation, through which starch molecules reorganize into a more resistant form [37,69,70]. Additionally, parboiled rice retains more resistant starch than does regular milled rice. Concentrations of resistant starch in different types of rice are shown in Table 3. There are some strategies to enhance RS, including genetic engineering (the knockout of *SBEIIb* or overexpression of *GBSSI* increases RS [37,71]) and mutagenesis (high-amylose mutants (e.g., "ae" mutants) contain elevated RS) [72,73]. Processing methods, including parboiling, cooling, and fermentation, can induce RS formation (RS3) [74,75]. Resistant starch in rice is primarily controlled by genes involved in amylose synthesis (*Wx*), starch branching (*SBEIIb*) [76], and starch structure modification (*SSIIa, SSIIIa*) [72,77]. Identifying and manipulating these key loci through breeding or biotechnology can enhance RS content, improving the nutritional quality of rice. Future research should focus on fine-mapping QTLs and developing high-RS rice varieties for better dietary health benefits.

Table 3. Resistant starch (RS) concentrations in rice (% of dry weight).

Grain Type	Processing Method	Resistant Starch Content (% DW)	Key Factors Influencing RS Levels	Refs.
White Rice (Cooked)	Normal cooking (boiled)	0.5–1.5%	High gelatinization reduces RS	[78–81]
Brown Rice (Cooked)	Normal cooking	1.0–2.5%	Higher fiber retains some RS	
Parboiled Rice	Pre-gelatinization and drying	2.0–4.0%	Retrogradation increases RS3	
Cooled Cooked Rice (Retrograded)	Cooking and refrigeration (4 °C, 12–24 h)	3.0–6.0%	Retrogradation forms RS3 (Type 3)	
High-Amylose Rice (e.g., RS Rice Varieties)	Raw preparation or cooking	5.0–15.0%	Genetically high amylose (25–30%) boosts RS2	

2.2.3. Resistant Starch Biosynthesis and Gene Loci

Like dietary fiber, RS is a form of dietary starch that evades digestion in the small intestine and offers health advantages such as better glycemic management, altered gut flora, and a lower risk of metabolic diseases. Although rice (*Oryza sativa* L.), a staple food for more than half of the world's population, normally has a low RS content, some genotypes and processing techniques can increase it. Resistant starch formation in rice depends on starch structure, which is determined by the interplay of multiple biosynthetic enzymes [70,82]. The key steps in RS biosynthesis include the following: increasing amylose content, as high-amylose starch is more resistant to digestion due to its linear structure and reduced enzymatic accessibility; starch branching, as reduced branching (controlled by starch branching enzymes, SBEs) leads to longer amylose chains, increasing RS; starch retrogradation, as upon cooling, the realignment of starch molecules (particularly amylose) forms resistant crystalline structures (RS3); and phosphorylation, as phosphate esters in starch (mediated by Glucan Water Dikinase, GWD) can influence RS formation.

Key Genes and Loci Controlling RS

Several QTLs and genes have been identified that influence RS content in rice [83,84].

A. Major genes involved in RS biosynthesis

The biosynthesis of RS is a complex process governed by multiple enzymes and regulatory genes involved in starch metabolism. Key genes, such as *GBSSI* (granule-bound starch synthase I), encoded by the Waxy (*Wx*) gene, determine amylose content, a major determinant of RS formation. Mutations in *Wx* lead to varying amylose levels, influencing RS content. Additionally, *SSIIa* (starch synthase IIa) and *SBEIIb* (starch branching enzyme IIb) are critical in amylopectin chain elongation and branching, with their suppression resulting in higher RS accumulation. Other important genes include *ISA1* (Isoamylase 1) and *PUL* (Pullulanase), which regulate starch debranching, and their downregulation increases RS by reducing starch digestibility. Furthermore, transcription factors such as *RSR1* (Rice Starch Regulator 1) modulate the expression of starch biosynthetic genes, indirectly affecting RS content. Recent advances in genome editing (e.g., CRISPR-Cas9) have enabled precise modifications in these genes to enhance RS levels in rice cultivars. Understanding the genetic and molecular mechanisms underlying RS biosynthesis provides valuable insights for developing high-RS rice varieties with potential nutritional and therapeutic benefits. The biosynthesis pathway is discussed in detail below:

1. Genes in the *Waxy* (*Wx*) group (e.g., GBSSI, granule-bound starch synthase I) are found at chromosome number six, and these genes encode the enzyme responsible for amylose synthesis [85,86].
2. Starch branching enzymes (SBEs) named *SBEI*, found at chromosome number five, and *SBEIIb*, found at chromosome number two, influence amylopectin branching, in addition to the knockout or suppression of *SBEIIb*, and they increase amylose and RS (e.g., sbe3-rs mutant) [87,88].
3. Pullulanase (*ISA1*, Isoamylase 1) is found at chromosome number eight and affects starch debranching; mutations can alter starch structure and RS content [89].
4. *SSIIa* (starch synthase IIa) is found at chromosome number six and potentially influences amylopectin chain length; its variants affect RS formation [77,90].
5. *SSIIIa* (starch synthase IIIa) is found at chromosome number eight and can modify starch granule morphology; its mutations can increase RS [90,91].

B. Key QTLs associated with RS

Understanding the genetic basis of RS content is crucial for developing high-RS rice varieties through molecular breeding. Key QTLs associated with RS in rice can be identified by integrating genomic approaches and biochemical characterization. Recent advances in QTL mapping and genome-wide association studies (GWASs) have revealed several genomic regions influencing RS accumulation. Major QTLs have been identified at chromosomes 6, 8, and 11, harboring genes involved in starch biosynthesis, such as *GBSSI* (*Waxy*), *SSIIa*, and *SBEIIb*. Polymorphisms in these genes significantly impact amylose content and starch structure, directly affecting RS levels. Additionally, novel QTLs linked to starch degradation and retrogradation processes have been discovered, highlighting the complex genetic regulation of RS. This review consolidates findings from multiple studies, emphasizing the role of QTLs in RS variation across diverse rice germplasms. By elucidating the genetic architecture of RS, this research provides a foundation for developing functional rice varieties with improved nutritional benefits. Several QTLs have been mapped in rice populations [92–94], including qRS7 (found at chromosome seven), associated with high RS in indica rice, and qRS6 (found at chromosome six), which co-localizes with *Wx* and *SSIIa*. Additionally, qAC2 (found at chromosome two) is linked to amylose content and RS.

2.2.4. Health Benefits

RS is a type of dietary starch that escapes digestion in the small intestine and ferments in the colon, functioning similarly to dietary fiber. It has gained significant attention due to its potential health benefits, including improved glycemic control, enhanced gut microbiota composition, and a reduced risk of chronic diseases such as obesity, type 2 diabetes, and colorectal cancer. Rice (*Oryza sativa*) is a major source of carbohydrates but typically contains low levels of RS. However, processing methods such as cooking and cooling, parboiling, and high-amylose rice breeding can increase RS content. This review explores the mechanisms by which RS exerts its health benefits, focusing on its role in promoting the production of short-chain fatty acids (SCFAs), particularly butyrate, which supports colonocyte health, reduces inflammation, and enhances insulin sensitivity. Additionally, RS modulates the gut microbiome by favoring beneficial bacteria such as Bifidobacterium and Lactobacillus while inhibiting pathogenic species. In rice, RS content can be enhanced through genetic modification, enzymatic treatments, and optimized cooking techniques, offering a practical approach to improving its nutritional value. The potential of RS-rich rice as a functional food is discussed, emphasizing its ability to lower postprandial glucose levels, improve satiety, and contribute to metabolic health. Future research should focus on optimizing RS levels in rice varieties and evaluating long-term health impacts in diverse populations. Enhancing RS in rice presents a promising strategy to addressing global health challenges related to metabolic disorders and gut health while accommodating cultural dietary preferences [14,95,96].

2.2.5. Resistant Starch in Rice and Barley

Rice (*Oryza sativa*) and barley (*Hordeum vulgare*) are two major cereal crops with distinct RS profiles due to differences in their genetic, biochemical, and structural properties. While rice is predominantly a low-RS staple, barley contains higher RS content, particularly in its waxy and high-amylose varieties. This review provides an in-depth comparison of RS in rice and barley, a focus on the following: the genetic and molecular basis—key genes such as *GBSSI* (Granule-Bound Starch Synthase I), *SSIIa* (Starch Synthase IIa), and *SBEIIb* (Starch Branching Enzyme IIb) regulate amylose-to-amylopectin ratios, which directly influence RS formation, while rice mutants such as high-amylose rice (e.g., RS111) and barley genotypes (e.g., Himalaya 292) exhibit elevated RS due to altered starch biosynthesis pathways; starch structure and composition—barley starch contains more amylose (up to 40% in high-amylose types) and a higher proportion of resistant fractions (RS2 and RS3) compared to most rice varieties, and the presence of β-glucan in barley further enhances its RS properties, whereas rice starch is more readily digestible unless chemically or physically modified; environmental and processing effects—post-harvest processing (e.g., parboiling in rice, roasting in barley) and cooking methods significantly impact RS retention, and barley's thicker cell walls and higher fiber content contribute to its superior RS stability under heat treatment compared to rice; health and nutritional implications—barley's higher RS content is linked to stronger prebiotic effects, improved insulin sensitivity, and greater satiety than rice; yet, biofortified rice varieties (e.g., through CRISPR-edited Waxy alleles) are emerging as promising alternatives for increasing RS in populations dependent on rice as a staple. This comprehensive analysis highlights the potential of leveraging barley's natural RS advantages while exploring genetic engineering and processing techniques to improve RS in rice. The findings provide a roadmap for developing next-generation high-RS cereals to address global nutritional challenges. Later we discuss in detail resistant starch as a functional component in barley.

2.3. Alkaloids in Rice

2.3.1. Types of Alkaloids in Rice

Alkaloids are a diverse group of nitrogen-containing secondary metabolites found in various plants, including rice (*Oryza sativa* L.). While rice is primarily known as a staple food crop, it also contains several biologically active alkaloids that contribute to its pharmacological and nutritional properties. Rice contains several alkaloids, although at lower levels compared to those in medicinal plants [97–99]. The primary alkaloids identified in rice include gramine, a tryptophan-derived alkaloid with insecticidal and antifungal properties; hordenine, a phenethylamine alkaloid known for its antioxidant and antimicrobial activities; and indole alkaloids, derived from tryptophan, as well as isoquinoline and diterpenoid alkaloids, which are predominantly found in different parts of the plant, such as the husk, bran, and leaves. Indole alkaloids, such as tryptamine and its derivatives, play a role in plant defense mechanisms. Isoquinoline alkaloids, including nuciferine and liensinine, are primarily found in rice husks. Additionally, diterpenoid alkaloids, such as oryzalides, are unique to rice and contribute to its resistance against pathogens. The presence of these alkaloids in rice not only influences its medicinal value but also impacts food safety, as some may exhibit toxicity at high concentrations. Understanding the types, distribution, and biological functions of alkaloids in rice is crucial for optimizing its health benefits while minimizing potential risks.

2.3.2. Biosynthesis of Alkaloids in Rice

The biosynthesis of alkaloids in rice (*Oryza sativa*) is governed by intricate metabolic pathways that are frequently upregulated in response to various stress conditions. In rice, the production of alkaloids involves multi-step enzymatic processes that originate from primary metabolic precursors, such as amino acids (e.g., tryptophan, tyrosine, and lysine) [100,101], and are regulated by a network of genes encoding key biosynthetic enzymes, including cytochrome P450 monooxygenases, methyltransferases, and oxidoreductases. Recent advances in genomics and metabolomics have unveiled stress-inducible transcription factors and signaling molecules, such as jasmonic acid and salicylic acid, which modulate alkaloid biosynthesis pathways. Furthermore, ecological and evolutionary perspectives suggest that the diversification of alkaloid structures in rice may enhance adaptive fitness under fluctuating environmental pressures [102]. Understanding the regulatory mechanisms and functional roles of alkaloid biosynthesis in rice not only provides insights into plant stress responses but also opens avenues for developing stress-resistant crop varieties through metabolic engineering. This review synthesizes current knowledge on the biochemical pathways, genetic regulation, and ecological significance of alkaloid production in rice, highlighting its potential applications in sustainable agriculture. The biosynthesis of alkaloids in rice involves complex pathways that are often induced under stress conditions [42,103–105]. The key steps include decarboxylation and methylation, enzymatic modifications that convert precursors into bioactive alkaloids, and regulation via environmental stress, such as pathogen attack, herbivory, and abiotic stress, which can upregulate alkaloid biosynthesis.

Major Classes of Alkaloids and Their Biosynthetic Pathways

A. Terpenoid indole alkaloids (TIAs; derived from tryptophan). The tryptophan pathway is a precursor for many alkaloids, including gramine and indole alkaloids (e.g., vinblastine, vincristine, strychnine) [106,107]. Key enzymes are tryptophan decarboxylase (TDC), which converts tryptophan to tryptamine; strictosidine synthase (STR), which condenses tryptamine with secologanin; and cytochrome P450s (e.g., CYP72A1), which modify strictosidine into complex TIAs.

B. Benzylisoquinoline alkaloids (BIAs; derived from tyrosine), e.g., morphine, codeine, and berberine [108–110]. Key enzymes are tyrosine decarboxylase (TYDC), which converts tyrosine to dopamine; norcoclaurine synthase (NCS), which condenses dopamine with 4-HPAA; berberine bridge enzyme (BBE), which forms (S)-reticuline; and CYP719B1 (canadine synthase), which produces berberine precursors.

C. Tropane alkaloids (derived from ornithine/arginine), e.g., atropine, scopolamine, and cocaine [111]. Key enzymes are ornithine decarboxylase (ODC), which produces putrescine; putrescine N-methyltransferase (PMT), which methylates putrescine; and hyoscyamine 6β-hydroxylase (H6H), which converts hyoscyamine to scopolamine.

D. Purine alkaloids (derived from xanthosine), e.g., caffeine and theobromine [112,113]. Key enzymes are xanthosine methyltransferase (XMT), which methylates xanthosine, and caffeine synthase (CS), which converts theobromine to caffeine.

2.3.3. Biotechnological Application

Alkaloids have garnered significant attention in biotechnology due to their pharmacological and agrochemical properties. In rice (*Oryza sativa* L.), alkaloids play crucial roles in defense mechanisms against pathogens, pests, and environmental stressors, making them promising candidates for biotechnological applications. Recent advances in metabolic engineering and synthetic biology have enabled the manipulation of alkaloid biosynthetic pathways to enhance rice resilience, nutritional quality, and medicinal value. This review explores the biotechnological potential of alkaloids in rice, focusing on their biosynthesis, genetic regulation, and metabolic engineering strategies. Additionally, CRISPR-Cas9 and RNA interference (RNAi) technologies have been employed to modulate endogenous alkaloid production, improving stress tolerance, and reducing dependency on chemical pesticides. Furthermore, alkaloid-enriched rice varieties hold promise for nutraceutical applications, offering bioactive compounds with antioxidant, anti-inflammatory, and anticancer properties. Challenges such as yield penalties, metabolic trade-offs, and regulatory hurdles are also discussed. By integrating multi-omics approaches and precision breeding, the development of alkaloid-biofortified rice could revolutionize sustainable agriculture and functional food industries. Alkaloid biosynthesis is governed by multi-enzyme pathways encoded by specific gene loci, with key enzymes including STR, NCS, PMT, and XMT playing pivotal roles [114]. Understanding these genetic and biochemical mechanisms enables metabolic engineering for enhanced alkaloid production, benefiting pharmaceuticals and agriculture [115,116]. Future research should focus on pathway elucidation in non-model species and synthetic biology applications.

2.3.4. Functional Roles of Alkaloids in Rice

Alkaloids play crucial functional roles in plant defense mechanisms, particularly in rice (*Oryza sativa* L.). As a staple crop, rice faces persistent biotic stressors, including herbivores, pathogens, and competing plants, which threaten yield and quality. Alkaloids contribute to rice defense through direct toxicity, antifeedant effects, and signaling modulation, deterring herbivores and inhibiting microbial proliferation. Notable rice alkaloids, such as gramine, hordenine, and avenanthramides, exhibit broad-spectrum antimicrobial and insecticidal properties, disrupting cellular processes in pests and pathogens. Additionally, alkaloids function as inducible defenses, with their biosynthesis often upregulated in response to herbivory or pathogen attack via jasmonic acid (JA) and salicylic acid (SA) signaling pathways. Recent studies highlight their role in allelopathy [117,118], suppressing competing weeds through root exudation. However, the metabolic cost of alkaloid production may produce a trade-off with growth, necessitating precise regulatory mechanisms. Advances in metabolomics and genetic engineering have unveiled key biosynthetic genes (e.g., cytochrome P450s and methyltransferases), enabling the development of alkaloid-enhanced rice varieties for sustainable pest management [110,119].

These bioactive compounds contribute to plant defense against biotic stressors, such as herbivores [120], pathogens [121], and competing plants, through their antimicrobial, antifeedant, and allelopathic properties. Additionally, alkaloids participate in abiotic stress responses, including those to drought [122], salinity, and heavy metal toxicity, by modulating oxidative stress through antioxidant activity and regulating stress-signaling pathways. Certain rice-associated alkaloids, such as gramine and hordenine, have been implicated in enhancing resistance to insect herbivory and microbial infections. Furthermore, alkaloids may interact with phytohormones, such as jasmonic acid and salicylic acid, to fine-tune defense mechanisms. Recent advances in metabolomics and genetic engineering have uncovered key biosynthetic pathways responsible for alkaloid production, offering potential for developing stress-resilient rice varieties [123].

Alkaloids play crucial functional roles in rice (*Oryza sativa*), particularly in relation to human health. While primarily known for their defensive properties against herbivores and pathogens in plants, certain rice-derived alkaloids exhibit significant bioactive potential, influencing human physiology in both beneficial and detrimental ways. Positive aspects include some rice alkaloids' possession of antioxidant, anti-inflammatory, and anticancer properties, contributing to disease prevention and health promotion. For instance, compounds such as tricin and gramine have demonstrated chemo-preventive effects by modulating cellular signaling pathways linked to oxidative stress and carcinogenesis. Conversely, certain alkaloids may act as antinutritional factors, interfering with nutrient absorption or exhibiting mild toxicity at high concentrations. Additionally, the presence of alkaloids in rice can influence metabolic health, with some studies suggesting their role in glucose regulation and neuroprotection. However, the mechanisms underlying these effects remain understudied, necessitating further research to elucidate their pharmacokinetics, optimal bioactive concentrations, and potential interactions with other dietary components. Understanding the dual nature of rice alkaloids both as health-promoting agents and as presenting possible risks can guide dietary recommendations, biofortification strategies, and pharmacological applications, ultimately enhancing the nutritional and medicinal value of rice as a global staple food.

2.3.5. Comparisons of Alkaloids Between Rice and Barley

While alkaloid diversity has been extensively studied in dicots, its presence and biosynthetic pathways in monocots, particularly cereals such as rice (*Oryza sativa*) and barley (*Hordeum vulgare*), remains less characterized. This review provides an in-depth comparative analysis of the alkaloid biosynthesis, diversity, and functional significance in rice and barley, highlighting key genetic, enzymatic, and metabolic differences between these two agriculturally important crops. Rice and barley exhibit distinct alkaloid profiles due to divergent evolutionary pressures and ecological adaptations. Barley, a temperate crop, produces gramine and hordenine—two major alkaloids with demonstrated roles in herbivore deterrence and fungal resistance. These compounds are derived from tryptophan and tyrosine, respectively, via specialized cytochrome *P450* (*CYP*) and methyltransferase enzymes. In contrast, rice primarily accumulates benzoxazinoid-related compounds and simple indole alkaloids, which contribute to pest resistance and allelopathy. The presence of benzoxazinoids in certain wild rice species, but their absence in cultivated rice suggests domestication-related loss, whereas barley has retained its alkaloid biosynthesis pathways due to strong selection for stress resilience. Genomic comparisons reveal that key alkaloid biosynthetic genes, particularly those encoding CYPs, decarboxylases, and O-methyltransferases, have undergone lineage-specific expansions in barley. For instance, the tryptamine N-methyltransferase (*HvTMT*) gene cluster is unique to barley and underlies gramine biosynthesis. Rice, however, lacks this pathway but possesses alternative detoxification-related *CYPs* (e.g., *CYP79A1*) that contribute to indole alkaloid diversification. Additionally, comparative transcriptomic studies indicate that barley alkaloid synthesis is highly inducible under biotic stress, whereas rice relies more on flavonoid and terpenoid defenses. The regulatory networks controlling alkaloid production also differ significantly. Barley employs jasmonate (JA)- and abscisic acid (ABA)-mediated signaling to upregulate alkaloid biosynthesis, whereas rice uses a broader array of phytohormones, including salicylic acid (SA), to modulate secondary metabolism. Furthermore, recent advances in metabolomics and genome-wide association studies (GWAS) have identified quantitative trait loci (QTLs) linked to alkaloid variation in barley (e.g., Alk1.1 on chromosome 2H), while analogous studies in rice remain limited. This comparative analysis underscores the evolutionary trade-offs between domestication and defense in cereals. Barley's retention of alkaloid pathways highlights its ecological adaptation to temperate climates with higher pest pressures, whereas rice's shift toward alternative defense mechanisms reflects its subtropical origins and human selection for palatability. Future research should explore metabolic engineering strategies to transfer beneficial alkaloid traits between these crops, potentially enhancing stress resilience without compromising agronomic performance. Understanding these differences provides a foundation for improving cereal crop defenses through targeted breeding and biotechnological approaches. Later we discuss alkaloids as a functional component in barley in detail.

2.4. Interactions Among GABA, RS, and Alkaloid Pathways

The interplay between GABA, resistant starch, and alkaloid biosynthesis pathways represents a complex and underexplored nexus in plant metabolism and human health. GABA, a key inhibitory neurotransmitter in mammals, also functions as a signaling molecule in plants, modulating stress responses and carbon–nitrogen balance. Resistant starch, a fermentable dietary fiber, influences gut microbiota composition and short-chain fatty acid (SCFA) production, indirectly affecting systemic GABAergic activity through the gut–brain axis. Meanwhile, alkaloids—nitrogen-containing secondary metabolites—often share biosynthetic precursors with GABA, particularly via glutamate and ornithine-derived pathways. Emerging evidence suggests that resistant starch fermentation may alter the availability of this compound by modulating microbial GABA metabolism or host glutamate decarboxylase (GAD) activity. Conversely, certain alkaloids can interact with GABA receptors, potentially competing with endogenous GABA or influencing its synthesis. This tripartite interaction has implications for metabolic health, neuroprotection, and stress adaptation in both plants and animals. For instance, GABA-enriched foods (e.g., fermented products or germinated grains) may synergize with resistant starch to enhance gut–brain signaling, while alkaloids from medicinal plants could either potentiate or inhibit GABAergic effects. This review synthesizes the current knowledge on the molecular crosstalk between these pathways, highlighting gaps in understanding how dietary components (resistant starch) and phytochemicals (alkaloids) collectively modulate GABA homeostasis.

The genetic loci controlling GABA, RS, and alkaloids in rice are distinct but may interact through shared precursors and stress responses [62,124]. Understanding these differences helps in breeding rice varieties with enhanced functional properties, such as high-GABA rice for relaxation, high-RS rice for diabetes management, and alkaloid-rich rice for medicinal applications (Table 4). Future research could explore multi-trait gene editing to develop rice with optimized levels of all three components.

Table 4. Key differences among the functional components in rice.

Feature	GABA	Resistant Starch	Alkaloids	Refs.
Primary Function	Neurotransmission, stress response	Producing dietary fiber, glycin control	Defense, producing medicinal effects	[125–127]
Key Genes	GAD1, GAD2	SSlla, SBEllb	TYDc, CYP450	
Main Chromosome QTLs	1, 5, 8	6, 7	4, 11	
Inducing Factors	Hypoxia, germination	High amylose, enzymatic defects	Pathogen attack, stress	

3. Interaction of and Differences in Gene Loci Among Barley Functional Components

Barley (*Hordeum vulgare*) contains several bioactive compounds, including GABA, RS, and alkaloids, which contribute to its nutritional and functional properties [128]. These components are regulated by distinct genetic loci and metabolic pathways, yet their interactions and differences in gene regulation remain underexplored [7,129,130]. In this review, we examines the key gene loci associated with GABA biosynthesis (e.g., *GAD* genes), RS formation (e.g., *SSIIa*, *SBEIIb*), and alkaloid production (e.g., *TYDC*, *CYP450s*), highlighting their chromosomal locations (e.g., GABA QTLs on chromosome 2H, RS-related genes on 7H) and metabolic crosstalk. We discuss

as germinated barley and fermented barley extracts, exhibit significant bioactive properties, including antioxidant, anti-inflammatory, and cardioprotective effects. Furthermore, GABA biosynthesis in barley has been linked to improved gut health via its modulation of the gut–brain axis and promotion of beneficial microbiota. The potential applications of GABA-enriched barley in functional foods and nutraceuticals are vast, offering a natural approach to mitigating chronic diseases such as hypertension, anxiety disorders, and metabolic syndrome. Additionally, GABA-rich barley is linked to improved cardiovascular health, stress reduction, and better sleep. It is also used in functional foods and beverages due to its potential to lower blood pressure and improve overall well-being [136,149].

3.1.5. Comparison of GABA in Rice and Barley

Rice (*Oryza sativa*) and barley (*Hordeum vulgare*) are two major cereal crops that accumulate GABA under different physiological conditions, yet their GABA biosynthesis, regulation, and accumulation patterns exhibit both similarities and differences. This comparative review explores the GABA metabolic pathways in rice and barley, focusing on the key enzymes glutamate decarboxylase (GAD) and GABA transaminase (GABA-T), as well as factors influencing GABA levels, such as germination, fermentation, and environmental stresses. Rice, particularly brown and germinated rice, is known for its high GABA content, which can be further enhanced through anaerobic treatment or microbial fermentation. Barley, on the other hand, significantly accumulates GABA during malting and sprouting, making it valuable for brewing and functional food applications. While both crops rely on the GABA shunt for stress adaptation, barley exhibits a stronger GABA response to drought and salinity compared to rice, possibly due to differences in GAD isoforms and regulatory mechanisms. Additionally, genetic variations between rice and barley influence GABA-enriched product development, with rice being more commonly utilized in GABA-fortified foods. Both rice and barley can be enhanced to increase their levels of GABA through germination or fermentation. The exact concentration depends on the variety, processing methods, and conditions (e.g., temperature, pH, and treatment duration). GABA-rich rice and barley are used in functional foods, such as GABA rice tea, GABA barley tea, and fermented products. These products are marketed for their health benefits, particularly in Asian countries. Additionally, each grain provides additional nutrients, such as fiber, vitamins, and minerals, alongside GABA. Brown rice and whole barley are considered more nutritious due to their higher fiber and nutrient contents.

3.2. *Resistant Starch in Barley*

A form of dietary fiber called RS ferments in the colon after evading digestion in the small intestine. It offers several health advantages, such as better glycemic management, gut health, and a lower risk of colorectal cancer [150–152]. Barley (*Hordeum vulgare*) is a rich source of RS, particularly high-amylose varieties. This review summarizes the biochemical pathways of RS biosynthesis in barley, identifies key genes involved in RS formation, and discusses genetic and biotechnological approaches to enhance RS content. Understanding the genetic regulation of RS barley can facilitate the development of functional foods with enhanced nutritional value.

3.2.1. Types of Resistant Starch in Barley

Barley (*Hordeum vulgare* L.) is a significant cereal grain known for its high content of dietary fiber, particularly RS, which confers various health benefits, including improved glycemic control, enhanced gut health, and a reduced risk of chronic diseases. Resistant starch, defined as the starch fraction that escapes digestion in the small intestine and ferments in the colon, is categorized into five main types (RS1–RS5), each with distinct structural and functional properties. In barley, resistant starch primarily exists as RS1 (starch that is physically inaccessible due to intact cell walls), RS2 (native granular starch with a high amylose content and a B- or C-type crystalline structure), and RS3 (retrograded starch formed after cooking and cooling). Additionally, RS4 (chemically modified starch) may be present in processed barley products, while RS5 (amylose–lipid complexes) is less common but can exist in barley due to interactions between starch and endogenous lipids. The proportion and bioavailability of these RS types in barley are influenced by genetic factors, environmental conditions, and processing methods. Recent advances in breeding and bioprocessing techniques aim to enhance RS content in barley cultivars, particularly RS2 and RS3, to maximize nutritional benefits. Understanding the distribution and behavior of different RS types in barley is crucial for developing functional foods with tailored health-promoting properties [153].

3.2.2. Factors Affecting Resistant Starch Content

There are many factors affecting resistant starch content, including barley variety; hulled barley (with the outer bran intact) has a higher resistant starch content compared to pearled barley (polished to remove the bran). Additionally, less processed barley retains more resistant starch [154,155]. Concentrations of starch in different types of barley are shown in Table 7 [156–159]. Other factors influence resistant starch; its content in cooked and cooled barley ranges between 3 and 5%, whereas hulled barley has a higher resistant starch content compared to pearled barley [160]. Abundant nitrogen reduces RS, while phosphorus has a minimal effect, and retrogradation (RS3 formation) is enhanced by cooking and cooling.

Table 7. Resistant starch (RS) concentrations in barley (% of dry weight).

Grain Type	Processing Method	Resistant Starch Content (% DW)	Key Factors Influencing RS Levels
Barley (Whole, Raw)	Unprocessed	3.0–8.0%	β-glucans and RS2 are at naturally high levels
Pearled Barley (Cooked)	Boiled	1.5–3.5%	Polishing reduces fiber and RS
Barley Flour (Baked)	Heat-treated (e.g., bread)	1.0–2.5%	Gelatinization lowers RS
Germinated Barley	Soaked/sprouted (48 h)	2.0–5.0%	Enzymes partially degrade RS
Barley Flakes (Rolled)	Steam-pressed and dried	4.0–7.0%	Processing preserves RS2

3.2.3. Resistant Starch Biosynthesis and Gene Loci

RS is classified into five types (RS1–RS5) based on its structure and resistance to digestion [161,162]. Barley naturally contains RS2 (granular starch in raw form) and RS3 (starch retrograded after cooking and cooling) [163]. High-RS varieties are nutritionally superior, making them valuable for diabetes management and digestive health [14,94,164]. Starch biosynthesis occurs in the amyloplast and involves four key enzymes [164–167]: ADP-glucose pyrophosphorylase (AGPase) produces ADP-glucose, the substrate for starch synthesis; granule-bound starch synthase (GBSS) synthesizes amylose (linear α-1,4-glucan chains); starch branching enzyme (SBE) introduces α-1,6-branches to form amylopectin; and starch debranching enzyme (DBE) trims excessive branches for proper crystallization. High amylose content (a key determinant of RS) is influenced by reduced SBEII activity, which reduces branches and makes amylose more linear, and increased GBSS activity, which increases amylose content.

Key Gene Loci Controlling RS in Barley

Several genes regulate starch structure and RS content in barley, the most significant of which are shown in Table 8. Resistant starch in barley is primarily controlled by *HvGBSSI*, *HvSBEIIa/b*, and *HvISA1*, with high-amylose genotypes containing the greatest RS contents.

Advances in gene editing and breeding are enabling the development of barley varieties with enhanced RS for improved human health. Future research should focus on optimizing agronomic practices and processing methods to maximize RS yield.

Table 8. Key gene loci controlling resistant starch in barley.

Gene	Enzyme	Function	Chromosomal Location (Morex V3)	Effect on RS	Refs.
HvGBSSI (Waxy)	Granule-bound starch synthase I	Amylose synthesis	HORVU.MOREX.r3.7HG0634210 (7H)	Increases RS	[168–171]
HvSBEIIa	Starch branching enzyme IIa	Amylopectin branching	HORVU.MOREX.r3.2HG0123400 (2H)	Increases RS	
HvSBEIIb	Starch branching enzyme IIb	Amylopectin branching	HORVU.MOREX.r3.2HG0123500 (2H)	Knocks out abundant amylose	
HvISA1 (DBE)	Isoamylase 1	Debranching enzyme	HORVU.MOREX.r3.4HG0387650 (4H)	Affects starch crystallinity	
HvAGP-L	ADP-glucose pyrophosphorylase (large subunit)	ADP-glucose synthesis	HORVU.MOREX.r3.1HG0056780 (1H)	Modulates starch yield	

3.2.4. Health Benefits

Barley, a nutrient-rich cereal grain, is a notable source of RS, particularly its whole and high-amylose varieties. This review explores the physiological effects of barley-derived resistant starch, emphasizing its role in improving metabolic health, gut microbiota composition, and disease prevention. Research indicates that barley RS enhances glycemic control by slowing glucose absorption, making it beneficial for individuals with insulin resistance or type 2 diabetes. Additionally, its fermentation in the colon produces short-chain fatty acids (SCFAs) such as butyrate, which supports colonocyte health, reduces inflammation, and may lower colorectal cancer risk. The prebiotic properties of barley RS promote the growth of beneficial gut bacteria (e.g., *Bifidobacterium* and *Lactobacillus*), improving gut barrier function and immune modulation. Furthermore, barley RS contributes to weight management by increasing satiety and reducing fat storage. Its ability to lower cholesterol levels through bile acid binding and SCFA production also highlights its cardiovascular benefits. Given these multifaceted advantages, incorporating barley-rich foods into diets could serve as a strategic approach to combating metabolic disorders, gastrointestinal diseases, and obesity [14,95,96].

3.3. *Alkaloids in Barley*

Alkaloids are nitrogen-containing secondary metabolites with diverse pharmacological and ecological roles, including defense against herbivores and pathogens [102,172]. While barley (*Hordeum vulgare*) is not a major alkaloid-producing crop compared to plants such as opium poppy or tobacco, it contains several bioactive alkaloids, such as gramine and hordenine, which contribute to its stress resilience and medicinal properties [173,174]. This review synthesizes current knowledge on alkaloid biosynthesis pathways in barley, identifies key gene loci involved in their production, and discusses genetic and biotechnological approaches to manipulating alkaloid content for improved crop resilience and nutraceutical applications.

3.3.1. Types of Alkaloids in Barley

Barley, a widely cultivated cereal grain, is known for its rich contents of alkaloids [175–177], which are structurally diverse and perform significant biological activities. In barley, the most studied alkaloids include hordenine, known for its stimulant and antimicrobial properties. Gramine also contributes to barley's defense mechanisms, and β-carboline alkaloids exhibit neuroprotective and antioxidant activities.

Alkaloids Biosynthesis and Gene Loci

- Alkaloids biosynthesis

The biosynthesis of alkaloids in barley has similarities with rice but has unique features [178,179].

- Gramine biosynthesis

Gramine, a bioactive indole alkaloid predominantly found in barley (*Hordeum vulgare*), exhibits significant defensive properties against herbivores and pathogens, contributing to plant stress resilience. Despite its ecological importance, the biosynthetic pathway and genetic regulation of gramine remain only partially elucidated [173,180]. Recent advances in genomics and metabolomics have identified key candidate genes involved in gramine biosynthesis, including putative tryptophan decarboxylases (TDCs) and methyltransferases (MTs), which likely catalyze the conversion of tryptophan to gramine via intermediate steps such as the formation of 3-aminomethylindole (AMI). Genetic mapping studies regarding barley have localized QTLs associated with gramine accumulation to chromosomes 2H, 5H, and 7H, implicating several cytochrome P450s (CYPs) and UDP-glycosyltransferases (UGTs) in its regulation. Furthermore, transcriptomic analyses under biotic stress conditions reveal the induction of specific biosynthetic genes, suggesting a jasmonate-mediated signaling pathway controlling gramine production. Comparative genomics with related Poaceae species highlights conserved enzymatic functions as well as species-specific gene duplications, which may explain differential gramine accumulation across cultivars [181,182].

- Hordenine biosynthesis

Hordenine (N,N-dimethyl-tyramine) is a bioactive alkaloid found in barley (*Hordeum vulgare*) and other related grasses, with potential pharmacological properties. Its biosynthesis involves the decarboxylation of tyrosine to tyramine via tyrosine decarboxylase (TyDC), followed by N-methylation via N-methyltransferases (NMTs), such as norcoclaurine-6-O-methyltransferase-like enzymes [183–185]. Recent advances in genomics and metabolomics have enabled the identification of candidate gene loci associated with hordenine biosynthesis in barley. Key QTLs have been mapped to specific chromosomal regions, particularly on chromosomes 2H and 5H, which harbor genes encoding enzymes involved in phenylpropanoid and alkaloid pathways. Transcriptomic analyses reveal the differential expression of these biosynthetic genes during barley germination, in which hordenine accumulation is highest. Furthermore, comparative genomics suggests the evolutionary conservation of hordenine-related genes in *Triticeae* species. Understanding the genetic regulation of hordenine biosynthesis could facilitate metabolic engineering for the enhanced production or the development of low-alkaloid barley varieties for improved food and feed safety [178,186,187].

- Avenanthramides (barley-specific phenolic alkaloids)

Avenanthramides (AVNs) are a unique class of phenolic alkaloids predominantly found in barley (*Hordeum vulgare*), known for their potent antioxidant, anti-inflammatory, and health-promoting properties. These compounds are synthesized via the phenylpropanoid pathway, via which hydroxycinnamoyl-CoA thioesters and anthranilic acid derivatives undergo condensation reactions catalyzed by hydroxycinnamoyl-CoA: hydroxy anthranilate N-hydroxycinnamoyl transferase (HHT) [188,189]. Recent genomic and transcriptomic studies have identified key gene loci associated with AVN biosynthesis in barley, including *HvHHT* and cytochrome P450 (*CYP*), genes involved in the modification of the AVN core structure. QTL mapping and genome-wide association studies (GWASs) have further revealed genetic variations influencing AVN accumulation, particularly on chromosomes 2H, 5H, and 7H. Additionally, the regulation of *AVN* biosynthesis is modulated by transcription factors such as those in the MYB and bHLH families, which respond to environmental stressors such as pathogen attack and UV radiation. Understanding the genetic and enzymatic mechanisms underlying AVN production provides valuable insights for biofortification strategies and the development of barley cultivars with enhanced nutraceutical value [190].

Key Gene Loci Controlling Alkaloids in Barley

Several genes regulate alkaloids in barley, the most significant of which are shown in Table 9. Barley produces several bioactive alkaloids, with gramine, hordenine, and avenanthramides being the most notable. Key genes (HvTDC, HvTyDC, HvHHT) have been partially characterized, but further research is needed to fully map their regulatory networks. Manipulating these pathways could enhance barley's stress resilience and nutritional value, offering opportunities for sustainable agriculture and functional food development.

Table 9. Key gene loci controlling alkaloids in barley.

Alkaloid	Key Gene	Enzyme	Chromosomal Location (Morex V3)	Function	Refs.
Gramine	HvTDC	Tryptophan decarboxylase	HORVU.MOREX.r3. 5HG0452100 (5H)	Converts tryptophan → tryptamine	[178,191–193]
	HvNMT	N-methyltransferase	Unknown (predicted 2H or 7H)	Methylates tryptamine	
Hordenine	HvTyDC	Tyrosine decarboxylase	HORVU.MOREX.r3. 3HG0278900 (3H)	Converts tyrosine → tyramine	
	HvBBE	Berberine bridge enzyme	HORVU.MOREX.r3. 6HG0523400 (6H)	Oxidizes N-methyltyramine → hordenine	
Avenanthramides	HvHHT	Hydroxycinn-amoyltransferase	HORVU.MOREX.r3. 4HG0391200 (4H)	Forms avenanthramide conjugates	

Barley (*Hordeum vulgare*), a key cereal crop, produces several bioactive alkaloids, including gramine, hordenine, and avenanthramides, which contribute to its resistance mechanisms and potential health benefits. Recent advances in genomics and metabolomics have identified key genes involved in alkaloid biosynthesis, enabling a deeper understanding of their regulation and applications in crop improvement as shown in Table 10. Gramine (indole alkaloid) functions as a defense compound toxic to insects and pathogens and biosynthesizes genes through tryptophan decarboxylase (TDC)—converting tryptophan to tryptamine and gramine synthase—catalyzing the methylation of tryptamine to form gramine. Hordenine (phenylethylamine alkaloid) has antimicrobial and allelopathic properties and biosynthesizes tyrosine decarboxylase (TYDC)—producing tyramine from tyrosine and N-methyltransferases (NMTs)—converting tyramine to hordenine. Avenanthramides (phenylalanine-derived alkaloids) functions as an antioxidant and anti-inflammatory compound and biosynthesizes genes through phenylalanine ammonia-lyase (PAL)—initiating the phenylpropanoid pathway and hydroxycinnamoyl-CoA transferases (HCTs)—synthesizing avenanthramide precursors. Alkaloid-related genes in barley govern critical defense and health-promoting compounds. Understanding their genetic regulation opens avenues for developing stress-resistant, nutritionally enhanced barley varieties with applications in sustainable agriculture and functional foods.

Table 10. Alkaloid-Related Genes in Barley and Their Functions.

Gene Name	Gene ID/Locus	Alkaloid Pathway	Function	Impact/Phenotype	Refs.
HvNMT (*N-methyltransferase*)	MLOC_11101 (Chromosome 2H)	Gramine biosynthesis	Catalyzes the methylation of tryptamine to form *N*-methyltryptamine (precursor to gramine)	High expression in leaves; contributes to aphid resistance	[194,195]
HvTDC (*Tryptophan decarboxylase*)	MLOC_61215 (Chromosome 5H)	Gramine & serotonin pathway	Converts tryptophan to tryptamine (first step in gramine synthesis)	Knockdown reduces gramine accumulation, increasing susceptibility to herbivores	[196]
HvCYP76M8 (*Cytochrome P450*)	MLOC_72190 (Chromosome 3H)	Hordenine biosynthesis	Hydroxylates tyramine to form *N*-methyltyramine (hordenine precursor)	Associated with fungal resistance and allelopathy	[197]
HvAOX1 (*Amine oxidase*)	MLOC_80922 (Chromosome 6H)	Hordenine catabolism	Oxidizes hordenine to "p"-hydroxyphenylacetaldehyde	Regulates hordenine levels in germinating barley	[198]
HvBBE (*Berberine bridge enzyme-like*)	MLOC_40114 (Chromosome 4H)	Benzoxazinoid pathway	Involved in DIBOA (2,4-dihydroxy-1,4-benzoxazin-3-one) biosynthesis	Linked to defense against pathogens and insects	[199]
HvOMT2 (*O-methyltransferase*)	MLOC_50233 (Chromosome 7H)	Avenanthramide-like compounds	Methylates hydroxycinnamoyl derivatives	Possible role in stress-induced phenolic alkaloid production	[200]
HvLDC (*Lysine decarboxylase*)	MLOC_30567 (Chromosome 1H)	Cadaverine-derived alkaloids	Converts lysine to cadaverine (precursor for quinolizidine alkaloids)	Low activity in barley; more prominent in lupins	[201]

3.3.2. Functional Roles of Alkaloids in Barley

Alkaloids provide defense mechanisms for barley, protecting it from pests and diseases, reducing the need for chemical pesticides. They also provide adaptation to abiotic stress by enhancing barley's resilience to environmental challenges [179]. Alkaloids also confer nutritional and health benefits, contributing to the grain's functional food properties, including antioxidant and anti-inflammatory effects [128].

3.3.3. Comparative Analysis of Alkaloids in Rice and Barley

While both rice cereals and barley contain alkaloids, their profiles and concentrations differ significantly. In terms of diversity, barley contains more diverse alkaloids compared to rice. In terms of concentration, alkaloids are generally present at higher concentrations in barley, making it a more potent source of bioactive compounds. In terms of functional roles, both grains utilize alkaloids for defense and stress adaptation, but barley's alkaloids are more extensively studied due to their pharmacological potential.

3.3.4. Application in Human Health and Agriculture

- Human health
  - Antioxidant properties: Alkaloids in rice and barley enhance the antioxidant capacity of these grains, potentially reducing oxidative stress in humans [202].
  - Anti-inflammatory effects: Some alkaloids exhibit anti-inflammatory properties, which may benefit chronic disease prevention [203].
  - Neuroprotective potential: β-carboline alkaloids in barley have shown promise in neuroprotection and mental health [204].
- Agriculture application

- Natural pesticides: Alkaloids can be harnessed for eco-friendly pest control in rice and barley cultivation [205].
- Stress-resilient crops: Understanding alkaloid biosynthesis can aid in developing stress-tolerant varieties of rice and barley [206].
- Functional foods: Enhancing alkaloid content through breeding or biotechnology can improve the nutritional value of these grains [207].

Rice and barley are not only staple foods but also rich sources of functional components that offer numerous health benefits. Incorporating whole-grain rice (e.g., brown rice, black rice) and barley into diets can provide essential nutrients, antioxidants, and bioactive compounds that support cardiovascular health, glycemic control, gut health, and overall well-being. Further research is needed to explore the full potential of these functional components and their applications in functional foods.

4. Conclusions

Rice (*Oryza sativa*) and barley (*Hordeum vulgare*) are two of the most important cereal crops globally, serving as staple foods for billions. Beyond their nutritional value, they are rich sources of bioactive compounds such as GABA, RS, and alkaloids, which contribute to their nutraceutical potential. A synergistic approach combining these compounds from both cereals could revolutionize functional food development, offering enhanced health benefits ranging from metabolic regulation to neuroprotection. GABA, a non-protein amino acid, acts as a major inhibitory neurotransmitter in the central nervous system, aiding stress reduction, blood pressure regulation, and improved sleep. In rice and barley, GABA accumulates under stress conditions (e.g., hypoxia, germination) via the glutamate decarboxylase (GAD) pathway. Key genes such as *OsGAD1/GAD2* (rice) and *HvGAD1* (barley) regulate GABA synthesis. Enhancing GABA content through targeted breeding or fermentation could amplify its anxiolytic and cardioprotective effects. Additionally, high-RS varieties of rice (e.g., high-amylose mutants such as *OsSSIIIa*) and barley (e.g., *HvGBSSI* mutants) have been developed by manipulating starch biosynthesis genes. Combining RS from both grains could enhance colonic fermentation, fostering a healthier gut microbiome. While alkaloids are less studied in cereals compared to medicinal plants, certain alkaloids in rice (e.g., tricin) and barley (e.g., gramine) exhibit antioxidant, anti-inflammatory, and anticancer properties. Biosynthetic pathways involve enzymes such as tryptophan decarboxylase (TDC) and cytochrome P450s, with genetic loci such as *OsTDC1* (rice) and *HvTDC* (barley) playing crucial roles. Optimizing alkaloid content could enhance the chemo-preventive potential of nutraceuticals derived from these grains. Future perspectives should focus on advancements in genomic editing (CRISPR-Cas9), fermentation biotechnology, and precision breeding, which can optimize the biosynthesis of these compounds. Identifying QTLs and regulatory genes will facilitate the development of highly nutraceutical rice and barley varieties. Furthermore, clinical trials validating synergistic effects will be essential for commercializing functional foods. In conclusion, integrating the nutraceutical properties of rice and barley presents a promising strategy for combating modern health challenges. By leveraging their complementary bioactive profiles, we can develop next-generation functional foods that promote holistic well-being, bridging the gap between agriculture and preventive medicine. Additionally, we discuss the health implications of these functional components, including their roles in reducing hypertension, managing diabetes, and exhibiting neuroprotective effects. Understanding the genetic differences between rice and barley in accumulating these compounds can guide biofortification strategies to enhance nutritional quality in cereal crops, ultimately benefiting human health and dietary outcomes. Our study has potentially wide implications, as we discuss the key functional components in both rice and barley and their potential health benefits. This review provides a detailed analysis of the types, structures, and functional roles of GABA, RS, and alkaloids in rice and barley, highlighting their nutritional significance, technological application, and potential for improving human health and agricultural sustainability. All the references were included in the manuscript related to our work, and some of them from our previous work at the department. We believe the results will be of interest to a wide readership including academics and researchers working in agriculture; genetics scientists, policy makers, and managers in government agencies; and agriculture and food security and policy makers.

Author Contributions: Conceptualization, E.E. and Y.Z.; methodology, X.Y.; software, J.Y.; validation, M.M.H., T.Y. and E.E.; formal analysis and investigation, E.E.; resources, X.Y.; data curation, Y.Z.; writing—original draft preparation, E.E.; writing—review and editing, E.E. and Y.Z.; visualization, Y.Z.; supervision, Y.Z.; project administration, Y.Z.; funding acquisition, Y.Z. All authors have read and agreed to the published version of the manuscript.

Funding: This work received funding from the Talented Young Scientist Program (no. P20R53019) and the China Agriculture Research System (no. CARS-05-01A-04).

Conflicts of Interest: The authors declare no conflicts of interest.

References

1. Kaur, A.; Purewal, S.S.; Phimolsiripol, Y.; Punia Bangar, S. Unraveling the Hidden Potential of Barley (*Hordeum vulgare*): An Important Review. *Plants* **2024**, *13*, 2421. [CrossRef] [PubMed]
2. Benjamin, J.; Idowu, O.; Babalola, O.K.; Oziegbe, E.V.; Oyedokun, D.O.; Akinyemi, A.M.; Adebayo, A. Cereal production in Africa: The threat of certain pests and weeds in a changing climate—A review. *Agric. Food Secur.* **2024**, *13*, 18. [CrossRef]
3. ElShamey, E.A.; Yang, X.; Yang, J.; Pu, X.; Yang, L.E.; Ke, C.; Zeng, Y. Occurrence, Biosynthesis, and Health Benefits of Anthocyanins in Rice and Barley. *Int. J. Mol. Sci.* **2025**, *26*, 6225. [CrossRef] [PubMed]
4. Guo, H.; Li, H.; Xiao, Y.; Wu, D.-T.; Gan, R.-Y.; Kang, Z.; Huang, Y.; Gao, H. Revisiting fermented buckwheat: A comprehensive examination of strains, bioactivities, and applications. *Crit. Rev. Food Sci. Nutr.* **2025**, 1–22. [CrossRef] [PubMed]
5. Park, Y.K.; Kim, J.; Ryu, M.S.; Yang, H.-J.; Jeong, D.-Y.; Shin, D.-H. The Health Benefits and Functional Properties of Gochujang: A Comprehensive Review of Fermentation and Bioactive Compounds. *Fermentation* **2025**, *11*, 67. [CrossRef]
6. Yoon, N.; Kim, Y.; Chin, J.H.; Lee, S. Comparative Analysis of Chemical Profiles and Bioactive Properties in Six Pigmented and Non-Pigmented Rice Varieties. *Chemistry* **2025**, *7*, 58. [CrossRef]
7. Shvachko, N.A.; Loskutov, I.G.; Semilet, T.V.; Popov, V.S.; Kovaleva, O.N.; Konarev, A.V. Bioactive components in oat and barley grain as a promising breeding trend for functional food production. *Molecules* **2021**, *26*, 2260. [CrossRef]
8. Loskutov, I.G.; Khlestkina, E.K. Wheat, barley, and oat breeding for health benefit components in grain. *Plants* **2021**, *10*, 86. [CrossRef]
9. Islam, M.T.; Al Hasan, M.S.; Ferdous, J.; Yana, N.T.; Mia, E.; Rakib, I.H.; Ansari, I.A.; Ansari, S.A.; Islam, M.A.; Bhuia, M.S. Caffeine and sclareol take the edge off the sedative effects of linalool, possibly through the GABAA interaction pathway: Molecular insights through in vivo and in silico studies. *Naunyn-Schmiedeberg's Arch. Pharmacol.* **2025**, 1–13. [CrossRef]
10. Aleebrahim-Dehkordi, E.; Jafari, A.; Bala, A.A.; Lagzian, A.; Ghoshouni, H.; Koochaki, P.; Hamedanchi, N.F.; Azadeh-Ghahfaroghi, S.Z.; Azadikhah, F.; Taei, N.; et al. Targeting the Receptor Complexes by Structure-based Natural Drug Compounds in Seizures: A New Dimension in Drug Discovery and Design. *Nat. Prod. J.* **2025**, *15*, E040624230683. [CrossRef]
11. Ahmad, R.; Pandit, C.; Yu, Y.-H.; Chen, W.-J.; Cheng, Y.-C.; Ali, I.; Cheng, Y.-H. The Impact of Fermented Gamma-Aminobutyric Acid on Poultry Growth Performance Through Insulin-like Growth Factor-1 Activation. *Fermentation* **2025**, *11*, 84. [CrossRef]
12. El Sayed, M.A.; Kheir, A.M.; Hussein, F.A.; Ali, E.F.; Selim, M.E.; Majrashi, A.; El Shamey, E.A. Developing new lines of Japonica rice for higher quality and yield under arid conditions. *PeerJ* **2021**, *9*, e11592. [CrossRef]
13. Thomas, S.; Aneena, E. Functional and Therapeutic Properties of Germinated Brown Rice (GBR). In *Traditional Foods: The Reinvented Superfoods*; Springer: Berlin/Heidelberg, Germany, 2024; pp. 225–244.
14. Niu, Y.; Wang, L.; Gong, H.; Jia, S.; Guan, Q.; Li, L.; Cheng, H. Nutrition and Gut Health: Preparation and Efficacy of Resistant Starch. *Foods* **2025**, *14*, 471. [CrossRef]
15. Carcea, M. Value of wholegrain rice in a healthy human nutrition. *Agriculture* **2021**, *11*, 720. [CrossRef]

16. Thilavech, T.; Suantawee, T.; Chusak, C.; Suklaew, P.O.; Adisakwattana, S. Black rice (*Oryza sativa* L.) and its anthocyanins: Mechanisms, food applications, and clinical insights for postprandial glycemic and lipid regulation. *Food Prod. Process. Nutr.* **2025**, *7*, 15. [CrossRef]
17. Zhang, X.; Jiang, A.; An, S.; Guo, C.; You, F.; Huang, Z.; Feng, S.; Zhang, Y.; Chang, X.; Yang, G.; et al. Dietary resistant starch supplementation improves the fish growth, lipid metabolism and intestinal barrier in largemouth bass (*Micropterus salmoides*) fed high-fat diets. *Int. J. Biol. Macromol.* **2025**, *306*, 141356. [CrossRef] [PubMed]
18. Hoque, M.; Emon, K.; Malo, P.; Hossain, M.; Tannu, S.; Roshed, M. Comprehensive guide to vitamin and mineral sources with their requirements. *Indiana J. Agric. Life Sci.* **2023**, *3*, 23–31.
19. Budzianowska, A.; Banaś, K.; Budzianowski, J.; Kikowska, M. Antioxidants to Defend Healthy and Youthful Skin—Current Trends and Future Directions in Cosmetology. *Appl. Sci.* **2025**, *15*, 2571. [CrossRef]
20. Newman, P.P.; Schmitt, B.L.; Maurmann, R.M.; Pence, B.D. Polysaccharides with Arabinose: Key Players in Reducing Chronic Inflammation and Enhancing Immune Health in Aging. *Molecules* **2025**, *30*, 1178. [CrossRef]
21. Santander, M.; Chica, V.; Correa, H.A.M.; Rodríguez, J.; Villagran, E.; Vaillant, F.; Escobar, S. Unravelling Cocoa Drying Technology: A Comprehensive Review of the Influence on Flavor Formation and Quality. *Foods* **2025**, *14*, 721. [CrossRef]
22. Subroto, E.; Djali, M.; Indiarto, R.; Lembong, E.; Baiti, N. Microbiological activity affects post-harvest quality of cocoa (*Theobroma cacao* L.) beans. *Horticulturae* **2023**, *9*, 805. [CrossRef]
23. Kumar, D.; Narwal, S.; Virani, S.; Verma, R.P.S.; Gyawali, S.; Singh, G.P. Barley grain beta glucan enrichment: Status and opportunities. In *Wheat and Barley Grain Biofortification*; Elsevier: Amsterdam, The Netherlands, 2020; pp. 295–308.
24. Li, L.; Pan, M.; Pan, S.; Li, W.; Zhong, Y.; Hu, J.; Nie, S. Effects of insoluble and soluble fibers isolated from barley on blood glucose, serum lipids, liver function and caecal short-chain fatty acids in type 2 diabetic and normal rats. *Food Chem. Toxicol.* **2020**, *135*, 110937. [CrossRef]
25. Ahmed, Z.; Tetlow, I.J.; Falk, D.E.; Liu, Q.; Emes, M.J. Resistant starch content is related to granule size in barley. *Cereal Chem.* **2016**, *93*, 618–630. [CrossRef]
26. Hamany Djande, C.Y.; Dubery, I.A. Hordatines, dimerised hydroxycinnamoylagmatine conjugates of barley (Hordeum vulgare L.): An appraisal of the biosynthesis, chemistry, identification and bioactivities. *Phytochem. Rev.* **2025**, *24*, 303–320. [CrossRef]
27. Mamta; Singh, D.; Gupta, B.; Bhardwaj, R.; Gopala Krishnan, S.; Bhowmick, P.K.; Nagarajan, M.; Vinod, K.K.; Ellur, R.K.; Singh, A.K.; et al. Genome-wide association study revealed novel MTAs for gamma-amino butyric acid in germinated brown rice. *J. Cereal Sci.* **2025**, *123*, 104189. [CrossRef]
28. Sun, H.; Yao, Q.; Hai, M.; Li, T.; Sun, J.; Liu, Z.; Ang, Y.; Zhao, Y.; Zhang, Y.; Cheng, X.; et al. Elite alleles of EPE1 identified via genome-wide association studies increase panicle elongation length in rice. *Rice* **2025**, *18*, 4. [CrossRef]
29. Alsharif, H. DeepRice6mA: A convolutional neural network approach for 6mA site prediction in the rice Genome. *PLoS ONE* **2025**, *20*, e0325216. [CrossRef]
30. Arif, M.; Mahmood, N.; Nawaz, A.F.; El-Sheikh, M.A.; Alansi, S.; Ahmad, P. Comprehensive genome-wide characterization of NAC transcription factors in Barley influence insights into stress tolerance and evolutionary dynamics. *Sci. Rep.* **2025**, *15*, 24257. [CrossRef]
31. Zaghum, M.J.; Ali, K.; Teng, S. Integrated genetic and omics approaches for the regulation of nutritional activities in rice (Oryza sativa L.). *Agriculture* **2022**, *12*, 1757. [CrossRef]
32. Kharabian-Masouleh, A.; Waters, D.L.; Reinke, R.F.; Ward, R.; Henry, R.J. SNP in starch biosynthesis genes associated with nutritional and functional properties of rice. *Sci. Rep.* **2012**, *2*, 557. [CrossRef]
33. Yao, S.; Zheng, L.; Zhao, Z.; Xiong, S. *Effect of Germination Conditions on γ-Aminobutyric Acid Content of Germinated Brown Rice*; Chinese Society of Agricultural Engineering: Beijing, China, 2006.
34. Hu, Y.; Huang, X.; Xiao, Q.; Wu, X.; Tian, Q.; Ma, W.; Shoaib, N.; Liu, Y.; Zhao, H.; Feng, Z. Advances in Plant GABA Research: Biological Functions, Synthesis Mechanisms and Regulatory Pathways. *Plants* **2024**, *13*, 2891. [CrossRef]
35. Li, X. Resistant starch and its applications. *Funct. Starch Appl. Food* **2018**, 63–90. [CrossRef]
36. ElShamey, E.A.Z.; Hamad, H.S.; Alshallash, K.S.; Alghuthaymi, M.A.; Ghazy, M.I.; Sakran, R.M.; Selim, M.E.; ElSayed, M.A.; Abdelmegeed, T.M.; Okasha, S.A.; et al. Growth regulators improve outcrossing rate of diverse rice cytoplasmic male sterile lines through affecting floral traits. *Plants* **2022**, *11*, 1291. [CrossRef]
37. Shen, L.; Li, J.; Li, Y. Resistant starch formation in rice: Genetic regulation and beyond. *Plant Commun.* **2022**, *3*, 100329. [CrossRef]
38. Goñi, I.; Garcia-Diz, L.; Mañas, E.; Saura-Calixto, F. Analysis of resistant starch: A method for foods and food products. *Food Chem.* **1996**, *56*, 445–449. [CrossRef]
39. Pusterla, N.; Lawton, K.; Barnum, S.; Ross, K.; Purcell, K. Investigation of an Outbreak of Equine Herpesvirus-1 Myeloencephalopathy in a Population of Aged Working Equids. *Viruses* **2024**, *16*, 1963. [CrossRef] [PubMed]
40. Idehen, E.; Tang, Y.; Sang, S. Bioactive phytochemicals in barley. *J. Food Drug Anal.* **2017**, *25*, 148–161. [CrossRef] [PubMed]

41. Matsuura, H.N.; Fett-Neto, A.G. Plant alkaloids: Main features, toxicity, and mechanisms of action. *Plant Toxins* **2015**, *2*, 1–15.
42. Jan, R.; Asaf, S.; Numan, M.; Lubna; Kim, K.-M. Plant secondary metabolite biosynthesis and transcriptional regulation in response to biotic and abiotic stress conditions. *Agronomy* **2021**, *11*, 968. [CrossRef]
43. Tsou, S.-F.; Hsu, H.-Y.; Chen, S.-D. Effects of different pretreatments on the GABA content of germinated brown rice. *Appl. Sci.* **2024**, *14*, 5771. [CrossRef]
44. Soni, P.; Sharma, M.M. Moringa oleifera Lam.: A Valuable Medicinal Plant, Boon of Nature. *Int. J. Ayurveda Pharma Res.* **2022**, *10*, 99–107. [CrossRef]
45. Patil, S.B.; Khan, M.K. Germinated brown rice as a value added rice product: A review. *J. Food Sci. Technol.* **2011**, *48*, 661–667. [CrossRef] [PubMed]
46. Almutairi, S.; Sivadas, A.; Kwakowsky, A. The effect of oral GABA on the nervous system: Potential for therapeutic intervention. *Nutraceuticals* **2024**, *4*, 241–259. [CrossRef]
47. Bai, Q.; Liu, S.; Xu, J.; Tufail, T.; Xu, B. Enrichment of GABA content in brown rice through heating and humidifying treatment: Quantification via TLC-ImageJ method. *Food Chem.* **2025**, *473*, 143033. [CrossRef] [PubMed]
48. Pannerchelvan, S.; Rios-Solis, L.; Wong, F.W.F.; Zaidan, U.H.; Wasoh, H.; Mohamed, M.S.; Tan, J.S.; Mohamad, R.; Halim, M. Strategies for improvement of gamma-aminobutyric acid (GABA) biosynthesis via lactic acid bacteria (LAB) fermentation. *Food Funct.* **2023**, *14*, 3929–3948. [CrossRef]
49. Hayat, F.; Khan, U.; Li, J.; Ahmed, N.; Khanum, F.; Iqbal, S.; Altaf, M.A.; Ahmad, J.; Javed, H.U.; Peng, Y.; et al. γ Aminobutyric acid (GABA): A key player in alleviating abiotic stress resistance in horticultural crops: Current insights and future directions. *Horticulturae* **2023**, *9*, 647. [CrossRef]
50. Sun, S.; Zhang, J.; Li, Y.; Xu, Y.; Yang, R.; Luo, L.; Xiang, J. Effects of Sodium Selenite on Accumulations of Selenium and GABA, Phenolic Profiles, and Antioxidant Activity of Foxtail Millet During Germination. *Foods* **2024**, *13*, 3916. [CrossRef]
51. Benincasa, P.; Falcinelli, B.; Lutts, S.; Stagnari, F.; Galieni, A. Sprouted grains: A comprehensive review. *Nutrients* **2019**, *11*, 421. [CrossRef]
52. Dokduang, H.; Katenak, A.; Kittirat, Y.; Tomacha, J.; Promraksa, B.; Taing, L.; Na Nakorn, S.; Klanrit, P.; Namwat, N.; Titapun, A.; et al. Chemical composition and anticancer potential of water extracts derived from ground powder of Thai germinated brown rice (*Oryza sativa* L.). *Nat. Prod. Res.* **2025**, 1–8. [CrossRef]
53. Thomas, S. Medicinal Properties and Process Optimisation for GABA Enrichment in Rice. Ph.D. Thesis, Department of Community Science, College of Agriculture, Vellanikkara, Kerala, India, 2023.
54. Karladee, D.; Suriyong, S. γ-Aminobutyric acid (GABA) content in different varieties of brown rice during germination. *Sci. Asia* **2012**, *38*, 13–17. [CrossRef]
55. Liyanaarachchi, G.; Mahanama, K.; Somasiri, H.; Punyasiri, P.; Kottawa-Arachchi, J. Total and free amino acid contents of popular rice varieties (*Oryza sativa* L.) consumed in the capital city of Sri Lanka. *J. Natl. Sci. Found. Sri Lanka* **2020**, *48*, 199–211. [CrossRef]
56. Castanho, A.; Pereira, C.; Lageiro, M.; Oliveira, J.C.; Cunha, L.M.; Brites, C. Improving γ-oryzanol and γ-aminobutyric acid contents in rice beverage amazake produced with brown, milled and germinated rices. *Foods* **2023**, *12*, 1476. [CrossRef]
57. Hou, D.; Tang, J.; Feng, Q.; Niu, Z.; Shen, Q.; Wang, L.; Zhou, S. Gamma-aminobutyric acid (GABA): A comprehensive review of dietary sources, enrichment technologies, processing effects, health benefits, and its applications. *Crit. Rev. Food Sci. Nutr.* **2024**, *64*, 8852–8874. [CrossRef] [PubMed]
58. Yogeswara, I.B.A.; Maneerat, S.; Haltrich, D. Glutamate decarboxylase from lactic acid bacteria—A key enzyme in GABA synthesis. *Microorganisms* **2020**, *8*, 1923. [CrossRef] [PubMed]
59. Ueno, H. Enzymatic and structural aspects on glutamate decarboxylase. *J. Mol. Catal. B Enzym.* **2000**, *10*, 67–79. [CrossRef]
60. Zhang, Q.; Zhu, L.; Li, H.; Chen, Q.; Li, N.; Li, J.; Zhao, Z.; Xiao, D.; Tang, T.; Bi, C.; et al. Insights and progress on the biosynthesis, metabolism, and physiological functions of gamma-aminobutyric acid (GABA): A review. *PeerJ* **2024**, *12*, e18712. [CrossRef]
61. Shelp, B.J.; Aghdam, M.S.; Flaherty, E.J. γ-Aminobutyrate (GABA) regulated plant defense: Mechanisms and opportunities. *Plants* **2021**, *10*, 1939. [CrossRef]
62. Sheteiwy, M.S.; Shao, H.; Qi, W.; Hamoud, Y.A.; Shaghaleh, H.; Khan, N.U.; Yang, R.; Tang, B. GABA-alleviated oxidative injury induced by salinity, osmotic stress and their combination by regulating cellular and molecular signals in rice. *Int. J. Mol. Sci.* **2019**, *20*, 5709. [CrossRef]
63. Sakamoto, T.; Miura, K.; Itoh, H.; Tatsumi, T.; Ueguchi-Tanaka, M.; Ishiyama, K.; Kobayashi, M.; Agrawal, G.K.; Takeda, S.; Abe, K.; et al. An overview of gibberellin metabolism enzyme genes and their related mutants in rice. *Plant Physiol.* **2004**, *134*, 1642–1653. [CrossRef]
64. Yuan, D.; Wu, X.; Gong, B.; Huo, R.; Zhao, L.; Li, J.; Lü, G.; Gao, H. GABA metabolism, transport and their roles and mechanisms in the regulation of abiotic stress (hypoxia, salt, drought) resistance in plants. *Metabolites* **2023**, *13*, 347. [CrossRef]

65. Zhao, G.-c.; Xie, M.-x.; Wang, Y.-c.; Li, J.-y. Molecular mechanisms underlying γ-aminobutyric acid (GABA) accumulation in giant embryo rice seeds. *J. Agric. Food Chem.* **2017**, *65*, 4883–4889. [CrossRef]
66. Grewal, J. Gamma-aminobutyric acid (GABA): A versatile bioactive compound. *Eur. J. Mol. Clin. Med.* **2020**, *7*, 3068–3075.
67. ElShamey, E.A.; Sakran, R.M.; ElSayed, M.A.; Aloufi, S.; Alharthi, B.; Alqurashi, M.; Mansour, E.; Abd El-Moneim, D. Heterosis and combining ability for floral and yield characters in rice using cytoplasmic male sterility system. *Saudi J. Biol. Sci.* **2022**, *29*, 3727–3738. [CrossRef] [PubMed]
68. Ding, Y.; Huang, J.; Zhang, N.; Rasmussen, S.K.; Wu, D.; Shu, X. Physiochemical properties of rice with contrasting resistant starch content. *J. Cereal Sci.* **2019**, *89*, 102815. [CrossRef]
69. Yi, X.; Li, C. Main controllers for improving the resistant starch content in cooked white rice. *Food Hydrocoll.* **2022**, *122*, 107083. [CrossRef]
70. Sangwongchai, W.; Tananuwong, K.; Krusong, K.; Natee, S.; Thitisaksakul, M. Starch chemical composition and molecular structure in relation to physicochemical characteristics and resistant starch content of four Thai commercial rice cultivars differing in pasting properties. *Polymers* **2023**, *15*, 574. [CrossRef]
71. Kumar, A.; Lal, M.K.; Bagchi, T.B.; Sah, R.P.; Sharma, S.; Baig, M.; Nayak, A.K. *Glycemic Index of Rice: Role in Diabetics*; NRRI Research Bulletin No. 44; ICAR-National Rice Research Institute: Cuttack, Odisha, India, 2024.
72. Miura, S.; Crofts, N.; Koyama, N.; Hosaka, Y.; Morita, R.; Abe, M.; Nakamura, Y.; Fujita, N. Physiological and compensatory roles of three starch-branching enzymes in different rice organs. *Plant Mol. Biol.* **2025**, *115*, 1–20. [CrossRef]
73. Blennow, A. Source of Starch. In *Handbook of Starch Science and Technology*; Miao, M., Chen, L., BeMiller, J.N., Eds.; CRC: Boca Raton, FL, USA, 2025.
74. Song, E.; Gwon, H.; Choi, J.; Islam, M.A.; Chun, J.; Chang, Y.H.; Hwang, J. Enhancing Functional Properties and Shelf Life of *Oryza sativa* L. via Grain Stabilization Techniques. *Foods* **2025**, *14*, 596. [CrossRef]
75. Yang, F.; Cai, J.; Zhou, Y.; Wei, D.; Xie, C.Y. Thermal treatments modulate the multiscale structure of chestnut starch: A focus on the relationship with the gut microbiota composition. *Res. Sq.* **2025**. [CrossRef]
76. Yan, J.; Yu, J.; Shen, H.; Zhou, L.; Chen, Z.; Fan, X.; Li, Q.; Zhang, C.; Liu, Q.; Huang, L. Creating a Superior Wx Allele with Temperature-Responsive Amylose Regulation and a Novel Transcriptional Pattern in Rice via CRISPR/Cas9-Mediated Promoter Editing. *Foods* **2025**, *14*, 1330. [CrossRef]
77. Wang, A.; Cheng, Q.; Li, W.; Kan, M.; Zhang, Y.; Meng, X.; Guo, H.; Jing, Y.; Chen, M.; Liu, G.; et al. Creation of high-resistant starch rice through systematic editing of amylopectin biosynthetic genes in rs4. *Plant Biotechnol. J.* **2025**, *23*, 480–488. [CrossRef] [PubMed]
78. Walter, M.; da Silva, L.P.; Denardin, C.C. Rice and resistant starch: Different content depending on chosen methodology. *J. Food Compos. Anal.* **2005**, *18*, 279–285. [CrossRef]
79. Lee, K.Y.; Lee, H.G. Comparative effects of slowly digestible and resistant starch from rice in high-fat diet-induced obese mice. *Food Sci. Biotechnol.* **2016**, *25*, 1443–1448. [CrossRef] [PubMed]
80. Ordonio, R.L.; Matsuoka, M. Increasing resistant starch content in rice for better consumer health. *Proc. Natl. Acad. Sci. USA* **2016**, *113*, 12616–12618. [CrossRef] [PubMed]
81. Kumar, A.; Sahoo, U.; Baisakha, B.; Okpani, O.A.; Ngangkham, U.; Parameswaran, C.; Basak, N.; Kumar, G.; Sharma, S. Resistant starch could be decisive in determining the glycemic index of rice cultivars. *J. Cereal Sci.* **2018**, *79*, 348–353. [CrossRef]
82. Sahoo, B.; Roy, A. Structure–function relationship of resistant starch formation: Enhancement technologies and need for more viable alternatives for whole rice grains. *J. Food Process Eng.* **2023**, *46*, e14214. [CrossRef]
83. Yuan, J.; Wang, X.; Zhao, Y.; Khan, N.U.; Zhao, Z.; Zhang, Y.; Wen, X.; Tang, F.; Wang, F.; Li, Z. Genetic basis and identification of candidate genes for salt tolerance in rice by GWAS. *Sci. Rep.* **2020**, *10*, 9958. [CrossRef]
84. Kong, W.; Zhang, C.; Zhang, S.; Qiang, Y.; Zhang, Y.; Zhong, H.; Li, Y. Uncovering the novel QTLs and candidate genes of salt tolerance in rice with linkage mapping, RTM-GWAS, and RNA-seq. *Rice* **2021**, *14*, 1–12. [CrossRef]
85. Ahuja, G.; Jaiswal, S.; Hucl, P.; Chibbar, R.N. Genome-specific granule-bound starch synthase I (GBSSI) influences starch biochemical and functional characteristics in near-isogenic wheat (*Triticum aestivum* L.) lines. *J. Agric. Food Chem.* **2013**, *61*, 12129–12138. [CrossRef]
86. Vrinten, P.L.; Nakamura, T. Wheat granule-bound starch synthase I and II are encoded by separate genes that are expressed in different tissues. *Plant Physiol.* **2000**, *122*, 255–264. [CrossRef]
87. Han, Y.; Gasic, K.; Sun, F.; Xu, M.; Korban, S.S. A gene encoding starch branching enzyme I (SBEI) in apple (Malus × domestica, Rosaceae) and its phylogenetic relationship to Sbe genes from other angiosperms. *Mol. Phylogenetics Evol.* **2007**, *43*, 852–863. [CrossRef]
88. Qu, J.; Xu, S.; Zhang, Z.; Chen, G.; Zhong, Y.; Liu, L.; Zhang, R.; Xue, J.; Guo, D. Evolutionary, structural and expression analysis of core genes involved in starch synthesis. *Sci. Rep.* **2018**, *8*, 12736. [CrossRef]

89. Jain, R.; Dhaka, N.; Krishnan, K.; Yadav, G.; Priyam, P.; Sharma, M.K.; Sharma, R.A. Temporal Gene Expression Profiles From Pollination to Seed Maturity in Sorghum Provide Core Candidates for Engineering Seed Traits. *Plant Cell Environ.* **2025**, *48*, 2662–2690. [CrossRef] [PubMed]
90. Yu, G.; Gaoyang, Y.; Liu, L.; Shoaib, N.; Deng, Y.; Zhang, N.; Li, Y.; Huang, Y. The structure, function, and regulation of starch synthesis enzymes SSIII with emphasis on maize. *Agronomy* **2022**, *12*, 1359. [CrossRef]
91. Fahy, B.; Gonzalez, O.; Savva, G.M.; Ahn-Jarvis, J.H.; Warren, F.J.; Dunn, J.; Lovegrove, A.; Hazard, B.A. Loss of starch synthase IIIa changes starch molecular structure and granule morphology in grains of hexaploid bread wheat. *Sci. Rep.* **2022**, *12*, 10806. [CrossRef] [PubMed]
92. Wang, K.; Yan, T.-z.; Xu, S.-l.; Yan, X.; Zhou, Q.-f.; Zhao, X.-h.; Li, Y.-f.; Wu, Z.-x.; Qin, P.; Fu, C.-j.; et al. Validating a segment on chromosome 7 of japonica for establishing low-cadmium accumulating indica rice variety. *Sci. Rep.* **2021**, *11*, 6053. [CrossRef]
93. Biselli, C.; Volante, A.; Desiderio, F.; Tondelli, A.; Gianinetti, A.; Finocchiaro, F.; Taddei, F.; Gazza, L.; Sgrulletta, D.; Cattivelli, L. GWAS for starch-related parameters in japonica rice (Oryza sativa L.). *Plants* **2019**, *8*, 292. [CrossRef]
94. Selvaraj, R.; Singh, A.K.; Singh, V.K.; Abbai, R.; Habde, S.V.; Singh, U.M.; Kumar, A. Superior haplotypes towards development of low glycemic index rice with preferred grain and cooking quality. *Sci. Rep.* **2021**, *11*, 10082. [CrossRef]
95. Wang, C.; Chao, C.; Sun, R.; Yu, J.; Yang, Y.; Copeland, L.; Wang, S. Increased Crystallite Stability Enhances Gut Microbial Fermentability of Type 5 Resistant Starch. *J. Agric. Food Chem.* **2025**, *73*, 2613–2622. [CrossRef]
96. Guo, J.; Szacilo, A.; Kong, L. Resistant starch, chronic kidney disease, and the gut microbiome: A mini-review. *Food Front.* **2025**, *6*, 5–14. [CrossRef]
97. Zeng, Y.; Sun, D.; Du, J.; Pu, X.; Yang, S.; Yang, X.; Yang, T.; Yang, J. Identification of QTLs for resistant starch and total alkaloid content in brown and polished rice. *Genet. Mol. Res.* **2016**, *15*, 15037268. [CrossRef]
98. Mohidem, N.A.; Hashim, N.; Shamsudin, R.; Che Man, H. Rice for food security: Revisiting its production, diversity, rice milling process and nutrient content. *Agriculture* **2022**, *12*, 741. [CrossRef]
99. Rathna Priya, T.; Eliazer Nelson, A.R.L.; Ravichandran, K.; Antony, U. Nutritional and functional properties of coloured rice varieties of South India: A review. *J. Ethn. Foods* **2019**, *6*, 1–11. [CrossRef]
100. D'Mello, J. An outline of pathways in amino acid metabolism. In *Amino Acids in Animal Nutrition*; CABI Publishing: Wallingford, UK, 2003; pp. 71–86.
101. Yang, Q.; Zhao, D.; Liu, Q. Connections between amino acid metabolisms in plants: Lysine as an example. *Front. Plant Sci.* **2020**, *11*, 928. [CrossRef]
102. Singh, S.K. Explorations of plant's chemodiversity: Role of nitrogen-containing secondary metabolites in plant defense. *Mol. Asp. Plant-Pathog. Interact.* **2018**, 309–332. [CrossRef]
103. Wang, W.; Li, Y.; Dang, P.; Zhao, S.; Lai, D.; Zhou, L. Rice secondary metabolites: Structures, roles, biosynthesis, and metabolic regulation. *Molecules* **2018**, *23*, 3098. [CrossRef] [PubMed]
104. Dubouzet, J.G.; Matsuda, F.; Ishihara, A.; Miyagawa, H.; Wakasa, K. Production of indole alkaloids by metabolic engineering of the tryptophan pathway in rice. *Plant Biotechnol. J.* **2013**, *11*, 1103–1111. [CrossRef] [PubMed]
105. Rabeh, K.; Hnini, M.; Oubohssaine, M. A comprehensive review of transcription factor-mediated regulation of secondary metabolites in plants under environmental stress. *Stress. Biol.* **2025**, *5*, 15. [CrossRef]
106. Soltani, N.; Nazarian-Firouzabadi, F.; Shafeinia, A.; Sadr, A.S.; Shirali, M. The expression of Terpenoid Indole Alkaloid (TIAs) pathway genes in *Catharanthus roseus* in response to salicylic acid treatment. *Mol. Biol. Rep.* **2020**, *47*, 7009–7016. [CrossRef]
107. Verma, P.; Mathur, A.K.; Srivastava, A.; Mathur, A. Emerging trends in research on spatial and temporal organization of terpenoid indole alkaloid pathway in *Catharanthus roseus*: A literature update. *Protoplasma* **2012**, *249*, 255–268. [CrossRef]
108. Liu, X.; Chen, J.; Ma, R.; Zhao, L.; Lian, C.; Chen, S.; Ma, Y. Metabolic profiling and transcriptome analysis of *Sinomenium acutum* provide insights into the biosynthesis of structurally diverse benzylisoquinoline alkaloids. *Sci. Rep.* **2025**, *15*, 5877. [CrossRef]
109. Khaldari, I.; Naghavi, M.R.; Motamedi, E.; Zargar, M. The effects of green and chemically-synthesized copper oxide nanoparticles on the production and gene expression of morphinan alkaloids in Oriental poppy. *Sci. Rep.* **2024**, *14*, 6000. [CrossRef]
110. Bhambhani, S.; Kondhare, K.R.; Giri, A.P. Diversity in chemical structures and biological properties of plant alkaloids. *Molecules* **2021**, *26*, 3374. [CrossRef]
111. Ghirga, F.; Bonamore, A.; Calisti, L.; D'Acquarica, I.; Mori, M.; Botta, B.; Boffi, A.; Macone, A. Green routes for the production of enantiopure benzylisoquinoline alkaloids. *Int. J. Mol. Sci.* **2017**, *18*, 2464. [CrossRef]
112. Jiang, T.; Zuo, S.; Liu, C.; Xing, W.; Wang, P. Progress in Methylxanthine Biosynthesis: Insights into Pathways and Engineering Strategies. *Int. J. Mol. Sci.* **2025**, *26*, 1510. [CrossRef] [PubMed]
113. Čižmárová, B.; Kraus, V., Jr.; Birková, A. Caffeinated Beverages—Unveiling Their Impact on Human Health. *Beverages* **2025**, *11*, 18. [CrossRef]
114. Facchini, P.J. Regulation of alkaloid biosynthesis in plants. *Alkaloids Chem. Biol.* **2006**, *63*, 1–44. [PubMed]

115. Zhang, J.; Hansen, L.G.; Gudich, O.; Viehrig, K.; Lassen, L.M.; Schrübbers, L.; Adhikari, K.B.; Rubaszka, P.; Carrasquer-Alvarez, E.; Chen, L.; et al. A microbial supply chain for production of the anti-cancer drug vinblastine. *Nature* **2022**, *609*, 341–347. [CrossRef]
116. Nourozi, E.; Hedayati, A.; Madani, H.; Hosseini, B.; Hemmaty, S. In vitro synthetic polyploidization and enhancement of anticancer compounds in *Catharanthus Roseus* (L.) G. Don important cultivars. *Sci. Rep.* **2025**, *15*, 6563. [CrossRef]
117. El Shamey, E.A.Z.; El Sayed, M.A.A.; Elgamal, W.H.A. Genetic studies on the relationship between isozymes and allelopathic activity in hybrid rice. *Curr. Appl. Sci. Technol.* **2016**, *16*, 60–69.
118. Selim, M.E.; ELShamey, E.E.; ELkholy, N.A.; Abdelrahman, M.; Abo-Marzoka, E.A.; Elgamal, W.H.; Elsayed, M.A.; Attia, K.A.; Ghazy, A.-H.; Abdel-Haleem, H. Genetic determinants of weed control in rice (Oryza sativa L.) using allelopathy approach. *Cereal Res. Commun.* **2024**, *52*, 1789–1798. [CrossRef]
119. Leite Dias, S.; D'Auria, J.C. The bitter truth: How insects cope with toxic plant alkaloids. *J. Exp. Bot.* **2025**, *76*, 5–15. [CrossRef] [PubMed]
120. El-Denary, M.; Elshamey, E. Analysis of Gene Expressions Associated with Increased Allelopathy in Rice (*Oryza sativa* L.) Under Low Nitrogen Treatment. *Egypt. J. Genet. Cytol.* **2014**, *43*, 113–131. [CrossRef]
121. Selim, M.; Hassan, A.; El Sayed, M.; El Shamey, E. Determination of Resistant on New Highly Productivity Lines for Major Rice Diseases. *J. Plant Prot. Pathol.* **2020**, *11*, 385–394. [CrossRef]
122. ElShamey, E.A.; Ali, E.F.; Selim, M.E.; ElSayed, M.A.; Ahmed, M.; Alotaibi, F.M.; Kamara, M.M.; AL-Harbi, M.S.; Kheir, A.M. Water deficit induced physiological and amino acid responses in some rice varieties using NMR-metabolic analysis. *Agron. J.* **2021**, *113*, 4690–4704. [CrossRef]
123. Yong Jin, C.; Sun-Hwa, H. Metabolic Engineering in Rice for Functional Metabolite Production. *Rice Sci.* **2025**, 12. [CrossRef]
124. Qaderi, M.M.; Martel, A.B.; Strugnell, C.A. Environmental factors regulate plant secondary metabolites. *Plants* **2023**, *12*, 447. [CrossRef]
125. Zhou, Z.; Robards, K.; Helliwell, S.; Blanchard, C. Composition and functional properties of rice. *Int. J. Food Sci. Technol.* **2002**, *37*, 849–868. [CrossRef]
126. Zeng, Y.-W.; Du, J.; Yang, S.-M.; Pu, X.-Y.; Wang, Y.-C.; Yang, T.; Sun, Z.-H.; Xin, P.-Y. The zonal characteristics and cultivated types difference of functional components in brown rice for core collection of Yunnan rice. *Guang Pu Xue Yu Guang Pu Fen Xi = Guang Pu* **2010**, *30*, 3388–3394.
127. Heuberger, A.L.; Lewis, M.R.; Chen, M.-H.; Brick, M.A.; Leach, J.E.; Ryan, E.P. Metabolomic and functional genomic analyses reveal varietal differences in bioactive compounds of cooked rice. *PLoS ONE* **2010**, *5*, e12915. [CrossRef]
128. Ramakrishna, R.; Sarkar, D.; Shetty, K. Functional bioactives from barley for human health benefits. In *Functional Foods and Biotechnology*; CRC Press: Boca Raton, FL, USA, 2019; pp. 61–85.
129. Huang, S.; Liu, H.; Xu, R.; Li, W.; Yang, H.; Bao, X.; Hang, Y.; Gong, Y.; Zhao, Y. RNA Editing Analysis Reveals Methyl Jasmonic Acid Regulation of Fucoxanthin and Fatty Acid Metabolism in *Phaeodactylum tricornutum*. *Mar. Drugs* **2025**, *23*, 66. [CrossRef]
130. Stitt, M.; Sulpice, R.; Keurentjes, J. Metabolic networks: How to identify key components in the regulation of metabolism and growth. *Plant Physiol.* **2010**, *152*, 428–444. [CrossRef]
131. Yin, X.; Wang, S.; Wang, Z.; Wen, H.; Bai, T.; Zhang, Y. Effects of Pretreatment Methods on Gamma-Aminobutyric Acid Enrichment and Quality Improvement in Highland Barley Beverages. *Foods* **2024**, *13*, 4053. [CrossRef]
132. Kihara, M.; Okada, Y.; Iimure, T.; Ito, K. Accumulation and degradation of two functional constituents, GABA and β-glucan, and their varietal differences in germinated barley grains. *Breed. Sci.* **2007**, *57*, 85–89. [CrossRef]
133. Waleed, A.-A.; Mahdi, A.A.; Al-Maqtari, Q.A.; Mushtaq, B.S.; Ahmed, A.; Karrar, E.; Mohammed, J.K.; Fan, M.; Li, Y.; Qian, H.; et al. The potential improvements of naked barley pretreatments on GABA, β-glucan, and antioxidant properties. *LWT* **2020**, *130*, 109698. [CrossRef]
134. Ma, Y.; Wang, P.; Wang, M.; Sun, M.; Gu, Z.; Yang, R. GABA mediates phenolic compounds accumulation and the antioxidant system enhancement in germinated hulless barley under NaCl stress. *Food Chem.* **2019**, *270*, 593–601. [CrossRef] [PubMed]
135. Liu, Y.; Wu, Y.; Jia, Y.; Ren, F.; Zhou, S. Effect of GABA combined with ultrasound stress germination treatment on phenolic content and antioxidant activity of highland barley. *J. Sci. Food Agric.* **2024**, *104*, 9023–9034. [CrossRef]
136. AL-Ansi, W.; Alkawry, T.A.A.; AL-Qupati, A.A.M.; Fadhl, J.A.; Abdullah, A.B.; Mahdi, A.A.; Al-Maqtari, Q.A.; Fan, M.; Li, Y.; Qian, H. Effect of germination time on in-vitro extractability, and bioavailability of GABA, total phenolic compounds, and antioxidant activity of highland barley bread. *Int. J. Food Sci. Technol.* **2025**, *60*, vvae079. [CrossRef]
137. Paluch-Lubawa, E.; Tanwar, U.K.; Stolarska, E.; Arasimowicz-Jelonek, M.; Mattoo, A.K.; Sobieszczuk-Nowicka, E. Increasing nitrogen use efficiency in agronomically important plants: An insight into gene characteristics on a genome-wide scale in barley. *Comput. Biol. Med.* **2024**, *183*, 109277. [CrossRef]
138. Abdullah; Wani, K.I.; Naeem, M.; Aftab, T. From neurotransmitter to plant protector: The intricate world of gaba signaling and its diverse functions in stress mitigation. *J. Plant Growth Regul.* **2025**, *44*, 403–418. [CrossRef]

139. Liu, R.; Gao, L.; Zhang, X.; Ge, P.; Wang, L.; Zhou, K.; Yang, C.; Wang, L.; Song, L. The Regulation of γ-Aminobutyric Acid on Antioxidative Defense Response of Pacific Oyster upon High-Temperature Stress. *Antioxidants* **2025**, *14*, 222. [CrossRef]
140. Ahmed, H.G.M.D.; Iqbal, R.; Yang, X.; Yang, J.; Pu, X.; Li, X.; Yang, L.E.; Zeng, Y. Bioactive Compounds in Barley Modulating Glycemic Response: Implications for Diabetes Management. *Nat. Prod. Commun.* **2025**, *20*, 1934578X251336985. [CrossRef]
141. Ding, W.; Liu, Y.; Liu, Y.; Wang, G.; Liu, X.; Peng, X.; Li, H.; Li, Z. Research Progress in Nutritional Components, Biological Activity, and Processing and Utilization of Chenopodium quinoa Willd. *ACS Food Sci. Technol.* **2025**, *5*, 411–427. [CrossRef]
142. Islam, S.N.u.; Kouser, S.; Hassan, P.; Asgher, M.; Shah, A.A.; Khan, N.A. Gamma-aminobutyric acid interactions with phytohormones and its role in modulating abiotic and biotic stress in plants. *Stress. Biol.* **2024**, *4*, 36. [CrossRef] [PubMed]
143. Sukhareva, B.; Mamaeva, O. Glutamate decarboxylase: Computer studies of enzyme evolution. *Biochemistry* **2002**, *67*, 1180–1188. [CrossRef]
144. Locy, R.D. 3 The Yeast g-Aminobutyrate (GABA) Shunt. *CABI* **2017**, 29–48. [CrossRef]
145. Zhang, Q.-F.; Hu, S.; Zhao, W.-R.; Huang, J.; Mei, J.-Q.; Mei, L.-H. Parallel strategy increases the thermostability and activity of glutamate decarboxylase. *Molecules* **2020**, *25*, 690. [CrossRef]
146. Al-Quraan, N.A.; Al-Ajlouni, Z.I.; Obedat, D.I. The GABA shunt pathway in germinating seeds of wheat (*Triticum aestivum* L.) and barley (*Hordeum vulgare* L.) under salt stress. *Seed Sci. Res.* **2019**, *29*, 250–260. [CrossRef]
147. Al-Ajlouni, Z.I.; AL-Quraan, N.A.; Al-Saqqar, H.Y.; Al-Ghzawi, A.L.; Al-Tawaha, A.R. GABA Metabolism and MDA Production in Barley (*Hordeum vulgare* L.) Seedlings under Salinity and Osmotic Stresses. *Open Agric. J.* **2024**, *18*, e18743315287408. [CrossRef]
148. Kabała, K.; Janicka, M. Relationship between the GABA pathway and signaling of other regulatory molecules. *Int. J. Mol. Sci.* **2024**, *25*, 10749. [CrossRef]
149. Barakat, H.; Aljutaily, T. Role of γ-Aminobutyric Acid (GABA) as an Inhibitory Neurotransmitter in Diabetes Management: Mechanisms and Therapeutic Implications. *Biomolecules* **2025**, *15*, 399. [CrossRef]
150. Englyst, H.N.; Trowell, H.; Southgate, D.; Cummings, J. Dietary fiber and resistant starch. *Am. J. Clin. Nutr.* **1987**, *46*, 873–874. [CrossRef]
151. Cummings, J.H.; Beatty, E.R.; Kingman, S.M.; Bingham, S.A.; Englyst, H.N. Digestion and physiological properties of resistant starch in the human large bowel. *Br. J. Nutr.* **1996**, *75*, 733–747. [CrossRef]
152. Farooq, M.A.; Yu, J. Recent Advances in Physical Processing Techniques to Enhance the Resistant Starch Content in Foods: A Review. *Foods* **2024**, *13*, 2770. [CrossRef] [PubMed]
153. Tao, J.; Bu, T.; Xie, Q.; Jia, Y.; Wang, Q.; Shen, N.; Zhang, X.; Xiao, Y.; Ye, L.; Chen, Z.; et al. Physicochemical Properties and Functional Characterization of Buckwheat Type 3 Resistant Starch Prepared by Various Treatment Methods. *LWT* **2025**, *218*, 117467. [CrossRef]
154. Xie, J.; Hong, Y.; Gu, Z.; Cheng, L.; Li, Z.; Li, C.; Ban, X. Highland barley starch: Structures, properties, and applications. *Foods* **2023**, *12*, 387. [CrossRef] [PubMed]
155. Farag, M.A.; Xiao, J.; Abdallah, H.M. Nutritional value of barley cereal and better opportunities for its processing as a value-added food: A comprehensive review. *Crit. Rev. Food Sci. Nutr.* **2022**, *62*, 1092–1104. [CrossRef] [PubMed]
156. Shu, X.; Backes, G.; Rasmussen, S.K. Genome-wide association study of resistant starch (RS) phenotypes in a barley variety collection. *J. Agric. Food Chem.* **2012**, *60*, 10302–10311. [CrossRef]
157. Emami, S.; Meda, V.; Pickard, M.D.; Tyler, R.T. Impact of micronization on rapidly digestible, slowly digestible, and resistant starch concentrations in normal, high-amylose, and waxy barley. *J. Agric. Food Chem.* **2010**, *58*, 9793–9799. [CrossRef]
158. Xie, J.; Cheng, L.; Li, Z.; Li, C.; Hong, Y.; Gu, Z. Effect of non-starch components on the structural properties, physicochemical properties and in vitro digestibility of waxy highland barley starch. *Int. J. Biol. Macromol.* **2024**, *255*, 128013. [CrossRef]
159. Wang, C.; Tian, X.; Zhang, X.; Zhang, Z.; Zhang, X.; Zeng, X. Physicochemical characterizations, digestibility, and lipolysis inhibitory effects of highland barley resistant starches prepared by physical and enzymatic methods. *Molecules* **2023**, *28*, 1065. [CrossRef]
160. Baptista, N.T.; Dessalles, R.; Illner, A.-K.; Ville, P.; Ribet, L.; Anton, P.M.; Durand-Dubief, M. Harnessing the power of resistant starch: A narrative review of its health impact and processing challenges. *Front. Nutr.* **2024**, *11*, 1369950. [CrossRef]
161. Zhong, H.; Chen, Z.; Han, Y.; Hu, K. The Molecular Mechanisms and New Classification of Resistant Starch–a Review. *Curr. Res. Food Sci.* **2025**, *10*, 101023. [CrossRef] [PubMed]
162. Haralampu, S. Resistant starch—A review of the physical properties and biological impact of RS3. *Carbohydr. Polym.* **2000**, *41*, 285–292. [CrossRef]
163. Roman, L.; Martinez, M.M. Structural basis of resistant starch (RS) in bread: Natural and commercial alternatives. *Foods* **2019**, *8*, 267. [CrossRef]
164. Yuan, D.; Zhang, Y.; Chen, X.; Xu, F.; Zhu, K.; Wang, J.; Zhang, Y. Physicochemical, Structural, and Digestive Properties of Green Banana Starch from Five Chinese Mutant Banana Species. *Foods* **2025**, *14*, 706. [CrossRef]

165. Beura, M.; Salman, C.; Rahaman, S.; Bollinedi, H.; Singh, A.; Ray, S.; Yeasin, M.; Kaur, R.; Krishnan, V. From data to digestibility: Prediction of resistant starch using machine learning for functional food development. *Int. J. Adv. Eng. Sci. Appl. Math.* **2025**, 1–15. [CrossRef]
166. Wenqi, Z. An Overview of UDP-Glucose Pyrophosphorylase in Plants. *Trop. Plant Biol.* **2025**, *18*, 1–10. [CrossRef]
167. Jiang, H.; Zhao, Y.; Chen, L.; Wan, X.; Yan, B.; Liu, Y.; Liu, Y.; Zhang, W.; Gao, J. Grain Weight and Taste Quality in Japonica Rice Are Regulated by Starch Synthesis and Grain Filling Under Nitrogen–Phosphorus Interactions. *Plants* **2025**, *14*, 432. [CrossRef]
168. Lu, X.; Wang, Y.; Zhen, X.; Che, Y.; Yu, H.; Ge, Y.; Wang, X.; Li, R.; Geng, M.; Zhou, B.; et al. Editing of the soluble starch synthase gene MeSSIII-1 enhanced the amylose and resistant starch contents in cassava. *Carbohydr. Polym.* **2025**, *348*, 122903. [CrossRef]
169. Li, Z.; Sun, F.; Xu, S.; Chu, X.; Mukai, Y.; Yamamoto, M.; Ali, S.; Rampling, L.; Kosar-Hashemi, B.; Rahman, S.; et al. The structural organisation of the gene encoding class II starch synthase of wheat and barley and the evolution of the genes encoding starch synthases in plants. *Funct. Integr. Genom.* **2003**, *3*, 76–85. [CrossRef]
170. Feng, X.; Rahman, M.M.; Hu, Q.; Wang, B.; Karim, H.; Guzmán, C.; Harwood, W.; Xu, Q.; Zhang, Y.; Tang, H.; et al. HvGBSSI mutation at the splicing receptor site affected RNA splicing and decreased amylose content in barley. *Front. Plant Sci.* **2022**, *13*, 1003333. [CrossRef] [PubMed]
171. Baysal, C.; He, W.; Drapal, M.; Villorbina, G.; Medina, V.; Capell, T.; Khush, G.S.; Zhu, C.; Fraser, P.D.; Christou, P. Inactivation of rice starch branching enzyme IIb triggers broad and unexpected changes in metabolism by transcriptional reprogramming. *Proc. Natl. Acad. Sci. USA* **2020**, *117*, 26503–26512. [CrossRef] [PubMed]
172. Wink, M. Evolution of secondary metabolites from an ecological and molecular phylogenetic perspective. *Phytochemistry* **2003**, *64*, 3–19. [CrossRef] [PubMed]
173. Zhang, J.; Jia, Q.; Li, N.; Gu, L.; Dan, W.; Dai, J. Recent developments of gramine: Chemistry and biological activity. *Molecules* **2023**, *28*, 5695. [CrossRef]
174. Lebecque, S.; Crowet, J.-M.; Lins, L.; Delory, B.M.; Du Jardin, P.; Fauconnier, M.-L.; Deleu, M. Interaction between the barley allelochemical compounds gramine and hordenine and artificial lipid bilayers mimicking the plant plasma membrane. *Sci. Rep.* **2018**, *8*, 9784. [CrossRef]
175. Robinson, T.; Robinson, T. Simple amino acid derivatives and protoalkaloids. *Biochem. Alkaloids* **1981**, *3*, 20–34.
176. Raj, R.; Shams, R.; Pandey, V.K.; Dash, K.K.; Singh, P.; Bashir, O. Barley phytochemicals and health promoting benefits: A comprehensive review. *J. Agric. Food Res.* **2023**, *14*, 100677. [CrossRef]
177. Hamany Djande, C.Y.; Steenkamp, P.A.; Piater, L.A.; Tugizimana, F.; Dubery, I.A. Metabolic reprogramming of barley in response to foliar application of dichlorinated functional analogues of salicylic acid as priming agents and inducers of plant defence. *Metabolites* **2023**, *13*, 666. [CrossRef]
178. Dias, S.L.; Chuang, L.; Liu, S.; Seligmann, B.; Brendel, F.L.; Chavez, B.G.; Hoffie, R.E.; Hoffie, I.; Kumlehn, J.; Bültemeier, A.; et al. Biosynthesis of the allelopathic alkaloid gramine in barley by a cryptic oxidative rearrangement. *Science* **2024**, *383*, 1448–1454. [CrossRef]
179. Yahyazadeh, M.; Meinen, R.; Hänsch, R.; Abouzeid, S.; Selmar, D. Impact of drought and salt stress on the biosynthesis of alkaloids in *Chelidonium majus* L. *Phytochemistry* **2018**, *152*, 204–212. [CrossRef]
180. Gower, B.G.; Leete, E. Biosynthesis of gramine: The immediate precursors of the alkaloid. *J. Am. Chem. Soc.* **1963**, *85*, 3683–3685. [CrossRef]
181. Jedrzejczak-Rey, N. Characterization of Specialized Metabolism Pathways Involved in Immune Responses of the Model Grass Brachypodium distachyon. Ph.D. Thesis, Institute of Bioorganic Chemistry of the Polish Academy of Science, Poznań, Poland, 2019.
182. Li, L.; Zheng, M.; Long, H.; Deng, G.; Ishihara, A.; Liu, F.; Liang, J.; Pan, Z.; Yu, M. Molecular cloning and characterization of two genes encoding tryptophan decarboxylase from *Aegilops variabilis* with resistance to the cereal cyst nematode (*Heterodera avenae*) and root-knot nematode (*Meloidogyne naasi*). *Plant Mol. Biol. Report.* **2016**, *34*, 273–282. [CrossRef]
183. Kim, S.-C.; Lee, J.-H.; Kim, M.-H.; Lee, J.-A.; Kim, Y.B.; Jung, E.; Kim, Y.-S.; Lee, J.; Park, D. Hordenine, a single compound produced during barley germination, inhibits melanogenesis in human melanocytes. *Food Chem.* **2013**, *141*, 174–181. [CrossRef] [PubMed]
184. Anwar, S.; Mohammad, T.; Shamsi, A.; Queen, A.; Parveen, S.; Luqman, S.; Hasan, G.M.; Alamry, K.A.; Azum, N.; Asiri, A.M.; et al. Discovery of Hordenine as a potential inhibitor of pyruvate dehydrogenase kinase 3: Implication in lung Cancer therapy. *Biomedicines* **2020**, *8*, 119. [CrossRef] [PubMed]
185. Frank, A.W.; Marion, L. The biogenesis of alkaloids: XVI. Hordenine metabolism in barley. *Can. J. Chem.* **1956**, *34*, 1641–1646. [CrossRef]
186. Karre, S.; Kumar, A.; Dhokane, D.; Kushalappa, A.C. Metabolo-transcriptome profiling of barley reveals induction of chitin elicitor receptor kinase gene (HvCERK1) conferring resistance against *Fusarium graminearum*. *Plant Mol. Biol.* **2017**, *93*, 247–267. [CrossRef]

187. Larsson, K.A.; Zetterlund, I.; Delp, G.; Jonsson, L.M. N-Methyltransferase involved in gramine biosynthesis in barley: Cloning and characterization. *Phytochemistry* **2006**, *67*, 2002–2008. [CrossRef]
188. Pretorius, C.J.; Dubery, I.A. Avenanthramides, distinctive hydroxycinnamoyl conjugates of oat, *Avena sativa* L.: An update on the biosynthesis, chemistry, and bioactivities. *Plants* **2023**, *12*, 1388. [CrossRef]
189. Li, Z.; Chen, Y.; Meesapyodsuk, D.; Qiu, X. The biosynthetic pathway of major avenanthramides in oat. *Metabolites* **2019**, *9*, 163. [CrossRef]
190. Bhullar, H.K.; Rajput, R.; Kaur, A.T.; Kaur, K.T. Cereal-Based Nutrients and Phytochemicals: Traditional and Novel Processing Techniques for Increasing Bioavailability. In *Cereals and Cereal-Based Foods*; Apple Academic Press: New York, NY, USA, 2021; pp. 209–235.
191. Gudys, K.; Guzy-Wrobelska, J.; Janiak, A.; Dziurka, M.A.; Ostrowska, A.; Hura, K.; Jurczyk, B.; Żmuda, K.; Grzybkowska, D.; Śróbka, J.; et al. Prioritization of candidate genes in QTL regions for physiological and biochemical traits underlying drought response in barley (*Hordeum vulgare* L.). *Front. Plant Sci.* **2018**, *9*, 769. [CrossRef]
192. Guptha, S.K.S. Resistance Gene Identification in Barley Against Fusarium Head Blight Based on Metabolo-Transcriptomics Approach. Ph.D. Thesis, McGill University, Montreal, QC, Canada, 2016.
193. Liu, Y.; Balcke, G.U.; Porzel, A.; Mahdi, L.; Scherr-Henning, A.; Bathe, U.; Zuccaro, A.; Tissier, A. A barley gene cluster for the biosynthesis of diterpenoid phytoalexins. *BioRxiv* **2021**, arXiv:2021.05.21.445084.
194. Grün, S.; Frey, M.; Gierl, A. Evolution of the indole alkaloid biosynthesis in the genus Hordeum: Distribution of gramine and DIBOA and isolation of the benzoxazinoid biosynthesis genes from *Hordeum lechleri*. *Phytochemistry* **2005**, *66*, 1264–1272. [CrossRef] [PubMed]
195. Matsuo, K. Recovery of RNA-dependent RNA polymerase 6 gene-knockout phenotypes in *Nicotiana benthamiana* via in vivo generation of inverted repeat construct of the trans-acting short interference RNA3 sequence. *Plant J.* **2025**, *123*, e70350. [CrossRef] [PubMed]
196. Matsuo, S.; Miyatake, K.; Endo, M.; Urashimo, S.; Kawanishi, T.; Negoro, S.; Shimakoshi, S.; Fukuoka, H. Loss of function of the Pad-1 aminotransferase gene, which is involved in auxin homeostasis, induces parthenocarpy in Solanaceae plants. *Proc. Natl. Acad. Sci. USA* **2020**, *117*, 12784–12790. [CrossRef]
197. Kong, C.-H.; Li, Z.; Li, F.-L.; Xia, X.-X.; Wang, P. Chemically mediated plant–plant interactions: Allelopathy and allelobiosis. *Plants* **2024**, *13*, 626. [CrossRef]
198. Nomura, Y.; Okada, M.; Tameshige, T.; Takenaka, S.; Shimizu, K.K.; Nasuda, S.; Nagano, A.J. Subgenome-informed statistical modeling of transcriptomes in 25 common wheat accessions reveals cis-and trans-regulation architectures. *Plant Cell Physiol.* **2025**, *66*, 347–357. [CrossRef]
199. Handrick, V.; Robert, C.A.; Ahern, K.R.; Zhou, S.; Machado, R.A.; Maag, D.; Glauser, G.; Fernandez-Penny, F.E.; Chandran, J.N.; Rodgers-Melnik, E.; et al. Biosynthesis of 8-O-methylated benzoxazinoid defense compounds in maize. *Plant Cell* **2016**, *28*, 1682–1700. [CrossRef]
200. Li, L.; Shi, Q.; Li, Z.; Gao, J. Genome-wide identification and functional characterization of the PheE2F/DP gene family in Moso bamboo. *BMC Plant Biol.* **2021**, *21*, 158. [CrossRef]
201. Bunsupa, S.; Katayama, K.; Ikeura, E.; Oikawa, A.; Toyooka, K.; Saito, K.; Yamazaki, M. Lysine decarboxylase catalyzes the first step of quinolizidine alkaloid biosynthesis and coevolved with alkaloid production in leguminosae. *Plant Cell* **2012**, *24*, 1202–1216. [CrossRef]
202. Kenjayevich, B.A.; Nizomiddinovich, T.F.; Shermamatovich, M.M. Extraction of Alkaloids and Processes Occurring in the Body. *Tadqiqotlar* **2025**, *57*, 247–250.
203. Das, A.; Ruhal, R. Potential of plants-based alkaloids, terpenoids and flavonoids as antibacterial agents: An update. *Process Biochem.* **2025**, *150*, 94–120. [CrossRef]
204. Heredia-Bátiz, J.M.; Manjarrez-Quintero, J.P.; Valdez-Baro, O.; Rivera-Salas, M.M.; Bayardo-Rosales, H.; Jiménez-Ortega, L.A.; Heredia, J.B. Role of Bioactive Compounds in Plant Disease Management. In *Biotechnological Intervention in Production of Bioactive Compounds: Biosynthesis, Characterization and Applications*; Springer: Berlin/Heidelberg, Germany, 2025; pp. 141–162.
205. Khan, S.U.; Nawaz, T.; Alam, O.; Khan, D.; Fahad, S.; Saud, S.; Lu, K. Quinoline: A Novel Solution for Next-Generation Pesticides, Herbicides, and Fertilizers. *Appl. Biochem. Biotechnol.* **2025**, *197*, 2097–2119. [CrossRef]
206. Varadharajan, V.; Rajendran, R.; Muthuramalingam, P.; Runthala, A.; Madhesh, V.; Swaminathan, G.; Murugan, P.; Srinivasan, H.; Park, Y.; Shin, H.; et al. Multi-Omics Approaches Against Abiotic and Biotic Stress—A Review. *Plants* **2025**, *14*, 865. [CrossRef]
207. Ayiecho, P.O.; Nyabundi, J.O. Speed Breeding, Breeding of Climate Resilient Crops and Functional Foods. In *Conventional and Contemporary Practices of Plant Breeding*; Springer: Berlin/Heidelberg, Germany, 2025; pp. 359–374.

International Journal of
Molecular Sciences

MDPI

Review

Current and Potential Use of Biologically Active Compounds Derived from *Cannabis sativa* L. in the Treatment of Selected Diseases

Bożena Bukowska

Department of Biophysics of Environmental Pollution, Faculty of Biology and Environmental Protection, University of Lodz, Pomorska Street141/143, 90-236 Lodz, Poland; bozena.bukowska@biol.uni.lodz.pl

Abstract: *Cannabis sativa* L. contains numerous compounds with antioxidant and anti-inflammatory properties, including the flavonoids and the cannabinoids, particularly Δ-9-tetrahydrocannabinol (THC) and cannabidiol (CBD). Cannabinoids have an effect on the endocannabinoid system (ECS), a cellular communication network, and are, hence, widely studied for medical applications. Epidiolex®, a 99% pure oral CBD extract, has been approved by the FDA for the treatment of epilepsy. Nabiximols (Sativex) is an oromucosal spray containing equal volume of THC and CBD, and it is commonly used as an add-on treatment for unresponsive spasticity in multiple sclerosis (MS) patients. Several in vitro and in vivo studies have also shown that cannabinoids can be used to treat various types of cancer, such as melanoma and brain glioblastoma; the first positive clinical trials on the anticancer effect of a THC:CBD blend with temozolomide (TMZ) in the treatment of highly invasive brain cancer are very promising. The cannabinoids exert their anticancer properties in in vitro investigations by the induction of cell death, mainly by apoptosis and cytotoxic autophagy, and the inhibition of cell proliferation. In several studies, cannabinoids have been found to induce tumor regression and inhibit angiogenic mechanisms in vitro and in vivo, as well as in two low-numbered epidemiological studies. They also exhibit antiviral effects by inhibiting *ACE2* transcription, blocking viral replication and fusion, and acting as anti-inflammatory agents; indeed, prior CBD consumption (a study of 93,565 persons in Chicago) has also been associated with a much lower incidence of SARS-CoV-2 infections. It is postulated that cannabis extracts can be used in the treatment of many other diseases such as systemic lupus erythematosus, type 1 diabetes, or various types of neurological disorders, e.g., Alzheimer's disease. The aim of this review is to outline the current state of knowledge regarding currently used medicinal preparations derived from *C. sativa* L. in the treatment of selected cancer and viral diseases, and to present the latest research on the potential applications of its secondary metabolites.

Keywords: anticancer properties; antioxidant effects; antiviral effects; anti-inflammatory properties; cannabidiol; cannabinoids; tetrahydrocannabinol

1. Introduction

Cannabis sativa L. belongs to the family *Cannabaceae* of the order Rosales [1]. The members of this small family are typically found in the temperate regions of the northern hemisphere. *C. sativa* L. is an annual herbaceous and dioecious plant characterized by considerable genetic variation [2,3].

Cannabis is often regarded as monotypic, i.e., comprising only one species, *C. sativa* L. [4]; however, the botanical nomenclature also identifies ssp. *indica* and ssp. *sativa* as subspecies. *C. sativa* and *C. indica* differ in their THC content, with this being less than 0.3% by weight in *C. sativa* but more than 0.3% in *C. indica*; in the latter, this level is sufficient to produce a psychoactive action. In contrast, *C. indica* has a lower CBD concentration than *C. sativa* [5].

C. sativa L. is an easy crop to grow and, due to its high growth rate and large green leaf area, it can achieve high yields compared to other cultivated plants without the need for plant protection products and fertilizers [3]. The plants also have roots that loosen and ventilate the soil, improve its water conditions, and generally provide a beneficial environment for neighboring plants. Most importantly, they can survive with very low soil water availability and could serve as promising subjects for designing sustainable and water-saving crops [6].

C. sativa L. seeds are a source of oil [7,8], and various valuable essential oils can be obtained from the plants inflorescences and leaves. While *C. sativa* L. seeds are used as feed for farm animals, there is growing interest in the use of cannabis oil, meal, flour, and protein powder made from processed seeds in human nutrition [9]. *C. sativa* L. oil has great potential as a nutraceutical due to its polyunsaturated fatty acid content and because it has an ideal α-linolenic acid (omega-6) to linoleic acid (omega-3) ratio of 2.5–3:1; this value is also enhanced by the presence of bioactive CBD. The unsaturated fatty acids found in *C. sativa* L. oil, and especially γ-linoleic acid, have industrial applications in cosmetic products such as moisturizers, shampoos, massage oils, conditioners, and balms. *C. sativa* L. products are non-allergenic and have skin rejuvenating properties. The cultivation of industrial *C. sativa* L. provides a source of various products used in textiles, personal care, food, nutraceuticals, and pharmaceuticals [10–12].

Approximately 500 types of chemical compounds have been determined in *C. sativa* L. [3], including primary metabolites (amino acids, fatty acids, and steroids) and secondary metabolites (cannabinoids, flavonoids, stilbenoids, terpenoids, and lignans) with significant health-promoting properties [13].

When evaluating its pharmacological effects, it is essential to note that *C. sativa* L. can be further subdivided in chemotypes based on its chemical composition. The plant is currently classified into five chemotypes depending on the levels of its main phytocannabinoids: THC, CBD, and cannabigerol (CBG). While cannabis strains and chemovars often show similar phytocannabinoid levels, they are characterized by different chemical profiles, and, hence, different pharmacological effects. Cerato et al. identified several new subgroups within traditional classifications resulting from the unique composition of minor phytocannabinoids that may be essential for the evaluation of the pharmacological effects of cannabis chemovars [14]. A recent analysis of industrial *C. sativa* L. based on untargeted cannabinomics found its chemical composition to vary according to the variety and geographical location of the crop; the data also confirm that phytochemicals can play an important role in distinguishing plant samples [15].

Due to the medical potential of its cannabinoids, the medical use of *C. sativa* L. is now legal in -21 European Union countries, Canada, and 47 U.S. states [16,17]. Cannabis-based drugs are also permitted for use in treating numerous illnesses and symptoms, including cancer, AIDS, multiple sclerosis, chronic pain, nausea, anorexia, and muscle spasms, in the European Union [16]. Sales of *C. sativa* L. products in the United States have increased dramatically in recent years: in 2018, the sale of *C. sativa* L.-derived CBD products amounted to $390 million. Globally, the market for *C. sativa* L. products was estimated at $2.6 billion in 2022, and is expected to reach $3.9 billion by 2031 [18].

The aim of this article is to present the current state of knowledge regarding the current and potential use of biologically active compounds in *C. sativa* L. in the treatment of selected diseases. It takes a methodical approach to select and assess the recent literature, with most papers having been published between 2019 and 2024. Around 250 articles were initially identified in the databases, of which 157 were considered suitable for this review. The publications were identified using the PubMed, MDPI, Frontiers, Elsevier, and Springer databases, as well as Google Scholar. The review includes original research papers and reviews related to the presented topic. The search was performed with the following keywords: *Cannabis sativa;* cannabinoids; "cannabinoids and disease"; "cannabinoids and cancer"; "cannabinoids and inflammation"; "antiviral properties of cannabinoids"; "cannabinoids and SARS-CoV-2"; 9-tetrahydrocannabinol; cannabidiol; dronabinol; nabiximols; nabilone; Epidiolex®; Sativex and cannflavin.

This review discusses the antioxidant, anti-inflammatory, and anticancer properties of the compounds present in *C. sativa* L. It also describes currently used medicinal preparations derived from *C. sativa* L. in the treatment of selected diseases, and presents the latest research on the potential applications of its secondary metabolites.

2. Biologically Active Compounds in *Cannabis sativa* L.

C. sativa L. is rich in secondary compounds, including 12 fatty acids, 200 terpenes, and 20 heterocyclic compounds with nitrogen atoms in their ring structure. These are accompanied by more than 50 hydrocarbons and 125 cannabinoids [19,20]. While the most well-known cannabinoids in the species are THC and CBD, various other compounds have been isolated, including 9-tetrahydrocannabinolic acid (Figure 1) [21].

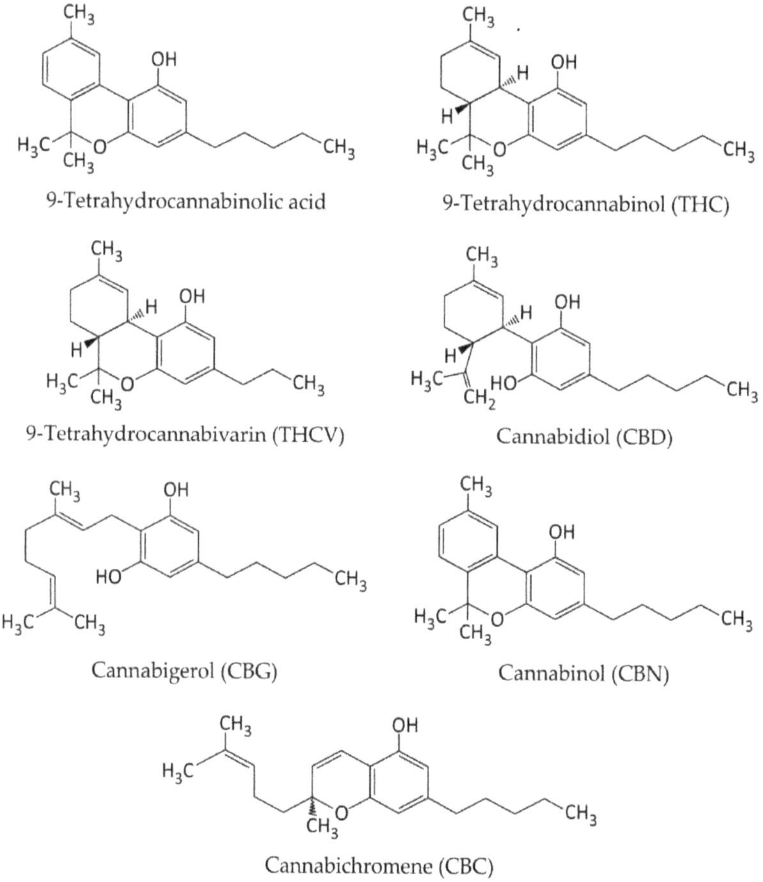

Figure 1. Structural formulas of cannabinoids occurring in *C. sativa* L.

C. sativa L. oil extract is dominated by the cannabinoids CBD (59.1%) and THC (20.2%). Various monoterpenes (α-pinene, camphene, β-myrcene, and D-limonene) and sesquiterpenes (β-caryophyllene, α-bergamotene, α-humelene, humulene II epoxide, and caryophyllene oxide) have also been identified. *C. sativa* L. oil has been found to significantly inhibit tumor necrosis factor α (TNF-α) release in lipopolysaccharide (LPS)-stimulated rat monocytes [7]. Shebaby et al. [7] also demonstrated that cannabis oil extract blocked cyclooxygenase-2 (*COX-2*) and nitric oxide synthase (*iNOS*) expression induced by LPS, as well as the phosphorylation of mitogen-activated protein kinases (MAPKs). Additionally, *C. sativa* L. oil extract decreased paw edema in rats, and histopathological analysis showed reduction in inflammation and swelling.

The biologically active content of the oil was also studied in two varieties of industrial *C. sativa* L.: Kompolti and Tisza. The findings identified three terpenes, viz. isocaryophyllene, α-humulene, and β-caryophyllene β-oxide, and two terpenophenols, viz. CBD and cannabigerolic acid. Both essential oils caused cancer cell death in vitro, with that of the Tisza cultivar showing particularly marked cytotoxicity in cholangiocarcinoma cells [22].

C. sativa L. seeds consist of 20 to 25% protein, 20 to 30% carbohydrate, 25 to 35% oil, and 10 to 15% insoluble fiber, as well as chemical elements such as potassium, magnesium, calcium, iron, and zinc. Furthermore, they contain vitamins with highly antioxidative potential, such as A, C, E, and beta-carotene [23].

Oil obtained from hemp seeds is characterized by its high antioxidant potential. For example, studies on quail found those given feed with the addition of hemp seeds showed increased antioxidant activity in the blood [24]. Other laboratory studies have indicated that hempseed oil has an antiradical capacity corresponding to 1.87 α-tocopherol equivalents pel ml of oil [25].

C. sativa L. sprouts contain more biologically active compounds than their seeds. They are characterized by a higher total content of polyphenols, including flavonoids and flavanols, as well as high levels of omega-3 acids [26].

Twenty secondary metabolites have been identified in cannabis roots, including β-amyrone, glutinol, fucosterol, stigmastanol, stigmasta-3,5-diene, stigmasta-3,5,22-triene, and oleamide. The ethanol extracts of three chemovars were found to have antioxidant effects in vitro, and were identified by FRAP and ABTS. The authors propose that the antioxidant effect of the chemovars is dependent on their component compounds and is associated with external factors, such as harvest time [27].

C. sativa L. waxes are rich in alkanes, monoterpenes, sesquiterpenes, terpenoids, and cannabinoids, and have been attributed various healing and antioxidant effects. They are obtained from whole plants, as well as their individual fibers, seeds, and leaves, by various technological processes, with one important source being the dust [28].

2.1. Cannabinoids—Major Natural Ligands of CB1 and CB2 Receptors, with Anti-Inflammatory Effects

Among the 125 known cannabinoids, the most represented are THC and CBD due to their high content and biological properties [20]. Recent years have shown a growth in interest in legalizing cannabis as a source of cannabinoids for treating epilepsy, multiple sclerosis, pain (migraine), or tic disorders [29–31]. These compounds may also be of value in treating viral diseases, e.g., SARS-CoV-2 [32]; inflammation, e.g., Crohn's disease [33] and rheumatoid arthritis [34]; cardiovascular disorders [35]; and in cancers [36,37] (Figure 2).

The endocannabinoid system (ECS) acts as a cellular communication network that maintains many biological functions and body homeostasis. Cannabinoids themselves are known to influence numerous biological processes such as memory, pain, reproduction, bone repair, and immunity [38], and have therapeutic potential in many diseases.

The cannabinoid receptors CB1 and CB2 mediate a variety of physiological processes including the activation of MAPK phosphorylation and A-type potassium channels, and the suppression of adenylyl cyclase activity, D-type potassium, and calcium channels; these are all regarded as desirable drug targets. Both receptors are activated by endogenous endocannabinoids and the psychoactive components of marijuana [39,40].

Of the two primary cannabinoids, THC has a direct impact on cannabinoid receptors, while CBD does not. In addition, CBD does not bind to CB1 and, hence, does not have any psychoactive effect; it also does not directly trigger any receptors. However, CBD plays an important role in the endocannabinoid system. It changes the ability of receptors to bind to cannabinoids, affects other types of receptor, and increases the natural level of endocannabinoids.

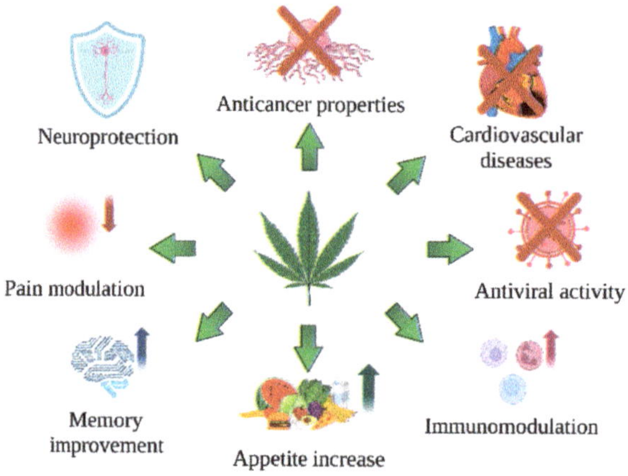

Figure 2. The therapeutic potential of *C. sativa* L. compounds. The up arrows indicate improved memory and increased immunity and appetite, while the down arrow indicates a pain reduction. The crossed red lines underline the potential protective role of *C. sativa* L. compounds in cancer, viruses, and cardiovascular diseases. Created in BioRender. Bukowska, B. (2024) BioRender.com/k65p423 (accessed on 1 September 2024).

THC is a partial agonist of the CB1 and CB2 receptor. In line with classical pharmacology, the response it generates appears to be strongly influenced by the expression and signaling efficiency of the cannabinoid receptors, as well as the ongoing endogenous release of cannabinoids. CBD shows an unexpectedly high potential as an antagonist of CB1/CB2 receptor agonists in cells and tissues [41].

It was reported that CBD can act as an antagonist or inverse agonist at concentrations below which it binds to either the CB1 or CB2 orthosteric sites [42]. Recently, several studies have found that CBD can act as a negative allosteric modulator of CB1: although it alters the potency and efficacy of orthosteric ligands, it does not activate the receptor itself [43–45]. CBD was also found to act as a partial agonist for CB2 [43].

CBD is considered to be of greater clinical interest than THC, partly due to its less addictive, hallucinogenic, and toxic side effects, and partly due to its anti-inflammatory, analgesic, anxiolytic, and anticancer properties, which may offer therapeutic potential in several disorders [46]. Cannabinoids exert their psychoactive and analgesic effects through CB1 receptors [41], and immunomodulatory ones through CB2 [41]. CB2 is mainly present in immune cells and plays an important role in controlling humoral and cell immune responses [47]. Targeted activation of CB2 could limit inflammation, while avoiding the psychoactive effects associated with CB1. In a clinical trial, daily cannabis treatment was associated with lower levels of pro-inflammatory biomarkers in the cerebral fluid of HIV patients [48].

THC binds with high affinity to both cannabinoid receptors as an agonist [49]. It is also an agonist of many other receptors, such as G-protein-coupled receptors and the peroxisome proliferator-activated receptor; it also serves as an antagonist of the TRPM8 transient receptor potential channel and 5-HT3 receptor A, and can increase the level of anandamide and adenosine [50]. THC suppresses the release of pro-inflammatory factors, including tumor necrosis factor α (TNF-α), interleukin-1β (IL-1β), IL-6, and IL-8, decreases nuclear factor-κB (*NF-κB*) expression, and inhibits the upregulation of cofilin-1 protein [51]. CBD and THC were found to be effective in alleviating the LPS-induced cytokine storm in human macrophages and primary HBECs, at least via modulation of NLRP3 inflammasome and STAT3 signaling pathways [52].

CBD has demonstrated anti-inflammatory effects in several conditions characterized by inflammation, including arthritis [53], neurodegenerative changes [54], and HIV [55], as well as various others [56]. Lowin et al. report that CBD increases intracellular calcium levels and reduces cell viability and IL-6/IL-8/MMP-3 production of rheumatoid arthritis synovial fibroblasts [53]. Recent studies have suggested that CBD is able to reduce the formation of lipid inflammation precursors (e.g., arachidonic acid) and cytokines in rat cerebral cortices, this being a primary event in the development of neurodegenerative diseases [54]. CBD has also been found to reduce the production of various inflammatory cytokines and chemokines in HC69.5 human microglial cells following HIV infection compared with THC. Treatment resulted in the reduction in inter alia MIF and SERPIN E1, IL-6, IL-8, and IL-1 β, as well as GM-CSF, MCP-1, CXCL1, and CXCL10 [55].

Cannabinoids inhibit pro-inflammatory cytokine activity by the effect of the CB1 and CB2 receptors. As such, CBD offers potential for reducing the inflammation associated with Crohn's disease and ulcerative colitis [33]. Cannabinoids can also be used to alleviate rheumatoid arthritis by regulating the immune response [34]. CBD and THC ameliorate autoimmune effect on the nervous system, and are, hence, used to treat severe spasticity and pain in MS [57]. They have also demonstrated potential in the treatment of systemic lupus erythematosus [58], type 1 diabetes [59], and various types of neurological disorders, e.g., Alzheimer's disease [60].

2.2. Antioxidant Compounds

Antioxidant substances, like phenolic compounds, tocopherols, and phytosterols, have been identified in *C. sativa* L. Those compounds have been long recognized to exhibit antioxidant properties in plants [6]. One study based on targeted metabolomics found that the metabolomes of seven cannabis varieties consisted of a total of 1001 metabolites, including 201 flavonoids, 86 alkaloids, and 149 phenolic acids [61].

Among polyphenols, a variety of flavonoids have been detected, such as apigenin, luteolin, kaempferol, quercetin, rutin, and methylated flavones like cannflavin A (CFL-A), B (CFL-B), and C (CFL-C), as well as the structural isomer of CFL-B, isocanflavin B (IsoB) [62,63]. Polyphenols are known to be powerful antioxidants that neutralize free radicals by donating an electron or hydrogen atom, and by chelating transition metals [64].

Barrett et al. found cannflavins A and B to inhibit the release of prostaglandin E2 (PGE2) from human rheumatoid synovial cells 30 times more effectively than aspirin under ex vivo conditions. Both cannflavins A and B demonstrate potent anti-inflammatory activity via the inhibition of PGE2 release induced by 12-o-tetradecanoylphorbol 13-acetate and microsomal prostaglandin E synthase-1 [65,66]. A review by Erridge found cannflavins (CFL-A) to have antioxidant properties in preclinical study models (carried out on laboratory animals or cell cultures), and to have antiparasitic (CFL-A and CFL-C), neuroprotective (CFL-A), and anticancer (isocannflavin B and CFL-B isomer) properties. In silico screening found CFL-A to have binding affinity for viral proteins [67] and to inhibit transforming growth factor beta-activated kinase 1, an important therapeutic target in inflammation-related diseases [68]. In vivo experiments showed a significant increase in the survival of male and female mice with pancreatic cancer compared to control cohorts after the use of the cannflavin B isomer (FBL-03G) [69].

Hemp seeds are rich in several phenolic compounds, including phenolic acids, lignamides, phenolic amides, and flavonoids [70]. The most abundant phenolics present in hemp flour samples were found to be hydroxycinnamic and protocatechuic acids [71].

Furthermore, *C. sativa* L. seeds have been found to contain a number of phenylpropionamides and their random oxidative coupling derivatives, i.e., lignanamides which significantly inhibited proliferation in a U-87 glioblastoma cell line by inducing apoptosis and suppressing autophagic cell death [72]; they also protected against nerve inflammation in a rat model of streptozotocin-induced Alzheimer's disease, which was attributed to the regulation of primary bile acid biosynthesis and taurine and hypotaurine metabolism [73]. In a mouse model, an extract containing fourteen phenylpropionamides prevented the learning and spatial memory damage induced by lipopolysaccharide [74].

Other bioactive substances include tocopherols that exhibit antioxidant effects by scavenging lipid peroxide radicals, which promote lipid peroxidation and protect unsaturated fatty acids from oxidative damage [75].

Also, the peptides derived from cannabis proteins have been found to be strong antioxidants, which can be employed in both the phytomedicine and food industry [76]. Wang et al. found bioactive *C. sativa* L. protein hydrolysates to have both anti-inflammatory and antioxidant properties in LPS-stimulated BV2 microglial cells, inhibiting the progression of inflammation associated with the NF-κB and Keap1-Nrf2 signaling pathways [77]. *C. sativa* L. proteins also demonstrate various immunomodulatory effects. Rodriguez-Martin et al. found treatment with hemp protein products to decrease certain pro-inflammatory mediators (TNFα, IL-1β, and IL-6) and increase anti-inflammatory mediators (IL-10 and IL-4) in lipopolysaccharide (LPS)-activated primary human monocytes [78].

Extracts from the flowers [79,80], leaves [81,82], roots [82], and seeds [83] of *C. sativa* L., and their oil [84,85], contain a rich composition of antioxidants, or single compounds (e.g., CBD or THC) [86]; these have been found to counteract the production of free radicals (Table 1). Recent studies have also shown that the digestion of hulled *C. sativa* L. seeds increases the inhibition of radical scavenging by these compounds [83]. THC and CBD have similar antioxidant activity to vitamins E and C [87,88]. The phenolic groups readily oxidized to quinoid forms, and the unsaturated bonds found in non-olivetolic fragments of these cannabinoids could be responsible for their antioxidant properties [39,87].

Cannabinoids also act as indirect antioxidants, capable of modulating the redox balance by increasing the level of reduced glutathione (GSH) [89,90]; activating antioxidant enzymes such as superoxide dismutase (SOD) [91], catalase (CAT), glutathione peroxidase (GPx), and glutathione reductase (GR) [85,91]; and by suppressing pro-oxidant enzymes [56] (Table 1).

Table 1. Antioxidant activity of extracts, oil, and single compounds contained in *C. sativa* L.

Extract/Compounds	Tested System	Concentrations	References
	C. sativa L. extracts and cannabinoids decrease free radicals' level		
C. sativa L. essential oil (β-caryophyllene, α-humulene, β-myrcene, α-pinene)	Chemical system	IC_{50} = 0.98 mg/mL for DPPH EC_{50} = 1.74 mg/mL for FRAP 0.101 mg AAE/g EO for TAC	[84]
C. sativa L. extracts of flowers (flavonoids, terpenoids, saponins, anthocyanins, tannins, and reducing sugars)	Chemical system	for DPPH Ethanol extract—IC_{50} = 0.23 mg/mL Hexanic extract—IC_{50} = 0.38 mg/mL Chloroformic extract—IC50 = 0.77 mg/mL	[79]
C. sativa L. extract (β-caryophyllene, its oxide, CBD, THC, α-pinene, α-humulene, 2-monolinolein, methyl eicosatetraenoate and γ-sitosterol)	Chemical system	for DPPH (at different vegetative stages) Leaf—15.03 to 35.04 mmol/L extract Roots—0.29 to 1.56 mmol/L extract	[82]
Hulled *C. sativa* L. seeds and their compounds after digestion (in vitro)	Chemical system	Percentage inhibition of radical scavenging activity (Pi) 39.97 ± 0.71% after the oral phase 50.28 ± 0.62% of the intestinal phase	[83]

Table 1. *Cont.*

Extract/Compounds	Tested System	Concentrations	References
C. sativa L. extract	In vitro— neuro-2a cell line	For DPPH IC_{50}—0.06 mg/mL for lyophilized aqueous extract IC_{50}—0.10 mg/mL for the residual biomass hexane extract) IC_{50}—0.005 mg ml/L for ascorbic acid	[80]
C. sativa L. oil 1 mg oil/kg BW for 3 weeks	In vivo— rats with a sucrose-rich diet	For ROS levels in the liver Rats (SRD)—2.46 arbitrary units Rats (SRD + Ca)—1.45 arbitrary units	[85]
Cannabinoids prevented the oxidation of DHR and H_2DCF.			
THC and CBD	In vitro— aged pancreatic islet cells	0.1–100 μmol/L	[86]
Cannabinoids: CBG, CBD, Δ9-THC, CBN, CBGA, CBDA and Δ9-THCA	chemical system	EC_{50} CBD—11 mmol/L EC_{50} THC—18 mmol/L EC_{50} BHT—18 mmol/L	[92]
THC and CBD	In vitro— rat neuronal cell cultures	EC_{50} of 2–4 μmol/L	[39]
CBD	In vitro—rat PC12 cells with β-amyloid induced toxicity	0.1–100 μmol/L	[93]
C. sativa L. oil and cannabinoids reduce the lipid peroxidation			
CBD	In vitro— rat PC12 cells with β-amyloid-induced toxicity	Decrease in MDA level in PC12 caused by β-amyloid 0.1–100 μmol/L	[93]
C. sativa L. oil 1 mg oil/kg BW for 3 weeks	In vivo— rats with a sucrose-rich diet	For TBARs in serum Rats (SRD)—3.94 μmol/L Rats (SRD + Ca)—2.99 μmol/L	[85]
CBD intraperitoneal injection	In vivo— Rats with traumatic spinal cord injury	2.5–20 mg/kg	[89]
CBD	In vivo— rats	50, 100, 200 ng/rat	[91]
C. sativa L. oil and cannabinoids increase the level of GSH			
CBD	In vivo— rats chronically irradiated with UV	120 mg/kg BW	[90]
CBD intraperitoneal injection	In vivo— rats with traumatic spinal cord injury	2.5–20 mg/kg	[89]
Cannabis oil 1 mg oil/kg BW for 3 weeks	In vivo— rats with sucrose-rich diet	For GSH levels in the liver Rats (SRD)—45 ηmol/mg protein Rats (SRD + Ca)—61 ηmol/mg protein	[85]
C. sativa L. oil and cannabinoids elevated the activity of antioxidative enzymes: SOD, CAT or/and GPX and GR			
CBD	In vivo— rats	Injected at 50, 100, and 200 ng/rat for five consecutive days	[91]
Cannabis oil 1 mg oil/kg BW for 3 weeks	In vivo— rats	For GPx and GR activities in the liver Rats (SRD)—85 and 28 mU/mg Rats (SRD + Ca)—147 and 44 mU/mg	[85]

AAE: ascorbic acid equivalents; CAT—catalase; DPPH—2,2-difenylo-1-pikrylohydrazyl; BW—body weight; FRAP—ferric reducing antioxidant power; DHR—dihydrorhodamine; EO—essential oil; GSH—reduced glutathione; GPX—glutathione peroxidase; GR—glutathione reductase H_2DCF—2′,7′ dichlorofluorescein; SOD—superoxide dismutase; SRD—rats with sucrose-rich diet; SRD + Ca—rats with SRD and oil; TAC—total antioxidant capacity.

2.3. Potential Use of Biologically Active Compounds Derived from Cannabis sativa L. in the Treatment of Cancers

Oxidative stress, and the consequent imbalance in the redox system and inflammation, plays an important role in many conditions, including cancer, inflammatory diseases, and neurodegenerative diseases. The various antioxidant components of *C. sativa* L. interrupt the chain reactions of free radicals by capturing them or converting them to less active forms; they also ameliorate oxidative conditions, preventing the formation of radicals [94].

As some conditions, including cancer, can alter the endocannabinoid system, the use of THC and CBD as anticancer agents has become a promising research area. Although it is widely accepted that conventional therapies, such as chemotherapy and radiotherapy, have more side effects in adult cancer patients then cannabinoid treatment, these results have not been translated into many clinical trials for adults. Nevertheless, interest in these compounds in Medicine and Oncology has been growing [95]. Cannabinoids are mostly used as support for patients receiving radiation or chemotherapy: they alleviate many of the common symptoms and side effects connected with the treatment of cancer [96].

Currently, little evidence exists regarding the effective treatment of cancer in animals, with only several epidemiological studies examining the effects of cannabinoid administration; as such, the preclinical model of cancer demands more solid evidence to confirm whether cannabinoids have anticancer effects in humans [36,97,98]. An in silico study docked six cannabinoids from *C. sativa* L. with MAPK-ERK signaling pathways to identify their possible binding interactions; among these, CBD and THC exerted the greatest binding affinities with proteins of the MAPK-ERK signaling pathways, and their pharmacokinetics and drug-likeness-related profiles suggest that these cannabinoids could be potential inhibitors in cancer treatment [99].

Studies have revealed that cannabinoids may have a different mechanism of anticancer potential in various cancer cell types in vitro and in animals in vivo (Figure 3).

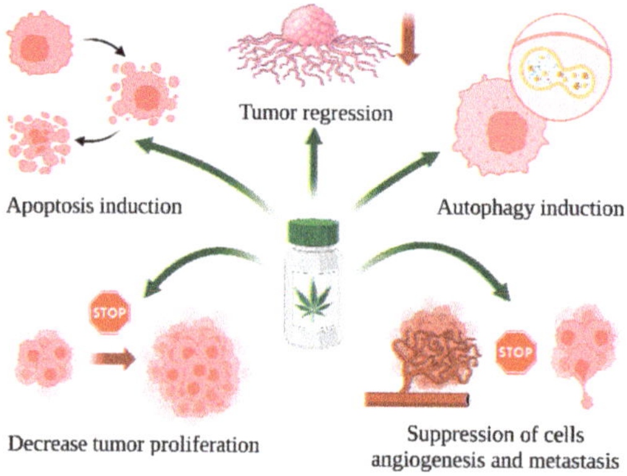

Figure 3. Mechanism of anticancer properties of *C. sativa* L. compounds by induction of apoptosis (e.g., by increase caspase 3/7 activity, upregulated *TP53* and *Bax*—expression [100]), induction of autophagy (e.g., upregulates LC3BII, downregulates p62,and inhibits p-PI3K, p-AKT, and p-mTOR pathways [101]), induction of tumor regression (e.g., decreases colony-stimulating factor 1 (CSF-1) level [102], induces DNA fragmentation, and arrests cell progression at the G1 cell cycle checkpoint) [103], decrease in tumor proliferation (e.g., by inhibiting ERK1/2 signaling pathway phosphorylation [104]), and suppression of cell invasion (e.g., by increased expression of *TIMP-1*, which mediates an anti-invasive effect [105]) in cells and animals. Created in BioRender. Bukowska, B. (2024) BioRender.com/p40b597 (accessed on 1 September 2024).

The in vitro research has included breast cancer cells [106], melanoma cells [96], lung cancer cells [97], and prostate cancer cells [107], while the in vivo studies include brain glioblastoma in animals [108,109], malignant melanoma tumors in mice [96], and metastatic melanoma in NSG mice [110]. Clinical trials have also been conducted [36,37].

Most importantly, THC and CBD, and other phytocannabinoids, have been revealed to trigger apoptosis, inhibit cancer cell proliferation, and modulate angiogenesis and metastasis [38,111,112] (Table 2).

Table 2. Different mechanisms of anticancer activity of extracts and single compounds contained in *C. sativa* L.

Extract, Compounds /Concentrations	Tested System	Mechanism of Action	References
C. sativa L. extracts and cannabinoids induction of apoptosis			
CBD 3 μmol/L	A549, H460 human lung cancer cells	Upregulates *COX-2* and *PPAR-γ*	[97]
C. sativa L. extracts—30 μmol/L CBD—10 μmol/L	PC3 prostate cancer cells	Increase caspase 3/7 activity, upregulate *TP53* and *Bax* expression, and induce silencing of *RBBP6*.	[100]
CBD 3.125–50.0 μmol/L	HCC—Hepatocellular carcinoma cells	Enhances p53 activation via ER stress	[113]
Extract from *C. sativa* 50% and 100% of their corresponding IC_{50} values	Melanoma cell lines MM418-C1, MM329, MM96L	Increases pro-apoptotic markers (Bax) expression Decreases anti-apoptotic markers (*Bcl-2-*) expression	[103]
CBD and THC 6, 10, 15 μmol/L	A375, A2058 and SK-Mel-28 melanoma cell lines	Release of mitochondrial cytochrome c and activate caspase-3/7	[110]
Cannabinol (CBN)	A172 glioblastoma, HepG2 liver cancer and HCC1806 breast cancer cell lines	Downregulates p21 and p27 arrested cell cycle in G1 or S phase	[114]
CBD induction of autophagy			
CBD 2–20 μmol/L	Human cholangiocarcinoma cells (KKU-213B, KKU-100, KKU-055)	Upregulates LC3BII, downregulates p62, inhibits p-PI3K, p-AKT, and p-mTOR pathways	[101]
CBD 3, 6, or 10 μmol/L	Head and neck squamous cell carcinoma (FaDu, SNU899, SCC15, Hep2)	Increases in Beclin- and LC3II-coding gene expression	[115]
C. sativa L. extracts and cannabinoids induction of tumor regression			
C. sativa L. essential oil extracted from flowers and leaves	Lewis lung cancer grafted mice model	Inhibits tumor growth, decreases of TNF-α and IL-6, increases in CD4+, CD8+ T lymphocytes count	[116]
CBD 5 mg/kg BW	Mice C57BL/6 with B16F10 murine melanoma tumor	Reduces tumor size	[96]
CBD—10 mg/kg BW THC—10 mg/kg BW	NSG mice	Deplete tumor growth	[110]
Cannabigerrol (CBG) 2.5 mg/kg or 3.75 mg/kg	Tumor-bearing mice	Decreases tumor progression by combining CBG and αPD-L1, decreases colony-stimulating factor 1 (CSF-1) level	[102]
PHEC-66 extract from *C. sativa* L.	MM418-C1, MM329, and MM96L melanoma cell lines	Induces DNA fragmentation, and arrests cell progression at the G1 cell cycle checkpoint.	[103]

Table 2. *Cont.*

Extract, Compounds /Concentrations	Tested System	Mechanism of Action	References
Cannabinoids decrease tumor proliferation.			
CBD and THC 3.9–500 μg/mL	A375 human melanoma cells	Inhibit ERK1/2 signaling pathway phosphorylation, which is responsible for the regulation of cell proliferation.	[104]
Cannabinol (CBN)	A172 glioblastoma, HepG2 liver cancer and HCC1806 breast cancer cell lines	Decreases cancer cell proliferation	[114]
Cannabinoids induce suppression of cell invasion/inhibition of angiogenesis			
R(+)-methanandamide analog 0.1 μmol/L THC—0.01 μmol/L	HeLa, C33A human cervical carcinoma, A549 human lung cancer cells	Increase expression of *TIMP-1*, which mediates an anti-invasive effect of cannabinoids.	[105]
Cannabinoid JWH-133 8 days at 50 μg/day	Mice with malignant gliomas	Inhibits expression of proangiogenic factors: vascular endothelial growth factor, and angiopoietin 2	[36]

2.3.1. Lung Cancer Cells/Lung Cancer

Ramer et al. [97] evaluated the role of COX-2 and PPAR-γ in the pro-apoptotic and regressive effects of cannabinoids in human lung cancer cells (A549, H460). They noticed that cannabinoids decreased the viability of cancer cells by triggering apoptosis, *COX-2* and *PPAR-γ* genes upregulation and their proteins expression. Cancer cell lines treated with cannabidiol demonstrated elevated levels of COX-2-dependent prostaglandins; in particular, PGD(2) and 15-deoxy-Δ(12,14)-PGJ(2) translocated PPAR-γ to the nucleus, and triggered PPAR-γ-dependent apoptotic cell death. Additionally, in A549-xenografted nude mice, cannabidiol induced upregulation of *COX-2* and *PPAR-γ* genes and their proteins in tumor tissue and tumor regression. Studies on A549 cells from human lung cancer have found CBD to decrease tumor invasion and metastasis, and inhibit cell growth [97,117,118].

CBD also slowed tumor invasion by human lung carcinoma cells (A549), human cervical carcinoma (C33A), and HeLa [105]. The authors of this study assessed the effect of THC and the stable anandamide analog R(+)-methanandamide (MA) on h expression of matrix metalloproteinases and endogenous tissue MMP inhibitors (TIMPs) in A549 cells. The role of TIMP-1 was determined by transfecting HeLa, C33A, or A549 cells with siRNA targeting TIMP-1. It was found that MA and THC induced a time- and concentration-dependent suppression of HeLa cell invasion through Matrigel, and that this was assisted by elevated *TIMP-1* expression; such elevated *TIMP-1* expression appeared to mediate the anti-invasive effects of cannabinoids. Ramer and Hinz, thus, recommend cannabinoids in the treatment of highly invasive cancers [105].

An in vivo study of a Lewis lung cancer grafted mouse model by Xu et al. [116] found that *C. sativa* L. essential oil (HEO) significantly inhibited tumor growth; it also downregulated the levels of tumor inflammatory cytokines TNF-α and IL-6, modified the immune organ index, and increased the numbers of immune-related CD4+ and CD8+ T lymphocytes and the CD4+/CD8+ ratio.

2.3.2. Hepatocellular Carcinoma Cells

Recently, Lee et al. studied the effects of a combination of cabozantinib and CBD therapy on hepatocellular carcinoma (HCC) cells. Their data indicate that cannabidiol could increase the death of apoptotic cells by cabozantinib (a multi kinase inhibitor) through phosphorylation of p53, which is regulated by ER stress in HCC cells. The addition of CBD increased the anticancer activity of cabozantinib, which the authors attribute to increased apoptosis and cytotoxicity [113].

2.3.3. Prostate Cancer Cells/Prostate Cancer

In other research, cannabis extract, cannabidiol, and cisplatin was found to cause antiproliferation of PC3 cancer cells by increasing caspase 3/7 activities. The study also showed apoptotic changes in studied cells when retinoblastoma binding protein 6 (RBBP6) was silenced, with upregulation of *TP53* and *Bax* expression and downregulation of *Bcl2* expression. In other experiments, it was observed that the tumors in the mice had decreased in size after treatment with cisplatin and CBD. CBD, hence, appears to be a viable therapy for treating prostate cancer cells when combined with the silencing of RBBP6 [100].

CBD seems to induce cell death mostly through the intrinsic apoptotic pathway. This is connected with the dysregulation of the Bcl-2 family of proteins, resulting in mitochondrial dysfunction (i.e., depolarization and ROS production). Cannabinoids can target various mitochondrial processes by modulating multiple complex signaling pathways; the targets include intracellular calcium regulation, the bioenergetic metabolism, and mitochondrial apoptosis and dynamics, such as mitochondrial fission and fusion, transport, mitophagy, and biogenesis [119].

An open-label, single-center Phase I/Ib trial was performed to determine the safety and initial efficacy of CBD using Epidiolex® in 18 patients with biochemically recurrent prostate cancer. Epidiolex® 800 mg/day was observed to be well tolerated, and to have an acceptable safety profile in these patients [107]. Unfortunately, there are inherent limitations to this study, such as the short duration of treatment with Epidiolex, the small sample size, and the lack of a comparator group.

2.3.4. Melanoma Cancer Cells/Malignant Melanoma Tumors

A study on a mouse model found cannabiberrol (CBG) to inhibit tumor progression and reduce the amount of macrophages associated with the tumor [102]. Combining CBG with the anti-programmed cell death ligand 1 (αPD-L1) further reduced tumor progression, increased survival, and increased the infiltration of active toxic T cells. Furthermore, the cannabis extract treatment also depleted colony-stimulating factor 1 (CSF-1) secretion by melanoma cells. The authors propose that CBG employs a novel mechanism to modulate the tumor microenvironment and increase immune checkpoint blockade therapy; as such, it offers promising therapeutic potential in various tumors treated with elevated expression of colony-stimulating factor 1.

Another study found *C. sativa* L. (PHEC-66) extract to trigger apoptosis in MM418-C1, MM329, and MM96L melanoma cell lines [103]. This was associated with increased expression of anti-apoptotic markers (i.e., *Bax*) and a decrease in anti-apoptotic markers (i.e., *Bcl-2-*). Additionally, PHEC-66 caused DNA fragmentation and stopped cell progression at the G1 cell cycle control point. These results suggest that PHEC-66 may have potential as an adjunct therapy for the treatment of malignant melanoma.

Simmerman et al. [96] assessed the effects of cannabinoid derivatives on malignant B16F10 murine melanoma tumors. The tumors were subcutaneously implanted in C57BL/6 mice, which were then treated with 5 mg cisplatin and 5 mg CBD. The mice treated with CBD demonstrated significantly smaller tumor sizes compared to the controls. Mice treated with cisplatin exhibited the longest survival period, while those treated with CBD demonstrated improved quality of life and movement. While cisplatin induced the strongest tumor growth suppression in all experimental groups, CBD significantly delayed tumor growth compared to controls, and subjective observations suggested that CBD was better tolerated.

Recent research has examined the pro-apoptotic effects of cannabinoids in metastatic melanoma in vivo and in vitro. Various melanoma cell lines were treated with cannabinoids at various concentrations, and their efficacy in combination with trametinib was assessed in vivo in NSG mice. Cannabinoids were noticed to deplete cell viability in the multiple melanoma cell lines in a concentration-dependent manner, triggering apoptosis by releasing mitochondrial cytochrome c and activating various caspases. These substances substantially depleted tumor growth in NSG mice and were as potent as trametinib [110].

Another study evaluated the expression of genes in A375 human melanoma cells after treatment with *C. sativa* L. extracts. It was observed that a mixture of THC and CBD of *C. sativa* L. fractions triggered apoptosis in the cancer cells. During apoptosis, a number of genes were upregulated, including DNA damage-induced transcript 3, colony-stimulating factor 2, and E2F transcription factor 1. The mixture inhibited ERK1/2 signaling pathway phosphorylation, which is responsible regulating the proliferation of A375 melanoma cells. In addition, the mixture of THC and CBD disrupted melanoma cell migration [104].

2.3.5. Glioblastoma and Neuroblastoma Cells/Malignant Gliomas

Growing evidence indicates that cannabinoids can be successfully employed in glioblastoma (GBM) treatment [109]. A few of in vivo studies have shown that cannabinoids are able to substantially deplete tumor volume in orthotopic and subcutaneous animal models of glioblastoma [120,121]. Others indicate that the cytotoxic effects of CBD in human and canine glioblastoma cells resulted from dysregulation of calcium homeostasis and mitochondrial activity [122]. Synthetic cannabinoids (WIN55,212-2 or JWH133) induce autophagy by the accumulation of LC3-II and mitochondrial apoptotic pathways in human glioblastoma cells (T98G and LN229), irrespective of any deficiency in *TP53* or *PTEN* tumor suppressors [123].

CBD was found to trigger autophagy in a human neuroblastoma cell line (SH-SY5Y) by regulating the phosphorylation of ERK1/2 and AKT kinases; this process involved ULK1, but in an mTORC1-independent pathway. The activation of these signaling pathways involves the cannabinoid receptors CB1 and CB2, as well as TRPV1 receptors, which are essential for promoting neuronal cell survival and proliferation [124].

Zhong et al. [114] assessed the anticancer effect of cannabinol on A172 glioblastoma, HepG2 liver cancer, and HCC1806 breast cancer cell lines. Cannabinol (CBN) induced apoptosis by p21 and p27 downregulation and an arrested cell cycle in the G1 or S phase by reducing CDK1, CDK2, and cyclin E1 levels. It was also revealed that CBN decreased the proliferation of the studied cancer cells and modulated the levels of cannabinoid receptors including GPR18, CB2, and GPR55.

Cannabinoids are characterized by the ability to inhibit tumor angiogenesis. Blázquez et al. [36] observed that local administration of the nonpsychotic cannabinoid JWH-133 in mice inhibited the angiogenesis of malignant gliomas. It was revealed that while the treated tumors were characterized by small and impermeable blood vessels, those of the untreated tumors were large and porous. Hence, tumor size may be influenced by the biological activities of the cannabinoids.

2.3.6. Studies on Humans—Clinical Trials

Although *C. sativa* L. compounds have been demonstrated to have clinical efficacy in preclinical studies, clinical trials with oncology patients are lacking.

An analysis of 207 preclinical articles, including 77 unique case reports, found no strong clinical trial data to confirm that *C. sativa* L. compounds may have benefits against various cancers: breast, central nervous system, gynecological, leukemia, lung, prostate, and pancreatic cancer [125]. Further prospective and randomized studies on this clinically important topic are needed to draw clear conclusions.

Two clinical trials have examined the effect of cannabinoids on gliomas. The first, a pilot study, covered nine patients [36], and the second, a phase II study, was realized by GW Pharmaceuticals on 21 adult patients (clinical study NCT01812603) [37]. Patients taking THC:CBD plus temozolomide (TMZ) demonstrated an 83% one-year survival rate, with a median survival of over 662 days (369 days in the control group). The patients in the control group received only TMZ and achieved a 44% one-year survival rate.

2.4. Antiviral Effects of Cannabinoids

The antiviral properties of cannabinoids [126–128] have been tested in diseases such as SARS-CoV-2 [32] and HIV [129]. The tested cannabinoids were found to inhibit severe viral infection of the SARS-CoV-2 virus by different mechanisms (Figure 4).

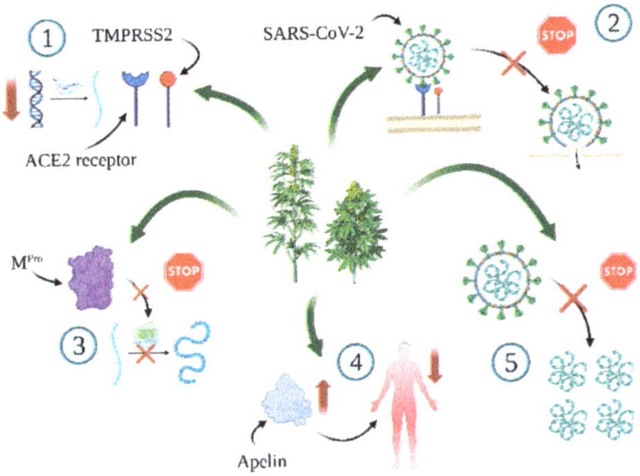

Figure 4. Molecular mechanism of antiviral effects of *C. sativa* L. compounds. (1) Downregulating serine protease TMPRSS2 and lowering *ACE2* transcription levels [130], (2) preventing SARS-CoV-2 spike protein-mediated membrane fusion [131], (3) blocking the virus translation and replication by suppressing enzyme of SARS-CoV-2 i.e., M^{pro} protease (in silico and in vitro experiments) [132], (4) upregulation of apelin level [133], and (5) blocking viral replication, in part by upregulating the host IRE1α RNase endoplasmic reticulum (ER) stress response and interferon signaling pathways [32]. Created in BioRender. Bukowska, B. (2024) BioRender.com/k571772 (accessed on 1 September 2024).

In SARS-CoV-2 infection, the virus enters the host via ACE2, which is present in several tissues, including the oral and nasal membranes, lung tissue, renal tissue, and gastrointestinal tract. The expression of *ACE2* in the nose epithelium is age dependent, being more highly expressed in elderly people and men, which would provide more opportunities for viruses to enter cells. As a result, the risk of serious clinical manifestation is higher in older adults than in the younger population [134]. CBD and THC reduce disease sensitivity by modulating ACE2 protein levels. Wang et al. [130] investigated the anti-inflammatory effects of cannabis extracts on tissues (Human 3D models: Epioral, Epi-Airway, and EpiIntestinal) initially treated with pro-inflammatory cytokines. The extracts from *C. sativa* L., especially CBD, demonstrated downregulated *ACE2* gene expression.

A study of 93,565 patients in Chicago with SARS-CoV-2 found those who had previously taken CBD had a significantly lower incidence of SARS-CoV-2 infections compared to matched pairs or the general population [32]. Only 1.2% of patients prescribed with CBD contracted SARS-CoV-2 compared to 12.2% of matched, non-cannabinoid patients. Hence, CBD and its 7-OH-CBD metabolite appear to be potential preventive and treatment agents for SARS-CoV-2 in the initial stages of infection. The findings indicate that CBD blocked SARS-CoV-2 infection early in its course and reduced the risk of SARS-CoV-2 infection in humans. CBD and 7-OH-CBD strongly blocked replication of SARS-CoV-2 in the A549 human lung cancer cell line. CBD also blocked the expression of viral genes. Moreover, treatment also triggered interferon expression and activated the antiviral signaling pathway [32]. These results are in agreement with previous research works showing the ability of cannabinoids to inhibit the formation of cytokine in macrophages [135].

Additionally, the experiments have shown the positive action of CBD, which may help during COVID-19 infection [133]. Studies on mice have shown that CBD increases the level of apelin, which reduces inflammation and protects the lungs. It was found that CBD intake increased the blood apelin level by as much as 20 times and, importantly, this was accompanied by in an increase in oxygen level in the lungs.

Cannabinoids may possibly prevent SARS-CoV-2 infections. Cannabinoid acids isolated from *C. sativa* L. were revealed to be allosteric and orthosteric ligands with micromolar affinity for the SARS-CoV-2 spike protein. Cannabigerolic acid and cannabidiolic acid protected human epithelial cells from infection with pseudoviruses expressing the SARS-CoV-2 spike protein and prevented SARS-CoV-2 from entering the cells. Importantly, cannabigerolic acid and cannabidiolic acid were equally effective against the SARS-CoV-2 alpha variant B.1.1.7 and beta variant B.1.351 [136]. Classen et al. [131] showed that cannabigerol and cannabicyclol (20 μM), other cannabinoids of *C. sativa* L., blocked SARS-CoV-2 fusion in Vero cells. Both compounds inhibit SARS-CoV-2 spike protein-mediated membrane fusion, as shown by a virus-free reporter fusion inhibition assay and by monitoring syncytia formation in Vero reporter cells.

2.5. Current Cannabinoid-Based Drugs Approved by the FDA

Cannabinoids are used in medical applications. The modern pharmacological armamentarium now includes synthetic CBD and THC (dronabinol), nabiximols (THC:CBD = 1.08:1.00), and synthetic nabilone [137].

Cannabinoid-containing preparations are used to treat neurological diseases such as epilepsy (drug Epidiolex®/Epidyolex®), multiple sclerosis (Sativex® i.e., nabiximols), pain (migraine), or tic disorders. Cannabinoids are also being trialed in neurodegenerative diseases such as Parkinson's disease (PD), Huntington's disease, amyotrophic lateral sclerosis (ALS), dementia, and dystonia [138].

2.5.1. Epidiolex® CBD Extract as a Drug for Epilepsy

For several years, cannabidiol (Epidiolex®/Epidyolex®) has been approved in the US and in Europe as an add-on therapy for the treatment of two forms of epilepsy that are often drug-resistant, Lennox–Gastaut (LGS) and Dravet syndrome [138]. The starting dosage of Epidiolex® is 2.5 mg/kg by mouth twice daily (5 mg/kg/day). After one week, the dosage can be increased to a maintenance dosage of 5 mg/kg twice daily (10 mg/kg/day); the maximum recommended maintenance dosage is 10 mg/kg twice daily (20 mg/kg/day) [139].

Randomized, controlled, and open-label research works have been performed to assess the effect of Epidiolex (99% pure CBD oral extract) in patients suffering from resistant epilepsy. The findings have led to FDA approval and the inclusion of Epidiolex by the Drug Enforcement Agency in Schedule V of the Controlled Substances Act [140].

A retrospective review of 108 patients with refractory epilepsy taking Epidiolex was executed in a single center. Patients received Epidiolex as 100 mg/mL oral solution, from 2 to 10 mg/kg/day to tolerance or a maximum 25–50 mg/kg/day dose for up to 192 weeks, depending on the single center. In the final assessment, 75% of the patients continued to use Epidiolex. The most usual causes of termination were lack of effectiveness, raised seizure activity, deterioration of behavior, and sedation. At baseline, 47.2% of patients were also taking clobazam, and 39.2% of these patients were able to lower their clobazam dose. Additionally, 53% could stop or decrease the dose of at least one other antiseizure drug [141]. Another extensive research work, executed between 2014 and 2019 with the participation of 892 patients, found CBD to have a positive effect in patients with treatment-resistant epilepsies. All patients received 20 mg/kg oral cannabidiol daily or matched placebo for 14 weeks. Cannabidiol administration was connected with a sustained reduction in seizures for up to 192 weeks with an acceptable safety profile, and the achieved results confirmed that CBD can be employed in long-term treatment [142].

Some findings indicate that cannabidiol use has a risk of side effects [143,144]. A recent meta-analysis of nine studies found CBD use was connected with an elevated risk of several adverse events in epilepsy patients, like somnolence, diarrhea, decreased appetite, and elevation of alanine and aspartate aminotransferases [145].

2.5.2. Sativex CBD Extract as a Drug for Treating Muscle Stiffness and Spasms Caused by MS

Sativex is the trademark of a cannabis drug called Nabiximols consisting of THC (2.7 mg per dose) and CBD (2.5 mg per dose). It has already been approved in Canada and most European countries for the treatment of moderate to severe spasticity in MS [57]. The license authorizes its use as a complement for treating MS. The dose requires titration until optimum symptom relief is achieved. This may take up to two weeks. The maximum licensed dose is 12 sprays per day [146].

Sativex users report dizziness and fatigue, however psychiatric symptoms, such as anxiety, mood changes, and paranoia, are rare [147]. Target symptoms were sleep disturbances, pain, and spasticity, with the most commonly reported side effects being drowsiness, emotional flattening, and difficulty concentrating [148].

However, at the doses generally used, Sativex did not affect driving ability and did not have any significant detrimental effects on cognition or mood; this was also noted in older people, as demonstrated by Rekand [149].

3. Limitations in *Cannabis sativa* L. Research and Use

Although the literature generally regards cannabis-based treatments as effective against neurological diseases [150–152], few studies have been conducted to assess the anti-inflammatory and anticancer properties of these compounds in humans: such research has been limited to in vitro and animal studies. While a few peer-reviewed clinical trials have been performed in humans, these have included only small numbers of patients [36,37].

It is important to note that translating preclinical research findings (from in vitro or animal studies) into clinical applications typically faces significant limitations.

Firstly, preclinical studies on cannabinoids are most commonly conducted on animals such as mice or rats, whose metabolism, immune systems, and physiology differ significantly from those of humans. It cannot be excluded that doses that are safe and effective in animals may be toxic or ineffective in humans. Additionally, the responses of the endocannabinoid system can vary between species, complicating the direct translation of results into clinical practice.

One of the primary challenges faced by researchers is to determine the optimal therapeutic dose for humans. In preclinical studies, very high doses of cannabinoids are often used to achieve therapeutic effects. However, the clinical applicability of such doses is limited by their adverse effects, such as the psychoactive impact of THC. The lack of precise data on effective and safe dosing poses a significant barrier in the planning of clinical trials or applications.

Another limitation is that many preclinical studies fail to account for the considerable variation in age, sex, lifestyle, diet, health status, genetics, and medications between patients. As such, it is difficult to predict the final effect of therapy.

Furthermore, little data exist on the long-term effects of cannabinoid treatment. While most preclinical studies focus on the short-term therapeutic effects of cannabinoids, it is crucial to understand their long-term impact on the human body, including the potential side effects of treatment, as well as interactions with other drugs, and the risk of developing tolerance or dependency.

Although Epidiolex® and Sativex® are approved as drugs for neurological diseases, both have potential side effects such as mental depression or suicidal ideation, which should also be taken into consideration [138,153,154].

It is important to mention the other challenges associated with the therapeutic use of *L. sativa*, such as the need to standardize extracts or interactions with drugs and other cannabinoids.

It is important to note that the composition of *C. sativa* extracts vary considerably in response to the variety and geographical location of the crop, as well as the method of isolation and standardization [14,15].

In addition, cannabinoids can inhibit efflux transporters and drug metabolizing enzymes and can possibly induce drug kinetic interactions with anticancer drugs. These interactions can enhance the cytostatic effect of the drugs and/or their side effects, or it may reduce the effective drug dose [155].

The interaction between cannabinoids themselves is also important. Indeed, a synergistic effect has been observed for the combination of CBD with cannabichromene (CBC) or THC. As a result, small concentrations can replicate much higher doses of the extract in inducing HNSCC cell death [156].

Also, the lack of appropriate legal regulations regarding the production and distribution of CBD and *C. sativa* L. oil products has led to considerable variation in the levels of CBD and tetrahydrocannabinol levels in the products [140].

Although the data presented in this article report the wide-ranging benefits of using *C. sativa* L. and cannabinoids, recreational use of marijuana continues to have negative consequences. Cannabis is the most commonly taken drug of natural origin, as evidenced by data from the report prepared by UNODC; about 269 million people used it in 2018 [157].

It is possible that the further relaxation of restrictions regarding the availability of medicinal marijuana may also result in an increase in health problems for users.

4. Conclusions

(1) *C. sativa* L. is an extraordinary plant that provides a valuable raw material for medical applications. Its secondary metabolites, cannabinoids, have attracted growing interest in the fight against illness, mainly due to their effect on CB1 and CB2 cannabinoid receptors.

(2) Cannabinoid-containing preparations are used to treat neurological diseases such as epilepsy, multiple sclerosis, and pain [150–152], and are also being trialed in neurodegenerative diseases e.g., in Parkinson's disease or amyotrophic lateral sclerosis [138].

(3) Cannabinoids and other antioxidant compounds of *C. sativa* L., such as cannflavin A and B, exert antioxidant potential by neutralizing free radicals, supporting metal chelation, increasing GSH level, and influencing the activity of antioxidative enzymes (Table 1).

(4) They also possess anti-inflammatory properties, inhibiting the progression of inflammation associated with the NF-κB and Keap1-Nrf2 signaling pathways [78], and inducing T-regulatory cells [51–53,55].

(5) In addition, their antiapoptotic, antiproliferative, antiangiogenic, and antimetastatic properties make them attractive potential cancer medications (Table 2)

(6) Cannabinoids have demonstrated antiviral potential (Figure 4) by inter alia blocking of SARS-CoV-2 translation and replication by suppressing M^{pro} protease, and by downregulating the serine protease TMPRSS2 and inhibiting *ACE2* transcription [130,132].

(7) There is a need to build a higher number of clinical trials on humans. These studies should include larger groups of patients and should use greater standardization of products and research methodology. There is also a pressing need to understand the long-term effects and action of cannabinoids in various diseases and their interactions with other drugs.

Funding: This work was funded by Research granted (B2011000000191.01) to the Department of Biophysics of Environmental Pollution, Faculty of Biology and Environmental Protection, University of Lodz.

Conflicts of Interest: The author declare no conflict of interest.

Abbreviations

AAE: ascorbic acid equivalents; ABTS: 2,2′-azinobis(3-ethylbenzothiaziline-6-sulfonate), BHT: butylated hydroxytoluene, CBC: cannabichromene, CBD: cannabidiol, CBDA: cannabidiolic acid, CBG: cannabigerol, CBGA: cannabigerolic acid, CBN: cannabinol, CFL-A: cannflavins A, CFL-B: cannflavins B, CFL-C: cannflavins C, COX-2: induced cyclooxygenase, CRP: C-reactive protein,

ECS: endocannabinoids system, FBL-03G: cannflavin B isomer, FDA: Food and Drug Administration, FRAP: fluorescence recovery after photobleaching, HCC: hepatocellular carcinoma cells, LPS: lipopolysaccharide, MA: R(+)-methanandamide; MAPKs: mitogen-activated protein kinase, PPAR-γ: peroxisome proliferator-activated receptor gamma, THC: tetrahydrocannabinol, TIMP-1: inhibitor of metalloproteinase-1, TMZ: temozolomide, TNF-α: tumor necrosis factor α; TRE: treatment-resistant epilepsies.

References

1. Strzelczyk, M.; Lochynska, M.; Chudy, M. Systematics and Botanical Characteristics of Industrial Hemp *Cannabis sativa* L. *J. Nat. Fibers* **2022**, *19*, 5804–5826. [CrossRef]
2. Pandey, Y.; Chaturvedi, T.; Swaroop, H.; Gupta, A.K.; Shanker, K.; Tiwari, G. Phytochemical and Genetic Marker (SCoT and CBDP) Based Study of Genetic Diversity and Population Structure in Natural Populations of *Cannabis sativa* L.: A High-Value Sustainable Biodiversity of North-Indian Himalaya. *Ind. Crops Prod.* **2023**, *200*, 116892. [CrossRef]
3. Shams, R.; Azizi, A.; Hamzei, J.; Noroozisharaf, A.; Moghadam, S.; Kordrostami, M. Genetic Structure and Diversity of Iranian Cannabis Populations Based on Phytochemical, Agro-Morphological and Molecular Markers. *Ind. Crops Prod.* **2020**, *158*, 112950. [CrossRef]
4. Erkelens, J.; Hazekam, A. That Which We Call Indica, by Any Other Name Would Smell as Sweet. *Cannabinoids* **2014**, *9*, 9–15.
5. Martínez, V.; Iriondo De-Hond, A.; Borrelli, F.; Capasso, R.; Del Castillo, M.D.; Abalo, R. Cannabidiol and Other Non-Psychoactive Cannabinoids for Prevention and Treatment of Gastrointestinal Disorders: Useful Nutraceuticals? *Int. J. Mol. Sci.* **2020**, *21*, 3067. [CrossRef]
6. Gill, A.R.; Loveys, B.R.; Cowley, J.M.; Hall, T.; Cavagnaro, T.R.; Burton, R.A. Physiological and Morphological Responses of Industrial Hemp (*Cannabis sativa* L.) to Water Deficit. *Ind. Crops Prod.* **2022**, *187*, 115331. [CrossRef]
7. Shebaby, W.; Saliba, J.; Faour, W.H.; Ismail, J.; El Hage, M.; Daher, C.F.; Taleb, R.I.; Nehmeh, B.; Dagher, C.; Chrabieh, E.; et al. In Vivo and in Vitro Anti-Inflammatory Activity Evaluation of Lebanese *Cannabis sativa* L. ssp. Indica (Lam.). *J. Ethnopharmacol.* **2021**, *270*, 113743. [CrossRef]
8. Eržen, M.; Čeh, B.; Kolenc, Z.; Bosancic, B.; Čerenak, A. Evaluation of Different Hemp (*Cannabis sativa* L.) Progenies Resulting from Crosses with Focus on Oil Content and Seed Yield. *Ind. Crops Prod.* **2023**, *201*, 116893. [CrossRef]
9. Burton, R.A.; Andres, M.; Cole, M.; Cowley, J.M.; Augustin, M.A. Industrial Hemp Seed: From the Field to Value-Added Food Ingredients. *J. Cannabis Res.* **2022**, *4*, 45. [CrossRef]
10. Singh, K.; Nassar, N.; Bachari, A.; Schanknecht, E.; Telukutla, S.; Zomer, R.; Piva, T.J.; Mantri, N. The Pathophysiology and the Therapeutic Potential of Cannabinoids in Prostate Cancer. *Cancers* **2021**, *13*, 4107. [CrossRef]
11. Liu, M.; Thygesen, A.; Summerscales, J.; Meyer, A.S. Targeted Pre-Treatment of Hemp Bast Fibres for Optimal Performance in Biocomposite Materials: A Review. *Ind. Crops Prod.* **2017**, *108*, 660–683. [CrossRef]
12. Crini, G.; Lichtfouse, E.; Gilles, C.; Crini, N. Applications of hemp in textiles, paper industry, insulation and building materials, horticulture, animal nutrition, food and beverages, nutraceuticals, cosmetics and hygiene, medicine, agrochemistry, energy production and environment: A review. *Environ. Chem. Lett.* **2020**, *18*, 1451–1476. [CrossRef]
13. Flores-Sanchez, I.J.; Choi, Y.H.; Verpoorte, R. Metabolite Analysis of *Cannabis sativa* L. by NMR Spectroscopy. In *Functional Genomics*; Kaufmann, M., Klinger, C., Eds.; Methods in Molecular Biology; Springer: New York, NY, USA, 2012; Volume 815, pp. 363–375, ISBN 978-1-61779-423-0.
14. Cerrato, A.; Citti, C.; Cannazza, G.; Capriotti, A.L.; Cavaliere, C.; Grassi, G.; Marini, F.; Montone, C.M.; Paris, R.; Piovesana, S.; et al. Phytocannabinomics: Untargeted Metabolomics as a Tool for Cannabis Chemovar Differentiation. *Talanta* **2021**, *230*, 122313. [CrossRef]
15. Cerrato, A.; Biancolillo, A.; Cannazza, G.; Cavaliere, C.; Citti, C.; Laganà, A.; Marini, F.; Montanari, M.; Montone, C.M.; Paris, R.; et al. Untargeted Cannabinomics Reveals the Chemical Differentiation of Industrial Hemp Based on the Cultivar and the Geographical Field Location. *Anal. Chim. Acta* **2023**, *1278*, 341716. [CrossRef] [PubMed]
16. European Drug Report 2023. Cannabis—The Current Situation in Europe. Available online: https://www.euda.europa.eu/publications/european-drug-report/2023/cannabis_en (accessed on 1 September 2024).
17. State Medical Marijuana Laws. 2024. Available online: https://www.cdc.gov/cannabis/about/state-medical-cannabis-laws.html (accessed on 1 September 2024).
18. Mikulic, M. Statista. 2022. Available online: https://www.statista.com/statistics/760498/total-us-cbd-sales/ (accessed on 1 September 2024).
19. Tutek, K.; Masek, A. Hemp and Its Derivatives as a Universal Industrial Raw Material (with Particular Emphasis on the Polymer Industry)—A Review. *Materials* **2022**, *15*, 2565. [CrossRef]
20. Abdel-Kader, M.S.; Radwan, M.M.; Metwaly, A.M.; Eissa, I.H.; Hazekamp, A.; ElSohly, M.A. Chemistry and Biological Activities of Cannflavins of the Cannabis Plant. *Cannabis Cannabinoid Res.* **2023**, *8*, 974–985. [CrossRef] [PubMed]
21. Zagzoog, A.; Mohamed, K.A.; Kim, H.J.J.; Kim, E.D.; Frank, C.S.; Black, T.; Jadhav, P.D.; Holbrook, L.A.; Laprairie, R.B. In Vitro and in Vivo Pharmacological Activity of Minor Cannabinoids Isolated from *Cannabis sativa*. *Sci. Rep.* **2020**, *10*, 20405. [CrossRef]

22. Di Sotto, A.; Gullì, M.; Acquaviva, A.; Tacchini, M.; Di Simone, S.C.; Chiavaroli, A.; Recinella, L.; Leone, S.; Brunetti, L.; Orlando, G.; et al. Phytochemical and Pharmacological Profiles of the Essential Oil from the Inflorescences of the *Cannabis sativa* L. *Ind. Crops Prod.* **2022**, *183*, 114980. [CrossRef]
23. Orhan, I.; Sener, B. Fatty Acid Content of Selected Seed Oils. *J. Herb. Pharmacother.* **2002**, *2*, 29–33. [CrossRef]
24. Da Porto, C.; Natolino, A.; Decorti, D. Effect of Ultrasound Pre-Treatment of Hemp (*Cannabis sativa* L.) Seed on Supercritical CO2 Extraction of Oil. *J. Food Sci. Technol.* **2015**, *52*, 1748–1753. [CrossRef]
25. Konca, Y.; Cimen, B.; Yalcin, H.; Kaliber, M.; Beyzi, S.B. Effect of Hempseed (*Cannabis sativa* sp.) Inclusion to the Diet on Performance, Carcass and Antioxidative Activity in Japanese Quail (Coturnix Coturnix Japonica). *Korean J. Food Sci. Anim. Resour.* **2014**, *34*, 141–150. [CrossRef] [PubMed]
26. Cerino, P.; Buonerba, C.; Cannazza, G.; D'Auria, J.; Ottoni, E.; Fulgione, A.; Di Stasio, A.; Pierri, B.; Gallo, A. A Review of Hemp as Food and Nutritional Supplement. *Cannabis Cannabinoid Res.* **2021**, *6*, 19–27. [CrossRef] [PubMed]
27. Kornpointner, C.; Sainz Martinez, A.; Marinovic, S.; Haselmair-Gosch, C.; Jamnik, P.; Schröder, K.; Löfke, C.; Halbwirth, H. Chemical Composition and Antioxidant Potential of *Cannabis sativa* L. Roots. *Ind. Crops Prod.* **2021**, *165*, 113422. [CrossRef]
28. Filer, C.N. Acidic Cannabinoid Decarboxylation. *Cannabis Cannabinoid Res.* **2022**, *7*, 262–273. [CrossRef] [PubMed]
29. Cristino, L.; Bisogno, T.; Di Marzo, V. Cannabinoids and the Expanded Endocannabinoid System in Neurological Disorders. *Nat. Rev. Neurol.* **2020**, *16*, 9–29. [CrossRef]
30. Baratta, F.; Pignata, I.; Ravetto Enri, L.; Brusa, P. Cannabis for Medical Use: Analysis of Recent Clinical Trials in View of Current Legislation. *Front. Pharmacol.* **2022**, *13*, 888903. [CrossRef]
31. Troyer, J.; Tanco, K. Review of the Use of Medicinal Cannabis Products in Palliative Care. *Cancers* **2024**, *16*, 1412. [CrossRef]
32. Nguyen, L.C.; Yang, D.; Nicolaescu, V.; Best, T.J.; Gula, H.; Saxena, D.; Gabbard, J.D.; Chen, S.-N.; Ohtsuki, T.; Friesen, J.B.; et al. Cannabidiol Inhibits SARS-CoV-2 Replication through Induction of the Host ER Stress and Innate Immune Responses. *Sci. Adv.* **2022**, *8*, eabi6110. [CrossRef]
33. Kumar, R.; Singh, S.; Maharshi, V. Therapeutic Effects of Cannabinoids on Ulcerative Colitis: A Systematic Review and Meta-Analysis of Randomized Controlled Trials. *J. Health Sci. Med. Res.* **2024**, *2024*, 20241041. [CrossRef]
34. Paland, N.; Hamza, H.; Pechkovsky, A.; Aswad, M.; Shagidov, D.; Louria-Hayon, I. Cannabis and Rheumatoid Arthritis: A Scoping Review Evaluating the Benefits, Risks, and Future Research Directions. *Rambam Maimonides Med. J.* **2023**, *14*, e0022. [CrossRef]
35. Ghazi Eid, B. Cannabinoids for Treating Cardiovascular Disorders: Putting Together a Complex Puzzle. *J. Microsc. Ultrastruct.* **2018**, *6*, 171–176. [CrossRef]
36. Blázquez, C.; Casanova, M.L.; Planas, A.; Gómez del Pulgar, T.; Villanueva, C.; Fernández-Aceñero, M.J.; Aragonés, J.; Huffman, J.W.; Jorcano, J.L.; Guzmán, M. Inhibition of Tumor Angiogenesis by Cannabinoids. *FASEB J.* **2003**, *17*, 333–568. [CrossRef] [PubMed]
37. Schultz, S.; Beyer, M. GW Pharmaceuticals Achieves Positive Results in Phase 2 Proof of Concept Study in Glioma. Available online: http://Ir.Gwpharm.Com/Static-Files/Cde942fe-555c-4b2f-9cc9-F34d24c7ad27 (accessed on 1 September 2024).
38. Andradas, C.; Truong, A.; Byrne, J.; Endersby, R. The Role of Cannabinoids as Anticancer Agents in Pediatric Oncology. *Cancers* **2021**, *13*, 157. [CrossRef]
39. Kopustinskiene, D.M.; Masteikova, R.; Lazauskas, R.; Bernatoniene, J. *Cannabis sativa* L. Bioactive Compounds and Their Protective Role in Oxidative Stress and Inflammation. *Antioxidants* **2022**, *11*, 660. [CrossRef]
40. Brust, C.A.; Swanson, M.A.; Bohn, L.M. Structural and Functional Insights into the G Protein-Coupled Receptors: CB1 and CB2. *Biochem. Soc. Trans.* **2023**, *51*, 1533–1543. [CrossRef]
41. Pertwee, R.G. The Diverse CB_1 and CB_2 Receptor Pharmacology of Three Plant Cannabinoids: Δ^9-tetrahydrocannabinol, Cannabidiol and Δ^9-tetrahydrocannabivarin. *Br. J. Pharmacol.* **2008**, *153*, 199–215. [CrossRef]
42. Badal, S.; Smith, K.N.; Rajnarayanan, R. Analysis of Natural Product Regulation of Cannabinoid Receptors in the Treatment of Human Disease. *Pharmacol. Ther.* **2017**, *180*, 24–48. [CrossRef]
43. Tham, M.; Yilmaz, O.; Alaverdashvili, M.; Kelly, M.E.M.; Denovan-Wright, E.M.; Laprairie, R.B. Allosteric and Orthosteric Pharmacology of Cannabidiol and Cannabidiol-dimethylheptyl at the Type 1 and Type 2 Cannabinoid Receptors. *Br. J. Pharmacol.* **2019**, *176*, 1455–1469. [CrossRef] [PubMed]
44. Laprairie, R.B.; Bagher, A.M.; Kelly, M.E.M.; Denovan-Wright, E.M. Cannabidiol Is a Negative Allosteric Modulator of the Cannabinoid CB $_1$ Receptor. *Br. J. Pharmacol.* **2015**, *172*, 4790–4805. [CrossRef]
45. Chung, H.; Fierro, A.; Pessoa-Mahana, C.D. Cannabidiol Binding and Negative Allosteric Modulation at the Cannabinoid Type 1 Receptor in the Presence of Delta-9-Tetrahydrocannabinol: An In Silico Study. *PLoS ONE* **2019**, *14*, e0220025. [CrossRef]
46. An, D.; Peigneur, S.; Hendrickx, L.A.; Tytgat, J. Targeting Cannabinoid Receptors: Current Status and Prospects of Natural Products. *Int. J. Mol. Sci.* **2020**, *21*, 5064. [CrossRef] [PubMed]
47. Capozzi, A.; Caissutti, D.; Mattei, V.; Gado, F.; Martellucci, S.; Longo, A.; Recalchi, S.; Manganelli, V.; Riitano, G.; Garofalo, T.; et al. Anti-Inflammatory Activity of a CB2 Selective Cannabinoid Receptor Agonist: Signaling and Cytokines Release in Blood Mononuclear Cells. *Molecules* **2021**, *27*, 64. [CrossRef] [PubMed]
48. Watson, C.W.-M.; Campbell, L.M.; Sun-Suslow, N.; Hong, S.; Umlauf, A.; Ellis, R.J.; Iudicello, J.E.; Letendre, S.; Marcotte, T.D.; Heaton, R.K.; et al. Daily Cannabis Use Is Associated With Lower CNS Inflammation in People With HIV. *J. Int. Neuropsychol. Soc.* **2021**, *27*, 661–672. [CrossRef]

49. Shahbazi, F.; Grandi, V.; Banerjee, A.; Trant, J.F. Cannabinoids and Cannabinoid Receptors: The Story so Far. *iScience* **2020**, *23*, 101301. [CrossRef]
50. Almogi-Hazan, O.; Or, R. Cannabis, the Endocannabinoid System and Immunity—The Journey from the Bedside to the Bench and Back. *Int. J. Mol. Sci.* **2020**, *21*, 4448. [CrossRef]
51. Yang, L.; Li, F.-F.; Han, Y.-C.; Jia, B.; Ding, Y. Cannabinoid Receptor CB2 Is Involved in Tetrahydrocannabinol-Induced Anti-Inflammation against Lipopolysaccharide in MG-63 Cells. *Mediat. Inflamm.* **2015**, *2015*, 362126. [CrossRef]
52. Suryavanshi, S.V.; Zaiachuk, M.; Pryimak, N.; Kovalchuk, I.; Kovalchuk, O. Cannabinoids Alleviate the LPS-Induced Cytokine Storm via Attenuating NLRP3 Inflammasome Signaling and TYK2-Mediated STAT3 Signaling Pathways In Vitro. *Cells* **2022**, *11*, 1391. [CrossRef]
53. Lowin, T.; Tingting, R.; Zurmahr, J.; Classen, T.; Schneider, M.; Pongratz, G. Cannabidiol (CBD): A Killer for Inflammatory Rheumatoid Arthritis Synovial Fibroblasts. *Cell Death Dis.* **2020**, *11*, 714. [CrossRef] [PubMed]
54. Opęchowska, A.; Karpiuk, K.; Zahorodnii, A.; Harasim-Symbor, E.; Chabowski, A.; Konstantynowicz-Nowicka, K. Anti-Inflammatory Effects of Cannabidiol in Early Stages of Neuroinflammation Induced by High-Fat Diet in Cerebral Cortex of Rats. *Toxicol. Appl. Pharmacol.* **2024**, *484*, 116856. [CrossRef]
55. Yndart Arias, A.; Kolishetti, N.; Vashist, A.; Madepalli, L.; Llaguno, L.; Nair, M. Anti-Inflammatory Effects of CBD in Human Microglial Cell Line Infected with HIV-1. *Sci. Rep.* **2023**, *13*, 7376. [CrossRef]
56. Atalay, S.; Jarocka-Karpowicz, I.; Skrzydlewska, E. Antioxidative and Anti-Inflammatory Properties of Cannabidiol. *Antioxidants* **2019**, *9*, 21. [CrossRef] [PubMed]
57. Markovà, J.; Essner, U.; Akmaz, B.; Marinelli, M.; Trompke, C.; Lentschat, A.; Vila, C. Sative® as Add-on Therapy vs. Further Optimized First-Line ANTispastics (SAVANT) in Resistant Multiple Sclerosis Spasticity: A Double-Blind, Placebo-Controlled Randomised Clinical Trial. *Int. J. Neurosci.* **2019**, *129*, 119–128. [CrossRef] [PubMed]
58. Navarini, L.; Bisogno, T.; Mozetic, P.; Piscitelli, F.; Margiotta, D.P.E.; Basta, F.; Afeltra, A.; Maccarrone, M. Endocannabinoid System in Systemic Lupus Erythematosus: First Evidence for a Deranged 2-Arachidonoylglycerol Metabolism. *Int. J. Biochem. Cell Biol.* **2018**, *99*, 161–168. [CrossRef]
59. Ghasemi-Gojani, E.; Kovalchuk, I.; Kovalchuk, O. Cannabinoids and Terpenes for Diabetes Mellitus and Its Complications: From Mechanisms to New Therapies. *Trends Endocrinol. Metab.* **2022**, *33*, 828–849. [CrossRef] [PubMed]
60. Li, S.; Huang, Y.; Yu, L.; Ji, X.; Wu, J. Impact of the Cannabinoid System in Alzheimer's Disease. *Curr. Neuropharmacol.* **2023**, *21*, 715–726. [CrossRef] [PubMed]
61. Ning, K.; Hou, C.; Wei, X.; Zhou, Y.; Zhang, S.; Chen, Y.; Yu, H.; Dong, L.; Chen, S. Metabolomics Analysis Revealed the Characteristic Metabolites of Hemp Seeds Varieties and Metabolites Responsible for Antioxidant Properties. *Front. Plant Sci.* **2022**, *13*, 904163. [CrossRef]
62. Pollastro, F.; Minassi, A.; Fresu, L.G. Cannabis Phenolics and Their Bioactivities. *Curr. Med. Chem.* **2018**, *25*, 1160–1185. [CrossRef]
63. Rea, K.A.; Casaretto, J.A.; Al-Abdul-Wahid, M.S.; Sukumaran, A.; Geddes-McAlister, J.; Rothstein, S.J.; Akhtar, T.A. Biosynthesis of Cannflavins A and B from *Cannabis sativa* L. *Phytochemistry* **2019**, *164*, 162–171. [CrossRef]
64. Bhuyan, U.; Handique, J.G. Plant Polyphenols as Potent Antioxidants: Highlighting the Mechanism of Antioxidant Activity and Synthesis/Development of Some Polyphenol Conjugates. In *Studies in Natural Products Chemistry*; Elsevier: Amsterdam, The Netherlands, 2022; Volume 75, pp. 243–266, ISBN 978-0-323-91250-1.
65. Barrett, M.L.; Gordon, D.; Evans, F.J. Isolation from *Cannabis sativa* L. of Cannflavin—A Novel Inhibitor of Prostaglandin Production. *Biochem. Pharmacol.* **1985**, *34*, 2019–2024. [CrossRef]
66. Barrett, M.L.; Scutt, A.M.; Evans, F.J. Cannflavin A and B, Prenylated Flavones from *Cannabis sativa* L. *Experientia* **1986**, *42*, 452–453. [CrossRef]
67. Erridge, S.; Mangal, N.; Salazar, O.; Pacchetti, B.; Sodergren, M.H. Cannflavins—From Plant to Patient: A Scoping Review. *Fitoterapia* **2020**, *146*, 104712. [CrossRef] [PubMed]
68. Chuanphongpanich, S.; Racha, S.; Saengsitthisak, B.; Pirakitikulr, P.; Racha, K. Computational Assessment of Cannflavin A as a TAK1 Inhibitor: Implication as a Potential Therapeutic Target for Anti-Inflammation. *Sci. Pharm.* **2023**, *91*, 36. [CrossRef]
69. Moreau, M.; Ibeh, U.; Decosmo, K.; Bih, N.; Yasmin-Karim, S.; Toyang, N.; Lowe, H.; Ngwa, W. Flavonoid Derivative of Cannabis Demonstrates Therapeutic Potential in Preclinical Models of Metastatic Pancreatic Cancer. *Front. Oncol.* **2019**, *9*, 660. [CrossRef]
70. Izzo, L.; Pacifico, S.; Piccolella, S.; Castaldo, L.; Narváez, A.; Grosso, M.; Ritieni, A. Chemical Analysis of Minor Bioactive Components and Cannabidiolic Acid in Commercial Hemp Seed Oil. *Molecules* **2020**, *25*, 3710. [CrossRef] [PubMed]
71. Bonacci, S.; Di Stefano, V.; Sciacca, F.; Buzzanca, C.; Virzì, N.; Argento, S.; Melilli, M.G. Hemp Flour Particle Size Affects the Quality and Nutritional Profile of the Enriched Functional Pasta. *Foods* **2023**, *12*, 774. [CrossRef] [PubMed]
72. Nigro, E.; Crescente, G.; Formato, M.; Pecoraro, M.T.; Mallardo, M.; Piccolella, S.; Daniele, A.; Pacifico, S. Hempseed Lignanamides Rich-Fraction: Chemical Investigation and Cytotoxicity towards U-87 Glioblastoma Cells. *Molecules* **2020**, *25*, 1049. [CrossRef]
73. Wu, D.; Liu, Y.; Guan, W.; Pan, J.; Kuang, H.; Yang, B. UHPLC-Orbitrap-Fusion-TMS-Based Metabolomics Study of Phenylpropionamides in the Seed of *Cannabis sativa* L. against Alzheimer's Disease. *Chem. Biodivers.* **2023**, *20*, e202201047. [CrossRef]
74. Zhou, Y.; Wang, S.; Ji, J.; Lou, H.; Fan, P. Hemp (*Cannabis sativa* L.) Seed Phenylpropionamides Composition and Effects on Memory Dysfunction and Biomarkers of Neuroinflammation Induced by Lipopolysaccharide in Mice. *ACS Omega* **2018**, *3*, 15988–15995. [CrossRef]

75. Kriese, U.; Schumann, E.; Weber, W.E.; Beyer, M.; Brühl, L.; Matthäus, B. Oil Content, Tocopherol Composition and Fatty Acid Patterns of the Seeds of 51 *Cannabis sativa* L. Genotypes. *Euphytica* **2004**, *137*, 339–351. [CrossRef]
76. Zhang, J.; Griffin, J.; Li, Y.; Wang, D.; Wang, W. Antioxidant Properties of Hemp Proteins: From Functional Food to Phytotherapy and Beyond. *Molecules* **2022**, *27*, 7924. [CrossRef]
77. Wang, S.; Luo, Q.; Fan, P. Cannabisin F from Hemp (*Cannabis sativa*) Seed Suppresses Lipopolysaccharide-Induced Inflammatory Responses in BV2 Microglia as SIRT1 Modulator. *Int. J. Mol. Sci.* **2019**, *20*, 507. [CrossRef] [PubMed]
78. Rodriguez-Martin, N.M.; Montserrat-de La Paz, S.; Toscano, R.; Grao-Cruces, E.; Villanueva, A.; Pedroche, J.; Millan, F.; Millan-Linares, M.C. Hemp (*Cannabis sativa* L.) Protein Hydrolysates Promote Anti-Inflammatory Response in Primary Human Monocytes. *Biomolecules* **2020**, *10*, 803. [CrossRef] [PubMed]
79. Ahidar, N.; Labhar, A.; Benamari, O.; Ahari, M.; Salhi, A.; Elyoussfi, A.; Amhamdi, H. Phenolic Content and Antioxidant Activity of *Cannabis Sativa* L. Flowers from the Ketama Region in Northern Morocco. *Ecol. Eng. Environ. Technol.* **2024**, *25*, 209–215. [CrossRef]
80. Cásedas, G.; Moliner, C.; Maggi, F.; Mazzara, E.; López, V. Evaluation of Two Different *Cannabis sativa* L. Extracts as Antioxidant and Neuroprotective Agents. *Front. Pharmacol.* **2022**, *13*, 1009868. [CrossRef]
81. Stasiłowicz-Krzemień, A.; Sip, S.; Szulc, P.; Cielecka-Piontek, J. Determining Antioxidant Activity of Cannabis Leaves Extracts from Different Varieties—Unveiling Nature's Treasure Trove. *Antioxidants* **2023**, *12*, 1390. [CrossRef]
82. Judžentienė, A.; Garjonytė, R.; Būdienė, J. Phytochemical Composition and Antioxidant Activity of Various Extracts of Fibre Hemp (*Cannabis sativa* L.) Cultivated in Lithuania. *Molecules* **2023**, *28*, 4928. [CrossRef]
83. Frazzini, S.; Torresani, M.C.; Roda, G.; Dell'Anno, M.; Ruffo, G.; Rossi, L. Chemical and Functional Characterization of the Main Bioactive Molecules Contained in Hulled *Cannabis sativa* L. Seeds for Use as Functional Ingredients. *J. Agric. Food Res.* **2024**, *16*, 101084. [CrossRef]
84. El-Mernissi, R.; El Menyiy, N.; Moubachir, R.; Zouhri, A.; El-Mernissi, Y.; Siddique, F.; Nadeem, S.; Ibork, H.; El Barnossi, A.; Wondmie, G.F.; et al. *Cannabis sativa* L. Essential Oil: Chemical Composition, Anti-Oxidant, Anti-Microbial Properties, and Acute Toxicity: In Vitro, In Vivo, and In Silico Study. *Open Chem.* **2024**, *22*, 20230214. [CrossRef]
85. Degrave, V.; Vega Joubert, M.B.; Ingaramo, P.; Sedan, D.; Andrinolo, D.; D'Alessandro, M.E.; Oliva, M.E. Effects of Full-Spectrum Cannabis Oil with a Cannabidiol:Tetrahydrocannabinol 2:1 Ratio on the Mechanisms Involved in Hepatic Steatosis and Oxidative Stress in Rats Fed a Sucrose-Rich Diet. *Med. Cannabis Cannabinoids* **2023**, *6*, 170–183. [CrossRef]
86. Baeeri, M.; Rahimifard, M.; Daghighi, S.M.; Khan, F.; Salami, S.A.; Moini-Nodeh, S.; Haghi-Aminjan, H.; Bayrami, Z.; Rezaee, F.; Abdollahi, M. Cannabinoids as Anti-ROS in Aged Pancreatic Islet Cells. *Life Sci.* **2020**, *256*, 117969. [CrossRef]
87. Dawidowicz, A.L.; Olszowy-Tomczyk, M.; Typek, R. CBG, CBD, Δ9-THC, CBN, CBGA, CBDA and Δ9-THCA as Antioxidant Agents and Their Intervention Abilities in Antioxidant Action. *Fitoterapia* **2021**, *152*, 104915. [CrossRef] [PubMed]
88. Hacke, A.C.M.; Lima, D.; De Costa, F.; Deshmukh, K.; Li, N.; Chow, A.M.; Marques, J.A.; Pereira, R.P.; Kerman, K. Probing the Antioxidant Activity of Δ^9-Tetrahydrocannabinol and Cannabidiol in *Cannabis sativa* Extracts. *Analyst* **2019**, *144*, 4952–4961. [CrossRef] [PubMed]
89. Baron-Flores, V.; Diaz-Ruiz, A.; Manzanares, J.; Rios, C.; Burelo, M.; Jardon-Guadarrama, G.; Martínez-Cárdenas, M.D.L.Á.; Mata-Bermudez, A. Cannabidiol Attenuates Hypersensitivity and Oxidative Stress after Traumatic Spinal Cord Injury in Rats. *Neurosci. Lett.* **2022**, *788*, 136855. [CrossRef] [PubMed]
90. Biernacki, M.; Jastrząb, A.; Skrzydlewska, E. Changes in Hepatic Phospholipid Metabolism in Rats under UV Irradiation and Topically Treated with Cannabidiol. *Antioxidants* **2021**, *10*, 1157. [CrossRef]
91. Khaksar, S.; Bigdeli, M.; Samiee, A.; Shirazi-zand, Z. Antioxidant and Anti-Apoptotic Effects of Cannabidiol in Model of Ischemic Stroke in Rats. *Brain Res. Bull.* **2022**, *180*, 118–130. [CrossRef]
92. Hampson, A.J.; Grimaldi, M.; Axelrod, J.; Wink, D. Cannabidiol and $(-)\Delta^9$-Tetrahydrocannabinol Are Neuroprotective Antioxidants. *Proc. Natl. Acad. Sci. USA* **1998**, *95*, 8268–8273. [CrossRef]
93. Iuvone, T.; Esposito, G.; Esposito, R.; Santamaria, R.; Di Rosa, M.; Izzo, A.A. Neuroprotective Effect of Cannabidiol, a Non-psychoactive Component from *Cannabis Sativa*, on B-amyloid-induced Toxicity in PC12 Cells. *J. Neurochem.* **2004**, *89*, 134–141. [CrossRef]
94. Marinho, A.M.D.N.; Silva-Neto, R.W.G.D. Anti-Inflammatory Effects of Cannabinoids. *Braz. J. Pain* **2023**, *6*, S31–S37. [CrossRef]
95. Sawtelle, L.; Holle, L.M. Use of Cannabis and Cannabinoids in Patients With Cancer. *Ann. Pharmacother.* **2021**, *55*, 870–890. [CrossRef]
96. Simmerman, E.; Qin, X.; Yu, J.C.; Baban, B. Cannabinoids as a Potential New and Novel Treatment for Melanoma: A Pilot Study in a Murine Model. *J. Surg. Res.* **2019**, *235*, 210–215. [CrossRef]
97. Ramer, R.; Heinemann, K.; Merkord, J.; Rohde, H.; Salamon, A.; Linnebacher, M.; Hinz, B. COX-2 and PPAR-γ Confer Cannabidiol-Induced Apoptosis of Human Lung Cancer Cells. *Mol. Cancer Ther.* **2013**, *12*, 69–82. [CrossRef] [PubMed]
98. Solinas, M.; Massi, P.; Cinquina, V.; Valenti, M.; Bolognini, D.; Gariboldi, M.; Monti, E.; Rubino, T.; Parolaro, D. Cannabidiol, a Non-Psychoactive Cannabinoid Compound, Inhibits Proliferation and Invasion in U87-MG and T98G Glioma Cells through a Multitarget Effect. *PLoS ONE* **2013**, *8*, e76918. [CrossRef] [PubMed]
99. Iqbal, S.; Matsabisa, M. In Silico Investigation of Cannabinoids from Cannabis Sativa Leaves as a Potential Anticancer Drug to Inhibit MAPK-ERK Signaling Pathway and EMT Induction. *Silico Pharmacol.* **2024**, *12*, 41. [CrossRef] [PubMed]

100. Motadi, L.R.; Jantjies, Z.E.; Moleya, B. Cannabidiol and Cannabis Sativa as a Potential Treatment in Vitro Prostate Cancer Cells Silenced with RBBp6 and PC3 Xenograft. *Mol. Biol. Rep.* **2023**, *50*, 4039–4047. [CrossRef] [PubMed]
101. Pongking, T.; Intuyod, K.; Thongpon, P.; Thanan, R.; Sitthirach, C.; Chaidee, A.; Kongsintaweesuk, S.; Klungsaeng, S.; Hongsrichan, N.; Sakonsinsiri, C.; et al. Cannabidiol Suppresses Proliferation and Induces Cell Death, Autophagy and Senescence in Human Cholangiocarcinoma Cells via the PI3K/AKT/mTOR Pathway. *J. Tradit. Complement. Med.* **2024**, *14*, 622–634. [CrossRef]
102. Wyrobnik, I.; Steinberg, M.; Gelfand, A.; Rosenblum, R.; Eid Mutlak, Y.; Sulimani, L.; Procaccia, S.; Ofran, Y.; Novak-Kotzer, H.; Meiri, D. Decreased Melanoma CSF-1 Secretion by Cannabigerol Treatment Reprograms Regulatory Myeloid Cells and Reduces Tumor Progression. *OncoImmunology* **2023**, *12*, 2219164. [CrossRef]
103. Bachari, A.; Nassar, N.; Telukutla, S.; Zomer, R.; Piva, T.J.; Mantri, N. Evaluating the Mechanism of Cell Death in Melanoma Induced by the Cannabis Extract PHEC-66. *Cells* **2024**, *13*, 268. [CrossRef]
104. Poommarapan, K.; Rummaneethorn, P.; Srisubat, A.; Suwanpidokkul, N.; Leenutaphong, P.; Nararatwanchai, T.; Srihirun, S.; Phetchengkao, W.; Suriyachan, K.; Tancharoen, S. Gene Profiling of *Cannabis sativa*-Mediated Apoptosis in Human Melanoma Cells. *Anticancer Res.* **2023**, *43*, 1221–1237. [CrossRef]
105. Ramer, R.; Hinz, B. Inhibition of Cancer Cell Invasion by Cannabinoids via Increased Expression of Tissue Inhibitor of Matrix Metalloproteinases-1. *JNCI J. Natl. Cancer Inst.* **2008**, *100*, 59–69. [CrossRef]
106. Dobovišek, L.; Krstanović, F.; Borštnar, S.; Debeljak, N. Cannabinoids and Hormone Receptor-Positive Breast Cancer Treatment. *Cancers* **2020**, *12*, 525. [CrossRef]
107. Myint, Z.W.; St. Clair, W.H.; Strup, S.E.; Yan, D.; Li, N.; Allison, D.B.; McLouth, L.E.; Ellis, C.S.; Wang, P.; James, A.C.; et al. A Phase I Dose Escalation and Expansion Study of Epidiolex (Cannabidiol) in Patients with Biochemically Recurrent Prostate Cancer. *Cancers* **2023**, *15*, 2505. [CrossRef] [PubMed]
108. Velasco, G.; Carracedo, A.; Blázquez, C.; Lorente, M.; Aguado, T.; Haro, A.; Sánchez, C.; Galve-Roperh, I.; Guzmán, M. Cannabinoids and Gliomas. *Mol. Neurobiol.* **2007**, *36*, 60–67. [CrossRef] [PubMed]
109. Dumitru, C.A.; Sandalcioglu, I.E.; Karsak, M. Cannabinoids in Glioblastoma Therapy: New Applications for Old Drugs. *Front. Mol. Neurosci.* **2018**, *11*, 159. [CrossRef] [PubMed]
110. Richtig, G.; Kienzl, M.; Rittchen, S.; Roula, D.; Eberle, J.; Sarif, Z.; Pichler, M.; Hoefler, G.; Heinemann, A. Cannabinoids Reduce Melanoma Cell Viability and Do Not Interfere with Commonly Used Targeted Therapy in Metastatic Melanoma In Vivo and In Vitro. *Biology* **2023**, *12*, 706. [CrossRef]
111. Mashabela, M.D.; Kappo, A.P. Anti-Cancer and Anti-Proliferative Potential of Cannabidiol: A Cellular and Molecular Perspective. *Int. J. Mol. Sci.* **2024**, *25*, 5659. [CrossRef]
112. Le, T.Q.; Meesiripan, N.; Sanggrajang, S.; Suwanpidokkul, N.; Prayakprom, P.; Bodhibukkana, C.; Khaowroongrueng, V.; Suriyachan, K.; Thanasitthichai, S.; Srisubat, A.; et al. Anti-Proliferative and Apoptotic Effect of Cannabinoids on Human Pancreatic Ductal Adenocarcinoma Xenograft in BALB/c Nude Mice Model. *Sci. Rep.* **2024**, *14*, 6515. [CrossRef]
113. Jeon, Y.; Kim, T.; Kwon, H.; Kim, J.-K.; Park, Y.-T.; Ham, J.; Kim, Y.-J. Cannabidiol Enhances Cabozantinib-Induced Apoptotic Cell Death via Phosphorylation of P53 Regulated by ER Stress in Hepatocellular Carcinoma. *Cancers* **2023**, *15*, 3987. [CrossRef]
114. Zhong, N.; Li, D.; Wang, B.; Kovalchuk, O.; Kovalchuk, I. Cannabinol Inhibits Cell Growth and Triggers Cell Cycle Arrest and Apoptosis in Cancer Cells. *Biocatal. Agric. Biotechnol.* **2023**, *48*, 102627. [CrossRef]
115. Go, Y.Y.; Kim, S.R.; Kim, D.Y.; Chae, S.-W.; Song, J.-J. Cannabidiol Enhances Cytotoxicity of Anti-Cancer Drugs in Human Head and Neck Squamous Cell Carcinoma. *Sci. Rep.* **2020**, *10*, 20622. [CrossRef]
116. Xu, Y.; Luo, J.; Guo, Y.; Zhou, J.; Shen, L.; Gu, F.; Shi, C.; Yao, L.; Hua, M. Chemical Compounds, Anti-Tumor and Anti-Neuropathic Pain Effect of Hemp Essential Oil In Vivo. *Fitoterapia* **2024**, *177*, 106092. [CrossRef]
117. Ramer, R.; Merkord, J.; Rohde, H.; Hinz, B. Cannabidiol Inhibits Cancer Cell Invasion via Upregulation of Tissue Inhibitor of Matrix Metalloproteinases-1. *Biochem. Pharmacol.* **2010**, *79*, 955–966. [CrossRef] [PubMed]
118. Ramer, R.; Rohde, A.; Merkord, J.; Rohde, H.; Hinz, B. Decrease of Plasminogen Activator Inhibitor-1 May Contribute to the Anti-Invasive Action of Cannabidiol on Human Lung Cancer Cells. *Pharm. Res.* **2010**, *27*, 2162–2174. [CrossRef] [PubMed]
119. Malheiro, R.F.; Carmo, H.; Carvalho, F.; Silva, J.P. Cannabinoid-Mediated Targeting of Mitochondria on the Modulation of Mitochondrial Function and Dynamics. *Pharmacol. Res.* **2023**, *187*, 106603. [CrossRef] [PubMed]
120. Rocha, F.C.M.; dos Santos Júnior, J.G.; Stefano, S.C.; da Silveira, D.X. Systematic Review of the Literature on Clinical and Experimental Trials on the Antitumor Effects of Cannabinoids in Gliomas. *J. Neurooncol.* **2014**, *116*, 11–24. [CrossRef] [PubMed]
121. Khodadadi, H.; Salles, É.L.; Alptekin, A.; Mehrabian, D.; Rutkowski, M.; Arbab, A.S.; Yeudall, W.A.; Yu, J.C.; Morgan, J.C.; Hess, D.C.; et al. Inhalant Cannabidiol Inhibits Glioblastoma Progression Through Regulation of Tumor Microenvironment. *Cannabis Cannabinoid Res.* **2023**, *8*, 824–834. [CrossRef]
122. Gross, C.; Ramirez, D.A.; McGrath, S.; Gustafson, D.L. Cannabidiol Induces Apoptosis and Perturbs Mitochondrial Function in Human and Canine Glioma Cells. *Front. Pharmacol.* **2021**, *12*, 725136. [CrossRef]
123. Ellert-Miklaszewska, A.; Ciechomska, I.A.; Kaminska, B. Synthetic Cannabinoids Induce Autophagy and Mitochondrial Apoptotic Pathways in Human Glioblastoma Cells Independently of Deficiency in TP53 or PTEN Tumor Suppressors. *Cancers* **2021**, *13*, 419. [CrossRef]
124. Vrechi, T.A.M.; Leão, A.H.F.F.; Morais, I.B.M.; Abílio, V.C.; Zuardi, A.W.; Hallak, J.E.C.; Crippa, J.A.; Bincoletto, C.; Ureshino, R.P.; Smaili, S.S.; et al. Cannabidiol Induces Autophagy via ERK1/2 Activation in Neural Cells. *Sci. Rep.* **2021**, *11*, 5434. [CrossRef]

125. Guggisberg, J.; Schumacher, M.; Gilmore, G.; Zylla, D.M. Cannabis as an Anticancer Agent: A Review of Clinical Data and Assessment of Case Reports. *Cannabis Cannabinoid Res.* **2022**, *7*, 24–33. [CrossRef]
126. Sea, Y.L.; Gee, Y.J.; Lal, S.K.; Choo, W.S. Cannabis as Antivirals. *J. Appl. Microbiol.* **2023**, *134*, lxac036. [CrossRef]
127. Van Breemen, R.B.; Simchuk, D. Antiviral Activities of Hemp Cannabinoids. *Clin. Sci.* **2023**, *137*, 633–643. [CrossRef] [PubMed]
128. Esposito, G.; Pesce, M.; Seguella, L.; Sanseverino, W.; Lu, J.; Corpetti, C.; Sarnelli, G. The Potential of Cannabidiol in the COVID-19 Pandemic. *Br. J. Pharmacol.* **2020**, *177*, 4967–4970. [CrossRef] [PubMed]
129. Costiniuk, C.T.; Jenabian, M.-A. Cannabinoids and Inflammation: Implications for People Living with HIV. *AIDS* **2019**, *33*, 2273–2288. [CrossRef] [PubMed]
130. Wang, B.; Kovalchuk, A.; Li, D.; Rodriguez-Juarez, R.; Ilnytskyy, Y.; Kovalchuk, I.; Kovalchuk, O. In Search of Preventative Strategies: Novel High-CBD *Cannabis Sativa* Extracts Modulate ACE2 Expression in COVID-19 Gateway Tissues. *Aging* **2020**, *12*, 22425. [CrossRef]
131. Classen, N.; Pitakbut, T.; Schöfbänker, M.; Kühn, J.; Hrincius, E.R.; Ludwig, S.; Hensel, A.; Kayser, O. Cannabigerol and Cannabicyclol Block SARS-CoV-2 Cell Fusion. *Planta Med.* **2024**, *90*, 717–725. [CrossRef]
132. Raj, V.; Park, J.G.; Cho, K.-H.; Choi, P.; Kim, T.; Ham, J.; Lee, J. Assessment of Antiviral Potencies of Cannabinoids against SARS-CoV-2 Using Computational and in Vitro Approaches. *Int. J. Biol. Macromol.* **2021**, *168*, 474–485. [CrossRef] [PubMed]
133. Salles, É.L.; Khodadadi, H.; Jarrahi, A.; Ahluwalia, M.; Paffaro, V.A.; Costigliola, V.; Yu, J.C.; Hess, D.C.; Dhandapani, K.M.; Baban, B. Cannabidiol (CBD) Modulation of Apelin in Acute Respiratory Distress Syndrome. *J. Cell. Mol. Med.* **2020**, *24*, 12869–12872. [CrossRef]
134. Beyerstedt, S.; Casaro, E.B.; Rangel, É.B. COVID-19: Angiotensin-Converting Enzyme 2 (ACE2) Expression and Tissue Susceptibility to SARS-CoV-2 Infection. *Eur. J. Clin. Microbiol. Infect. Dis.* **2021**, *40*, 905–919. [CrossRef]
135. Muthumalage, T.; Rahman, I. Cannabidiol Differentially Regulates Basal and LPS-Induced Inflammatory Responses in Macrophages, Lung Epithelial Cells, and Fibroblasts. *Toxicol. Appl. Pharmacol.* **2019**, *382*, 114713. [CrossRef]
136. van Breemen, R.B.; Muchiri, R.N.; Bates, T.A.; Weinstein, J.B.; Leier, H.C.; Farley, S.; Tafesse, F.G. Cannabinoids Block Cellular Entry of SARS-CoV-2 and the Emerging Variants. *J. Nat. Prod.* **2022**, *85*, 176–184. [CrossRef]
137. Likar, R.; Nahler, G. The Use of Cannabis in Supportive Care and Treatment of Brain Tumor. *Neuro-Oncol. Pract.* **2017**, *4*, 151–160. [CrossRef] [PubMed]
138. Hidding, U.; Mainka, T.; Buhmann, C. Therapeutic Use of Medical Cannabis in Neurological Diseases: A Clinical Update. *J. Neural Transm.* **2024**, *131*, 117–126. [CrossRef] [PubMed]
139. EPIDIOLEX® (Cannabidiol) Offers Flexible Dosing for Tolerability and Response Optimization. 2024. Available online: https://www.epidiolexhcp.com/dosing/dosing-and-administration (accessed on 1 September 2024).
140. VanDolah, H.J.; Bauer, B.A.; Mauck, K.F. Clinicians' Guide to Cannabidiol and Hemp Oils. *Mayo Clin. Proc.* **2019**, *94*, 1840–1851. [CrossRef] [PubMed]
141. Georgieva, D.; Langley, J.; Hartkopf, K.; Hawk, L.; Margolis, A.; Struck, A.; Felton, E.; Hsu, D.; Gidal, B.E. Real-World, Long-Term Evaluation of the Tolerability and Therapy Retention of Epidiolex® (Cannabidiol) in Patients with Refractory Epilepsy. *Epilepsy Behav.* **2023**, *141*, 109159. [CrossRef] [PubMed]
142. Szaflarski, J.P.; Devinsky, O.; Lopez, M.; Park, Y.D.; Zentil, P.P.; Patel, A.D.; Thiele, E.A.; Wechsler, R.T.; Checketts, D.; Sahebkar, F. Long-term Efficacy and Safety of Cannabidiol in Patients with Treatment-resistant Epilepsies: Four-year Results from the Expanded Access Program. *Epilepsia* **2023**, *64*, 619–629. [CrossRef]
143. Devinsky, O.; Cross, J.H.; Laux, L.; Marsh, E.; Miller, I.; Nabbout, R.; Scheffer, I.E.; Thiele, E.A.; Wright, S. Trial of Cannabidiol for Drug-Resistant Seizures in the Dravet Syndrome. *N. Engl. J. Med.* **2017**, *376*, 2011–2020. [CrossRef]
144. Thiele, E.A.; Marsh, E.D.; French, J.A.; Mazurkiewicz-Beldzinska, M.; Benbadis, S.R.; Joshi, C.; Lyons, P.D.; Taylor, A.; Roberts, C.; Sommerville, K.; et al. Cannabidiol in Patients with Seizures Associated with Lennox-Gastaut Syndrome (GWPCARE4): A Randomised, Double-Blind, Placebo-Controlled Phase 3 Trial. *Lancet* **2018**, *391*, 1085–1096. [CrossRef]
145. Fazlollahi, A.; Zahmatyar, M.; ZareDini, M.; Golabi, B.; Nejadghaderi, S.A.; Sullman, M.J.M.; Gharagozli, K.; Kolahi, A.-A.; Safiri, S. Adverse Events of Cannabidiol Use in Patients With Epilepsy: A Systematic Review and Meta-Analysis. *JAMA Netw. Open* **2023**, *6*, e239126. [CrossRef]
146. Cannabis Extract (Sativex®). 2024. Available online: https://www.panmerseyapc.nhs.uk/media/2377/cannabis_support.pdf (accessed on 1 September 2024).
147. Syed, Y.Y.; McKeage, K.; Scott, L.J. Delta-9-Tetrahydrocannabinol/Cannabidiol (Sativex®): A Review of Its Use in Patients with Moderate to Severe Spasticity Due to Multiple Sclerosis. *Drugs* **2014**, *74*, 563–578. [CrossRef]
148. Santarossa, T.M.; So, R.; Smyth, D.P.; Gustavsen, D.S.; Tsuyuki, D.R.T. Medical Cannabis Use in Canadians with Multiple Sclerosis. *Mult. Scler. Relat. Disord.* **2022**, *59*, 103638. [CrossRef]
149. Rekand, T. THC:CBD Spray and MS Spasticity Symptoms: Data from Latest Studies. *Eur. Neurol.* **2014**, *71*, 4–9. [CrossRef] [PubMed]
150. Jazz Pharmaceuticals. *A Randomized Controlled Trial of Cannabidiol (GWP42003-P, CBD) for Seizures in Tuberous Sclerosis Complex (GWPCARE6)*; Jazz Pharmaceuticals: Dublin, Ireland, 2022.
151. Jazz Pharmaceuticals. *Announces Initiation of Phase 3 Trial Evaluating Epidiolex®/Epidyolex® (Cannabidiol) for Patients with Epilepsy with Myoclonic-Atonic Seizures*; Jazz Pharmaceuticals: Dublin, Ireland, 2022.

152. Pauli, C.S.; Conroy, M.; Vanden Heuvel, B.D.; Park, S.-H. Cannabidiol Drugs Clinical Trial Outcomes and Adverse Effects. *Front. Pharmacol.* **2020**, *11*, 63. [CrossRef] [PubMed]
153. Brown, J.; Winterstein, A. Potential Adverse Drug Events and Drug–Drug Interactions with Medical and Consumer Cannabidiol (CBD) Use. *J. Clin. Med.* **2019**, *8*, 989. [CrossRef] [PubMed]
154. Van Ours, J.C.; Williams, J.; Fergusson, D.; Horwood, L.J. Cannabis Use and Suicidal Ideation. *J. Health Econ.* **2013**, *32*, 524–537. [CrossRef] [PubMed]
155. Woerdenbag, H.J.; Olinga, P.; Kok, E.A.; Brugman, D.A.P.; Van Ark, U.F.; Ramcharan, A.S.; Lebbink, P.W.; Hoogwater, F.J.H.; Knapen, D.G.; De Groot, D.J.A.; et al. Potential, Limitations and Risks of Cannabis-Derived Products in Cancer Treatment. *Cancers* **2023**, *15*, 2119. [CrossRef]
156. Blal, K.; Besser, E.; Procaccia, S.; Schwob, O.; Lerenthal, Y.; Abu Tair, J.; Meiri, D.; Benny, O. The Effect of Cannabis Plant Extracts on Head and Neck Squamous Cell Carcinoma and the Quest for Cannabis-Based Personalized Therapy. *Cancers* **2023**, *15*, 497. [CrossRef]
157. United Nations Office on Drugs and Crime World Drug Report Drug Use and Health Consequences. 2020. Available online: https://wdr.unodc.org/wdr2020/field/WDR20_Booklet_2.pdf (accessed on 1 September 2024).

MDPI AG
Grosspeteranlage 5
4052 Basel
Switzerland
Tel.: +41 61 683 77 34

International Journal of Molecular Sciences Editorial Office
E-mail: ijms@mdpi.com
www.mdpi.com/journal/ijms

www.ingramcontent.com/pod-product-compliance
Lightning Source LLC
Chambersburg PA
CBHW061509210926
51850CB00008B/1863
9783725872640